Advanced Technology in Environmental Remediation and Resource Utilization

Advanced Technology in Environmental Remediation and Resource Utilization

Editors

Deling Yuan
Shoufeng Tang

MDPI

Basel • Beijing • Wuhan • Barcelona • Belgrade • Novi Sad • Cluj • Manchester

Editors
Deling Yuan
Yanshan University
Qinhuangdao
China

Shoufeng Tang
Yanshan University
Qinhuangdao
China

Editorial Office
MDPI
St. Alban-Anlage 66
4052 Basel, Switzerland

This is a reprint of articles from the Special Issue published online in the open access journal *Coatings* (ISSN 2079-6412) (available at: https://www.mdpi.com/journal/coatings/special_issues/advanced_technology_environment_remediation).

For citation purposes, cite each article independently as indicated on the article page online and as indicated below:

Lastname, A.A.; Lastname, B.B. Article Title. *Journal Name* **Year**, *Volume Number*, Page Range.

ISBN 978-3-7258-0535-8 (Hbk)
ISBN 978-3-7258-0536-5 (PDF)
doi.org/10.3390/books978-3-7258-0536-5

Contents

Deling Yuan, Kai Yang, Eryu Zhu, Xiongbo Li, Mengting Sun, Lichun Xiao, et al.
Peracetic Acid Activated with Electro-Fe^{2+} Process for Dye Removal in Water
Reprinted from: *Coatings* **2022**, *12*, 466, doi:10.3390/coatings12040466 1

Linzhou Zhuang, Shiyi Li, Jiankun Li, Keyu Wang, Zeyu Guan, Chen Liang and Zhi Xu
Recent Advances on Hydrogen Evolution and Oxygen Evolution Catalysts for Direct Seawater Splitting
Reprinted from: *Coatings* **2022**, *12*, 659, doi:10.3390/coatings12050659 16

Deling Yuan, Zhihui Zhai, Eryu Zhu, Huilin Liu, Tifeng Jiao and Shoufeng Tang
Humic Acid Removal in Water via UV Activated Sodium Perborate Process
Reprinted from: *Coatings* **2022**, *12*, 885, doi:10.3390/coatings12070885 34

Yong Han, Bingjia Wei, Xiaoqiang Guo and Tifeng Jiao
Influence of Electrostatic Field on Mixed Aqueous Solution of Calcium and Ferrous Ions: Insights from Molecular Dynamics Simulations
Reprinted from: *Coatings* **2022**, *12*, 1165, doi:10.3390/coatings12081165 45

Haiqiu Fang, Dongfang Yang, Zizhen Su, Xinwei Sun, Jiahui Ren, Liwei Li and Kai Wang
Preparation and Application of Graphene and Derived Carbon Materials in Supercapacitors: A Review
Reprinted from: *Coatings* **2022**, *12*, 1312, doi:10.3390/coatings12091312 57

Jiaqian Xu, Mengzhou Wang, Cong Li, Mengxing Han, Qi Wang and Qina Sun
Metakaolin-Reinforced Sulfoaluminate-Cement-Solidified Wasteforms of Spent Radioactive Resins—Optimization by a Mixture Design
Reprinted from: *Coatings* **2022**, *12*, 1466, doi:10.3390/coatings12101466 76

Lu Wang, Yutong Fu, Shuaijie Wang, Fei Ye, Enming Cui and Qina Sun
Domesticating a Halotolerant Bacterium of *Vibrio* sp. LY1024 with Heterotrophic Nitrification–Aerobic Denitrification Property for Efficient Nitrogen Removal in Mariculture Wastewater Treatment
Reprinted from: *Coatings* **2022**, *12*, 1786, doi:10.3390/coatings12111786 88

Yingying Han, Lichun Xiao and Hongrui Chen
Chemical Agglomeration to Enhance Blast Furnace Dust Capture Efficiency in Wet Electrostatic Precipitators
Reprinted from: *Coatings* **2022**, *12*, 1937, doi:10.3390/coatings12121937 100

Henrikas Cesiulis and Natalia Tsyntsaru
Eco-Friendly Electrowinning for Metals Recovery from Waste Electrical and Electronic Equipment (WEEE)
Reprinted from: *Coatings* **2023**, *13*, 574, doi:10.3390/coatings13030574 116

Anna Modrzejewska-Sikorska, Mariola Robakowska, Emilia Konował, Hubert Gojzewski, Łukasz Gierz, Bartosz Wieczorek, et al.
Lignin and Starch Derivatives with Selenium Nanoparticles for the Efficient Reduction of Dyes and as Polymer Fillers
Reprinted from: *Coatings* **2023**, *13*, 1185, doi:10.3390/coatings13071185 127

Saeed Alhawtali, Mohanad El-Harbawi, Abdulrhman S. Al-Awadi, Lahssen El Blidi, Maher M. Alrashed and Chun-Yang Yin
Enhanced Adsorption of Methylene Blue Using Phosphoric Acid-Activated Hydrothermal Carbon Microspheres Synthesized from a Variety of Palm-Based Biowastes
Reprinted from: *Coatings* **2023**, *13*, 1287, doi:10.3390/coatings13071287 . . . . **139**

Sana Gharsallah, Abdulrahman Alsawi, Abdulelah H. Alsulami, Clarence Charnay and Mahmoud Chemingui
Enhancing Magnesium Phosphate Cement Paste for Efficient Fluoride Adsorption
Reprinted from: *Coatings* **2023**, *14*, 9, doi:10.3390/coatings14010009 . . . . **158**

coatings

Article

Peracetic Acid Activated with Electro-Fe^{2+} Process for Dye Removal in Water

Deling Yuan [1,2], Kai Yang [1,2], Eryu Zhu [1,2], Xiongbo Li [1,2], Mengting Sun [1,2], Lichun Xiao [1,2,*], Qiga Hari [3,*] and Shoufeng Tang [1,2,4,*]

1 Hebei Key Laboratory of Heavy Metal Deep-Remediation in Water and Resource Reuse, School of Environmental and Chemical Engineering, Yanshan University, Qinhuangdao 066004, China; yuandl@ysu.edu.cn (D.Y.); yangkaiysu@163.com (K.Y.); zhueryu2022@163.com (E.Z.); lixbysu@163.com (X.L.); sunmt19@126.com (M.S.)
2 Hebei Key Laboratory of Applied Chemistry, School of Environmental and Chemical Engineering, Yanshan University, Qinhuangdao 066004, China
3 Heibei Province Low-Carbon and Clean Building Heating Technology Innovation Center, Hebei Key Laboratory of Green Construction and Intelligent Maintenance for Civil Engineering, School of Civil Engineering and Mechanics, Yanshan University, Qinhuangdao 066006, China
4 State Key Laboratory of Metastable Materials Science and Technology, School of Materials Science and Engineering, Yanshan University, Qinhuangdao 066004, China
* Correspondence: xlc@ysu.edu.cn (L.X.); hariqiga@hotmail.com (Q.H.); tangshf@ysu.edu.cn (S.T.)

Abstract: An electro-Fe^{2+}-activated peracetic acid (EC/Fe^{2+}/PAA) process was established for organic dye removal in water. The operation factors such as the PAA dosage, Fe^{2+} amount, current density, and pH were investigated on methylene blue (MB) removal for the synergistic EC/Fe^{2+}/PAA system. Efficient MB decolorization (98.97% and 0.06992 min^{-1}) was achieved within 30 min under 5.4 mmol L^{-1} PAA, 30 μmol L^{-1} Fe^{2+}, 15 mA cm^{-2} current intensity, and pH 2.9. Masking tests affirmed that the dominating radicals were hydroxyl radicals (OH), organic radicals ($CH_3CO_2\cdot$, $CH_3CO_3\cdot$), and singlet oxygen ($^1O_2$), which were generated from the activated PAA by the synergetic effect of EC and Fe^{2+}. The influence of inorganic ions and natural organic matter on the MB removal was determined. Moreover, the efficacy of the EC/Fe^{2+}/PAA was confirmed by decontaminating other organic pollutants, such as antibiotic tetracycline and metronidazole. The studied synergy process offers a novel, advanced oxidation method for PAA activation and organic wastewater treatment.

Keywords: peracetic acid; ferrous ions; electrolysis; synergistic effect; dye elimination

Citation: Yuan, D.; Yang, K.; Zhu, E.; Li, X.; Sun, M.; Xiao, L.; Hari, Q.; Tang, S. Peracetic Acid Activated with Electro-Fe^{2+} Process for Dye Removal in Water. *Coatings* **2022**, *12*, 466. https://doi.org/10.3390/coatings12040466

Academic Editor: Ioannis Pashalidis

Received: 10 March 2022
Accepted: 28 March 2022
Published: 29 March 2022

1. Introduction

The effluents discharged from industries contain large amounts of organics, which can degrade the aquatic environment and bring adverse impacts for ecosystems [1,2]. Dye wastewater, as one of the representative organic wastewaters, contains many refractory compounds with high color and high concentration [3,4]. Because the traditional biological technologies are insufficient to treat the refractory organic contaminants in water, many physical and chemical methods have been developed over recent decades [5–7]. Advanced oxidation technologies (AOTs) can generate powerful reactive species—for example, the strong oxidative and nonselective hydroxyl radicals ($\cdot OH$, E_0 = 1.9–2.7 V) [8], which can transform the macromolecular substances into small molecular substances and eventually mineralize them into CO_2 and H_2O [9]. Therefore, the application of AOTs in the treatment of organic dye wastewater has attracted widespread attention [10].

Peracetic acid (PAA) is a typical disinfector during water purification [11], which is a mixture of four ingredients, such as H_2O, PAA, H_2O_2, and acetic acid. As an electrophilic agent, PAA can attack electron-rich parts or functional groups in pollutants, so it can oxidize some organic substances, such as amino acids [12] and β-lactam antibiotics [13]. Owing to the mediocre redox potential of 1.76 V, PAA cannot directly degrade organic pollutants that are resistant and persistent [14]. However, the peroxide bond of PAA is easily broken and activated due to the lower bond energy (38 kcal mol^{-1}) compared with that of H_2O_2 (51 kcal mol^{-1}) [15]. Therefore, researchers are becoming increasingly concerned about PAA activation methods [16]. The activated PAA could generate nonselective ·OH and selective organic free radicals (R-C·) including CH_3CO_2·, CH_3CO_3·, CH_3O_2·, and ·CH_3, etc., which could be effective for the elimination of organics in aqueous circumstances [17].

Previous research has shown that PAA can be activated by heat [18], ultraviolet [19], ultrasound [20,21], and cobalt metal ions and oxides (Co^{2+}, Co_3O_4, and $CoFe_2O_4$) [22–25]. Nevertheless, the above activation methods may have respective limitations—for example, high energy consumption and potential leaching toxicity [16]. Apart from these, Fe^{2+}-activating PAA is the most promising method, because iron species are eco-friendly and omnipresent in nature [26]. The inherent shortcomings of Fe^{2+}-activating PAA systems have hindered their development, such as the slow regeneration of Fe^{2+} and easy generation of iron precipitation, which is also the restrictive factor affecting the traditional Fenton process [27].

A number of studies have shown that EC can effectively promote Fe^{2+}/Fe^{3+} conversion through the cathode reaction for Fe^{2+} regeneration [28,29], forming the electro-Fenton process. The electro-Fenton process can generate H_2O_2 in situ and then enhance the production of ·OH radicals [30], thereby tremendously improving the degradation of pollutants. Moreover, many scientists have found that adding various peroxides (peroxydisulfate and peroxymonosulfate) into the electrolysis system could form a cooperative effect for removing organics in water [31–33]. Our previous study proved that electrolysis (EC) could effectively excite PAA for dye removal, but the activation efficiency still has room for improvement [34]. Hence, we propose the coupling of EC and Fe^{2+}-activating PAA (EC/Fe^{2+}/PAA) processes in this work.

The ternary synergistic system was expected to realize the efficient Fe^{2+}/Fe^{3+} cycle and PAA activation simultaneously. Methylene blue (MB), as a typical cationic dye, has complex and stable structures, which was selected to verify the performance of the synergistic system. The systematical study had the following aims: firstly, to research the enhancement of MB removal in the EC/Fe^{2+}/PAA process; secondly, to assess the effects of operation conditions and water constituents on the MB decolorization in the synergistic system; thirdly, to affirm the dominant reactive species for MB elimination; lastly, to propose a possible reaction mechanism of MB degradation in the EC/Fe^{2+}/PAA process.

2. Experimental

2.1. Materials and Reagents

Details of materials and reagents are introduced in Text S1 in the Supplementary Materials (SM).

2.2. Equipment and Procedure

A power source output a direct current to drive the EC system. The anode and cathode were platinum and graphite plates, separately, which were spaced 2 cm apart. The detailed information about the EC system is provided in Text S2 of Supplementary Materials. A beaker (100 mL) was applied as the reactor for the batch test. The designated concentrations of PAA and Fe^{2+} were introduced simultaneously to the reactor, which contained the target pollutant. The reaction was conducted at a temperature of 25 °C, with stable stirring. The initial pH was regulated through the addition of NaOH and H_2SO_4. The EC reaction started after switching on the power. Sampling was performed at a fixed time node, and 1 mL $Na_2S_2O_3$ solution was added for quenching active species before the concentration was determined. Different reactive species scavengers were added to the reactor separately to investigate the reaction mechanism in the masking tests.

2.3. Analytical Methods

The concentrations of MB, methyl orange (MO), and rhodamine B (RhB) were determined by the spectrophotometric method at 664, 465, and 554, nm, respectively. The metronidazole (MTZ) and tetracycline (TCH) concentrations were measured by High-Performance Liquid Chromatography (HPLC). The degradation byproducts of MB were investigated by HPLC combined with mass spectrum (MS). The amounts of iron ions were determined by the phenanthroline spectrophotometric method [35]. The change in PAA was measured by the N,N-diethyl-p-phenylenediamine (DPD) colorimetric method [36]. The H_2O_2 concentration was quantified through titanium potassium oxalate colorimetry [35].

2.4. Energy Efficiency

The calculated electrical energy per order (EE/O) was denoted as electrical energy consumption by the following equation [37]:

$$\frac{\mathrm{EE}}{\mathrm{O}} = \frac{U \times I \times T}{V \times 60 \times \log(\mathrm{C}_0/\mathrm{C})} \tag{1}$$

where EE/O represents the electrical energy (kWh) consumed for removing the pollutant by an order of magnitude in a significant volume (V, L) of wastewater and a certain time (T, min); U and I stand for the voltage (V) and current (A), respectively; C_0 and C indicate the beginning and ending contaminant concentrations, respectively.

3. Results and Discussion

3.1. Performance of EC/Fe^{2+}/PAA Process

The decolorizations of MB in the PAA, Fe^{2+}, Fe^{2+}/PAA, EC/Fe^{2+}, EC, EC/PAA, and EC/Fe^{2+}/PAA systems are comprehensively compared in Figure 1a. After 60 min of reaction, the decolorization ratios of MB in the PAA alone and sole Fe^{2+} systems were only 5.13% and 7.26%, respectively. In the Fe^{2+}/PAA system, the decolorization of MB reached 35.15%. The removals of MB in the EC, EC/Fe^{2+}, and EC/PAA systems were 54.55%, 52.24%, and 66.64%, respectively. However, the decoloring performance was greatly improved for the EC/Fe^{2+}/PAA process, and the elimination efficiency of MB increased to 98.97%, showing a significant synergistic effect. In Figure 1b, the kinetic fitting results indicate that the reaction rate constant of EC/Fe^{2+}/PAA reached 0.06992 min^{-1}, which was obviously higher than other processes (the PAA and Fe^{2+} were both fitted by a zero-order kinetic model, and other systems were all fitted through a pseudo-first-order kinetic model). The above results demonstrate that the ternary synergistic process could remarkably promote the decolorization efficiency and rate of MB.

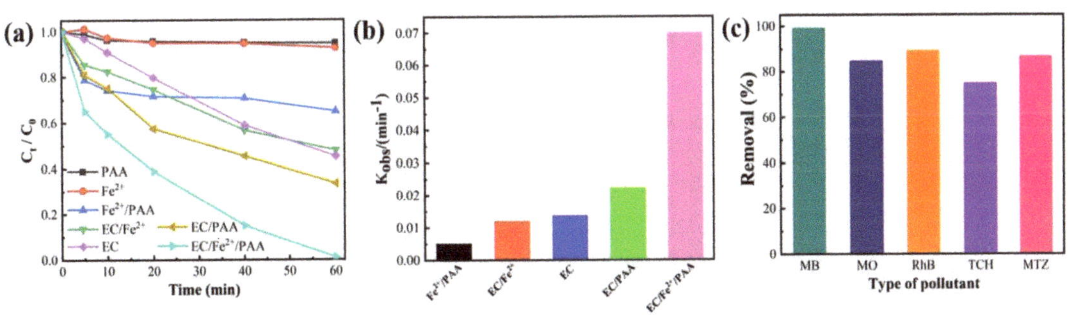

Figure 1. Decolorization ratio (**a**) and reaction rate constant (**b**) of MB in various systems. (**c**) Diverse organic pollutant removal in EC/Fe^{2+}/PAA system. Experimental conditions: $[MB]_0$ = $[MO]_0$ = $[RhB]_0$ = $[TCH]_0$ = $[MTZ]_0$ = 20 mg L^{-1}, $[Fe^{2+}]_0$ = 30 μmol L^{-1}, $[PAA]_0$ = 5.4 mmol L^{-1}, $[H_2O_2]_0$ = 11.25 mmol L^{-1}, $[NaNO_3]$ = 30 mmol L^{-1}, pH = 2.9, j = 15 mA cm^{-2}.

In the EC/PAA system, the electron-donating condition for the cathode could activate PAA (Equations (2) and (3)) [38]. Furthermore, the ·OH produced by the electron loss reaction on the anode not only could remove the MB (Equation (4)) [39], but also activate PAA to generate $CH_3CO_3\cdot$ and $CH_3CO_2\cdot$ for the oxidation of organic pollutants (Equations (5)–(7)) [20]. Meanwhile, the electron transfer on the surface of the electrode could directly oxidize MB. However, the main drawback of this system is that the reactions of PAA activation and radical generation are confined to the electrode surface. In the Fe^{2+}/PAA process, the transition metal ions of Fe^{2+} were often used as an activator, which could activate PAA to generate organic free radicals and ·OH (Equations (8) and (9)) [24]. However, the generated Fe^{3+} was difficult to reduce and easily produced iron sludge, leading to insufficient oxidation for MB removal.

$$CH_3CO_3H + e^- \rightarrow CH_3CO_2\cdot + OH^- \quad (2)$$

$$CH_3CO_3H + e^- \rightarrow CH_3CO_2^- + \cdot OH \quad (3)$$

$$H_2O \rightarrow \cdot OH + H^+ + e^- \quad (4)$$

$$CH_3CO_3H + \cdot OH \rightarrow CH_3CO_3\cdot + H_2O \quad (5)$$

$$CH_3CO_3\cdot + CH_3CO_3\cdot \rightarrow 2CH_3CO_2\cdot + O_2 \quad (6)$$

$$CH_3CO_3H + CH_3CO_2\cdot \rightarrow CH_3CO_3\cdot + CH_3CO_2H \quad (7)$$

$$CH_3CO_3H + Fe^{2+} \rightarrow CH_3CO_2\cdot + Fe^{3+} + OH^- \quad (8)$$

$$CH_3CO_3H + Fe^{2+} \rightarrow CH_3CO_2^- + Fe^{3+} + \cdot OH \quad (9)$$

Nevertheless, after adding the Fe^{2+} to the EC/PAA system, the combined effect of the electro-Fenton and PAA activation reactions could promote the oxidative radicals' generation and then enhance the dye elimination. Furthermore, the synergistic effect could extend the oxidation and activation reactions from the electrode surfaces to the entire solution, greatly improving the capability of this system. In addition, the generated Fe^{3+} could be reduced to Fe^{2+} by obtaining electrons at the cathode (Equation (10)) [32], realizing the reuse of Fe^{2+} for sustainable PAA activation.

$$Fe^{3+} + e^- \rightarrow Fe^{2+} \quad (10)$$

To prove that the EC/Fe^{2+}/PAA system is effective in removing diverse organic pollutants, three reactive dyes (MB, MO, and RhB) and two antibiotics (TCH and MTZ) were selected for the elimination tests [40]. As illustrated in Figure 1c, the significant abatement of MB, MO, RhB, TCH, and MTZ by the EC/Fe^{2+}/PAA process was observed, and their removal efficiencies reached 98.89%, 84.33%, 88.88%, 74.53%, and 86.15% during 60 min, respectively, certifying the effectiveness of the synergy process for the common organic contaminants' decontamination. These organic compounds contain methyl and hydroxyl groups, and these electro-rich functional groups could be susceptible to reaction with R-C· via electron transferring [41].

3.2. *Influencing Factors for EC/Fe^{2+}/PAA Process*

3.2.1. PAA Dosage

The effect of different PAA dosages for MB decolorization during the EC/Fe^{2+}/PAA process was researched (Figure 2a). The MB removal was enhanced from 61.26% to 91.80% at 60 min with the PAA dosage augmented from 0.9 to 5.4 mmol L^{-1}. Boosting the PAA dose could supply the substrate of active species generation for removing MB. Nevertheless, while the PAA amount was raised to 9.0 mmol L^{-1}, the MB elimination was inhibited slightly, demonstrating that the superfluous PAA could not continuously improve the MB oxidation. This is because the H_2O_2 in PAA has a negative impact on the ·OH radical through scavenging effects [18], which could impact the PAA activation and then inhibit the dye elimination. Therefore, the optimal concentration of PAA was set at 5.4 mmol L^{-1}.

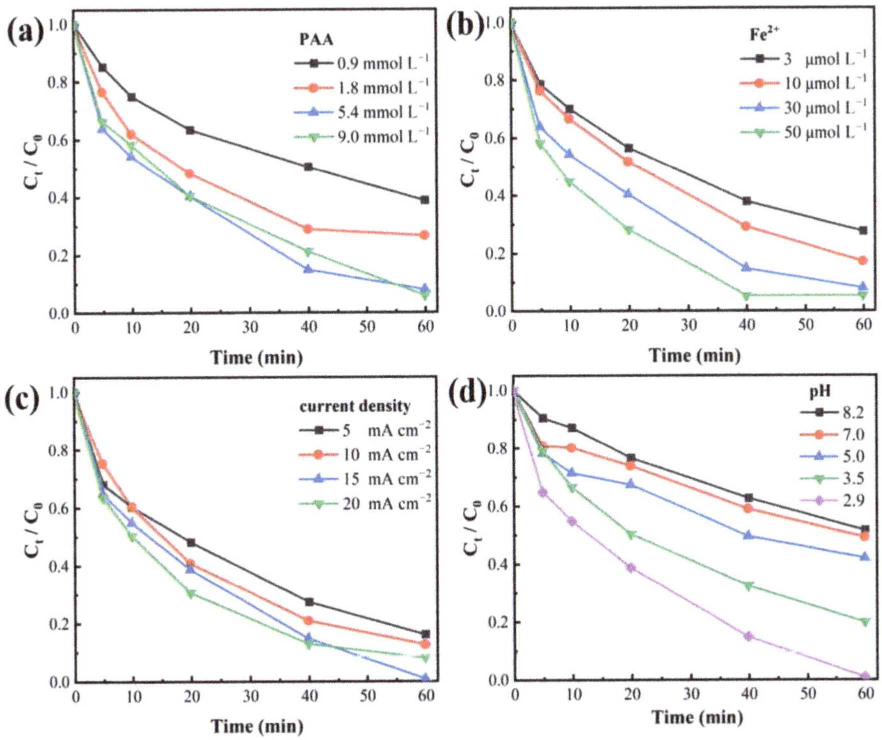

Figure 2. Decolorization ratio of MB under different reaction conditions: (**a**) PAA dosage, (**b**) Fe^{2+} dosage, (**c**) current density, and (**d**) solution pH. Experimental conditions: $[MB]_0$ = 20 mg L^{-1}, $[Fe^{2+}]_0$ = 30 μmol L^{-1}, $[PAA]_0$ = 5.4 mmol L^{-1}, $[H_2O_2]_0$ = 11.25 mmol L^{-1}, $[NaNO_3]$ = 30 mmol L^{-1}, pH = 2.9, j = 15 mA cm^{-2}.

3.2.2. Fe^{2+} Dosage

The effect of Fe^{2+} dosage on MB decolorization was evaluated (Figure 2b). With the Fe^{2+} concentration increasing from 3 to 30 μmol L^{-1}, the decolorization efficiency of MB was enhanced from 72.48% to 91.80% within 60 min, which indicated that an appropriate increase in Fe^{2+} dosage could catalyze PAA more efficiently and rapidly, hence strengthening the reactive radicals' formation and the MB removal. When the Fe^{2+} dosage was raised to 50 μmol L^{-1}, the MB removal ratio was not enhanced at 60 min, but the decolorization rate was improved obviously. Considering the cost saving, the suitable concentration of Fe^{2+} was chosen as 30 μmol·L^{-1}.

3.2.3. Current Intensity

As can be seen from Figure 2c, with the current intensity rising from 5 to 15 mA cm^{-2}, the decolorization ratio of MB was augmented from 84.74% to 93.73%. The stronger current is a benefit for the ·OH formation on the anode surface and Fe^{3+} reduction on the cathode surface [42,43], which could promote PAA activation and the oxidation of organics. Nevertheless, as the current was enhanced to 20 mA cm^{-2}, the decolorization of MB incurred a slight drop. This phenomenon reflects the fact that the exorbitant current density would improve the electrolysis of water, and the generated O_2 at the anode would accelerate the oxidation of Fe^{2+} to Fe^{3+} (Equation (11)) [44], then suppressing the PAA activation and dye removal. Hence, the suitable current density value was set to be 15 mA cm^{-2}.

$$2H_2O \rightarrow 4H^+ + O_2 + 4e^- \quad (11)$$

3.2.4. Initial pH

The pH condition could impact the form and redox potential of PAA. Thus, the effects of different solution pH on the MB removal were explored (Figure 2d). The result showed that acidic conditions were more conducive to the decolorization of MB. Under the alkaline condition (pH = 8.2), the removal of MB was only 56.03%. Nevertheless, the decolorization reached 98.97% at pH 2.9. The pKa value of PAA is approximately 8.2 [15]. The lower the pH, the more protonated PAA (PAA^0) was the dominant form in solution [16]. When the pH was near to 8.2, the PAA was prone to self-decomposition and hydrolysis, and the proportion of deprotonated PAA (PAA^-) would be enhanced in solution [16]. The oxidizing ability of PAA^0 is higher than that of PAA^- [12]. Moreover, compared to PAA^-, the PAA^0 more easily accepts electrons, and is more liable to be activated by Fe^{2+} [45]. In addition, the Fe^{3+}/Fe^{2+} cycle becomes more difficult as the pH increases [46]. Therefore, the acidic environment was conducive to the removal of MB in the EC/Fe^{2+}/PAA process.

3.3. Mechanism of EC/Fe^{2+}/PAA Process

3.3.1. Scavenging Test

Radical quenching experiments were employed to distinguish the various active substances in the EC/Fe^{2+}/PAA process. Five scavengers were employed, namely tert-butyl alcohol (TBA), methanol (MeOH), phenol, P-benzoquinone (p-BQ), and L-Histidine (L-His). TBA is a common scavenging agent for ·OH on account of the fast reaction rate (5×10^8 M^{-1} s^{-1}) with ·OH [36]. MeOH is a hydrophilic agent, which can quench ·OH, CH_3CO_2, and CH_3CO_3 at the same time [41]. Phenol is a hydrophobic chemical, which can react quickly with OH, $CH_3CO_2\cdot$, and $CH_3CO_3\cdot$ near the electrode [39]. p-BQ is an excellent masking agent for superoxide anions ($O_2^{\cdot-}$) [47]. L-Histidine is a typical inhibiting agent for singlet oxygen ($^1O_2$) [47].

A summary of the scavenging results is given in Figure 3a. Overall, 98.97% of MB was removed in the absence of a scavenger. The MB decay declined by approximately 6.04% with the addition of p-BQ, proving the inappreciable role of $O_2^{\cdot-}$ in this system. The suppression of MB decolorization was more prominent with the existence of L-His and the removal ratio of MB dropped to 80%, which proved that the $^1O_2$ plays a non-negligible role in MB elimination (Equations (12) and (13)) [48,49]. Adding excess MeOH and TBA cannot entirely restrain the MB decolorization, and the decolorization was reduced to 72.70% and 60.15%, respectively. However, when 50 mol L^{-1} of phenol was added, the inhibition effect on the MB removal became acute, and the decolorization was reduced to 28.49%. This result indicated that the free radicals on the electrode surface could play an important role in MB elimination. Hence, the major active species for MB removal were ·OH, $CH_3CO_2\cdot$, $CH_3CO_3\cdot$, and $^1O_2$ in the EC/Fe^{2+}/PAA system. Moreover, the presence of phenol had a more obvious suppression influence than TBA and MeOH, demonstrating that the PAA activation and related radicals' reactions for MB elimination mostly originated from the electrode surface. In addition, the above results confirmed that the PAA was primarily activated by the EC system, as observed in Figure 1a.

$$O_2^{\cdot-} + \cdot OH \rightarrow HO^- + {}^1O_2 \quad (12)$$

$$O_2^{\cdot-} + 2H^+ \rightarrow H_2O_2 + {}^1O_2 \quad (13)$$

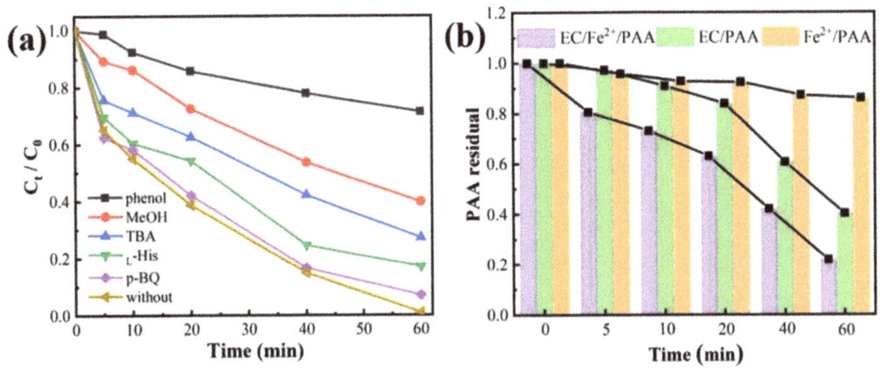

Figure 3. MB removal under various radical quenching agents in EC/Fe^{2+}/PAA (**a**). Remaining amount of PAA under different systems (**b**). Experimental conditions: $[MB]_0$ = 20 mg L^{-1}, $[Fe^{2+}]_0$ = 30 μmol L^{-1}, $[PAA]_0$ = 5.4 mmol L^{-1}, $[H_2O_2]_0$ = 11.25 mmol L^{-1}, TBA = 6 mol L^{-1}, MeOH = 10 mol L^{-1}, phenol = 50 mmol L^{-1}, p-BQ = 200 μmol L^{-1}, L-His = 200 μmol L^{-1}, $[NaNO_3]$ = 30 mmol L^{-1}, pH = 2.9, j = 15 mA cm^{-2}.

3.3.2. PAA Utilization Rate

As the major precursor of active radicals in the EC/Fe^{2+}/PAA is PAA, monitoring the residual PAA concentration can reflect its activation efficiency. As presented in Figure 3b, the remaining proportion of PAA (the remaining PAA to initial PAA) in the three systems (Fe^{2+}/PAA, EC/PAA, and EC/Fe^{2+}/PAA) decreased with the increase in time and the downward trend was enhanced significantly. After 60 min of reaction, only 13.87% of PAA was consumed in the Fe^{2+}/PAA process, and 59.62% of PAA was decomposed in the EC/PAA system, revealing that the PAA's activation ability for the EC process was stronger than that for the Fe^{2+} system. The highest decomposition ratio of PAA was significantly increased to 78.23% for the EC/Fe^{2+}/PAA process, which was higher than that for the other two systems. This result clearly shows that the EC/Fe^{2+}/PAA system yields an excellent synergistic effect for PAA activation. In summary, the above phenomena were in agreement with the outcomes of MB removal and radical quenching tests.

3.3.3. Trend of Fe Irons

Fe^{2+} is a crucial element in EC/Fe^{2+}/PAA, which could not excite PAA, but also plays the role of catalyst in the electro-Fenton process. The changes in the Fe^{2+} and Fe^{3+} concentration were compared in the Fe^{2+}/PAA and EC/Fe^{2+}/PAA systems (Figure 4). As displayed in Figure 4a, the Fe^{2+} content decreased rapidly with the increase in Fe^{3+} within 5 min in the Fe^{2+}/PAA process, and the trend tended towards becoming stable after 10 min. At the same time, the total Fe amount only fell by 2% during the 60 min reaction. The results were mainly ascribed to the fast depletion of Fe^{2+}, the activation of PAA, and the ineffective conversion of Fe^{3+} to Fe^{2+}, once again verifying the outcome of MB removal shown in Figure 1a. However, as illustrated in Figure 4b, in the EC/Fe^{2+}/PAA process, the concentration of Fe^{2+} remained at a higher level than that in Fe^{2+}/PAA, and the amount of Fe^{3+} also maintained a lower level than that of Fe^{2+}/PAA. These results prove that Fe^{3+} could be continuously converted into Fe^{2+} under the effect of cathode electron transfer, resulting in a higher concentration of Fe^{2+} in the system. Moreover, the total Fe content in the EC/Fe^{2+}/PAA process was decreased to 80%, which was more remarkable compared with the Fe^{2+}/PAA. In the electrolysis system, the Fe ions could undergo a reduction reaction by receiving electrons from the cathode, and then be deposited on the surface of the cathode, resulting in a decline in the total Fe concentration [39].

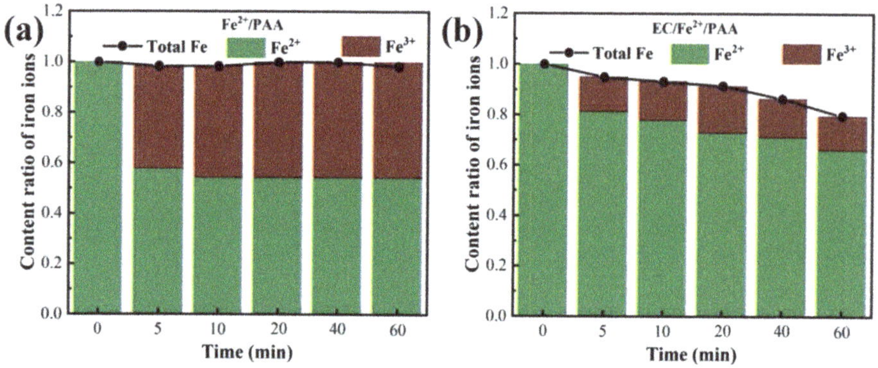

Figure 4. Changes in Fe content in Fe^{2+}/PAA process (**a**) and EC/Fe^{2+}/PAA (**b**). Experimental conditions: $[MB]_0$ = 20 mg L^{-1}, $[Fe^{2+}]_0$ = 30 μmol L^{-1}, $[PAA]_0$ = 5.4 mmol L^{-1}, $[H_2O_2]_0$ = 11.25 mmol L^{-1}, $[NaNO_3]$ = 30 mmol L^{-1}, pH = 2.9, j = 15 mA cm^{-2}.

3.4. MB Degradation Mechanism

The UV–visible spectrum was used to confirm the decolorization of MB in the EC/Fe^{2+}/PAA. As can be seen from Figure S1 of Supplementary Materials, the absorption peak at 664 nm gradually disappeared within 60 min [50], indicating that the chromogenic group of sulfhydryl (-S-) of MB was destroyed by the free radicals of EC/Fe^{2+}/PAA [51].

The TOC removal for MB was determined during the collaborative process. As seen in Figure S2 in Supplementary Materials, the TOC almost remained unchanged in the 120 min treatment. This result indicated that the application of activated PAA could increase the TOC, which is derived from the residual PAA and acetic acid [34]. Hence, the PAA-related process combined with the biological treatment could be a promising option because the increased TOC is a good carbon source for microorganisms [22,52].

The HPLC-MS spectra of MB before and after synergistic oxidation are presented in Figure S3 of Supplementary Materials. The peak of m/z = 284 was attributed to the MB, which was evidently decreased after 60 min reaction, corroborating the decomposition of MB. Based on the results of MS analyses, the possible degradation pathway of MB was suggested (Figure 5). Initially, the cation of MB (MB^+) could be formed via the electron transfer between MB and $CH_3CO_2\cdot$ or $CH_3CO_3\cdot$. Next, the MB^+ would rapidly react with H_2O or O_2 to produce hydroxylated byproduct No. 1 through the reactions of hydroxyl extraction or addition [53]. The MB could be demethylated to generate byproduct No. 2 under the action of ·OH. In addition, the central aromatic ring of MB could also be opened by ·OH to produce byproduct No. 3, which contained amino (-NH_2) and sulfoxide (C-S (=O)-C) functional groups [54]. The methylated byproduct No. 2 could also be further demethylated to form byproduct No. 4 after losing four methyl groups. After this, the N = C bond and sulfhydryl group of byproduct No. 4 could be broken down to form byproducts No. 5 and No. 6. The byproduct No. 4 could generate the hydroxylated byproduct No. 7 via the attack of ·OH. The benzene ring of byproduct No. 6 was cleaved to produce small-molecule byproduct No. 8, which was further split into small-molecule fragment byproduct No. 9. The hydroxylated byproduct No. 1 also could be broken to form byproducts No. 5 and No. 10 through the ring opening reaction under the function of ·OH. The byproduct No. 3 could be hydroxylated to form byproduct No. 11, and then demethylated to produce byproduct No. 12, and the sulfoxide was cracked to generate byproduct No. 13 [49,55]. Finally, these degradation intermediates could be further oxidized and converted to CO_2, H_2O, $SO_4{}^{2-}$, and $NO_3{}^-$ [56].

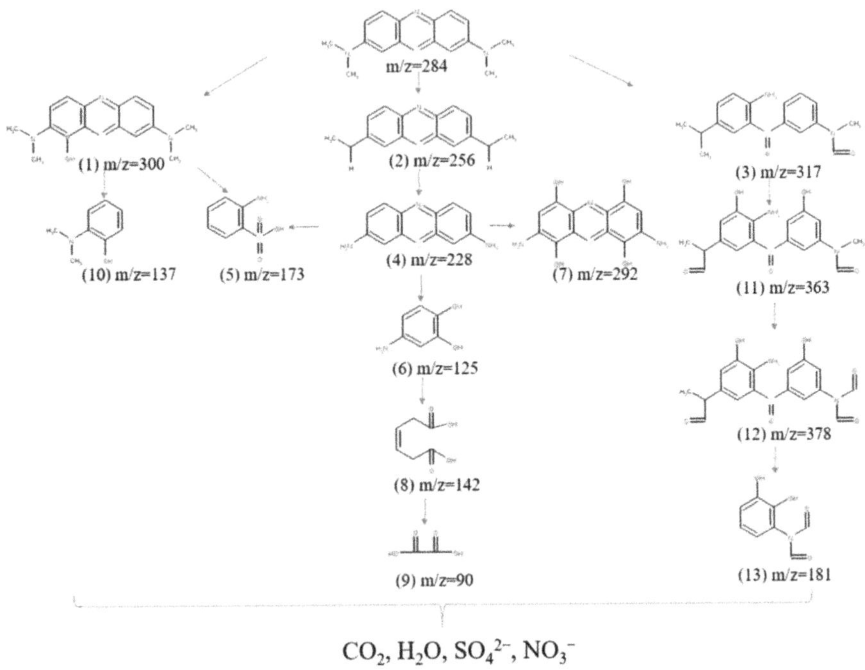

Figure 5. Possible degradation pathway in EC/Fe^{2+}/PAA.

3.5. Water Matrices Effects

The effect of Cl^- on the MB removal in the EC/Fe^{2+}/PAA was explored. As seen in Figure 6a, the existence of Cl^- (1–20 mmol L^{-1}) had a certain passive effect on the MB elimination, and the removal efficiency decreased from 98.97% to 83.47%. Chloride ions can scavenge free radicals, resulting in the generation of reactive chlorine species (Cl·, $Cl_2^{\cdot-}$, and $HClO^{\cdot-}$, Equations (14)–(18)) [36,57], whose oxidation capacities were lower than those of reactive species generated in the PAA activation. In addition, the coexisting iron and chloride ions could oxidize Fe^{2+} to Fe^{3+} and form $Fe(Cl)^{2+}$ and $FeCl_2^+$ complexes (Equation (19)) [14], harming the reaction of Fe^{2+} activating PAA.

$$CH_3CO_3\cdot + Cl^- + H^+ \rightarrow Cl\cdot + CH_3CO_3H \quad (14)$$

$$Cl^- + \cdot OH \rightarrow HClO\cdot^- \quad (15)$$

$$HClO\cdot^- + H^+ \rightarrow Cl\cdot + H_2O \quad (16)$$

$$Cl^- + Cl\cdot \rightarrow Cl_2^- \quad (17)$$

$$Cl\cdot + Cl\cdot \rightarrow Cl_2 \quad (18)$$

$$Fe^{2+} + Cl\cdot \rightarrow Fe^{3+} + Cl^- \quad (19)$$

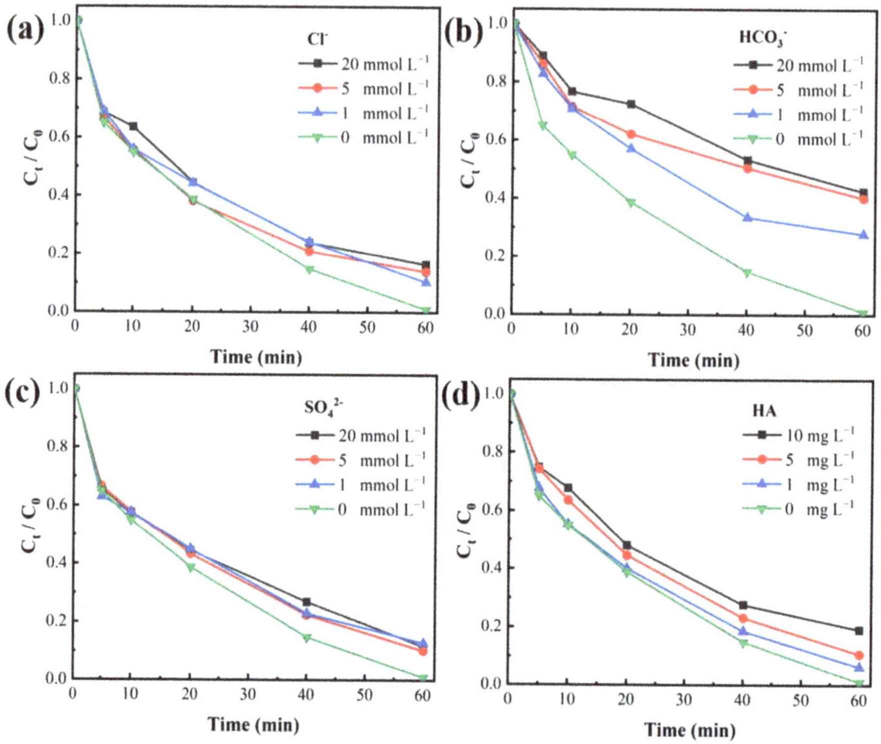

Figure 6. Effects of presence of (**a**) Cl^-, (**b**) HCO_3^-, (**c**) SO_4^{2-}, and (**d**) HA (0–10 mg L^{-1}) on MB decolorization in EC/Fe^{2+}/PAA process. Experimental conditions: $[MB]_0$ = 20 mg L^{-1}, $[Fe^{2+}]_0$ = 30 μmol L^{-1}, $[PAA]_0$ = 5.4 mmol L^{-1}, $[H_2O_2]_0$ = 11.25 mmol L^{-1}, $[NaNO_3]$ = 30 mmol L^{-1}, pH = 2.9, j = 15 mA cm^{-2}.

The effect of bicarbonate ions on MB removal in the EC/Fe^{2+}/PAA was investigated (Figure 6b). Increasing bicarbonate ions amount led to the strong inhibition of the MB removal, and when the HCO_3^- was increased to 20 mmol L^{-1}, the decolorization ratio was reduced to 57.51%. This inhibitory effect is due to the scavenging effect of HCO_3^- on ·OH. Moreover, the masking product is a carbonate radical (Equation (20) [27,58], which is a weak oxidant compared with ·OH. Meanwhile, the presence of HCO_3^- would increase the solution's pH [57], which is harmful to the activation of PAA by Fe^{2+}.

$$HCO_3^- + \cdot OH \rightarrow CO_3^{\cdot -} + H_2O \tag{20}$$

Figure 6c shows that the MB elimination in the EC/Fe^{2+}/PAA was restrained through the existence of SO_4^{2-}. When the SO_4^{2-} concentration was augmented from 1 to 20 mmol L^{-1}, the MB removal ratio dropped from 98.97% to 80.79%. SO_4^{2-} would react with ·OH in the acidic environment following the production of $SO_4^{\cdot -}$ (Equation (21)) [59,60]. $SO_4^{\cdot -}$ is also a strongly oxidizing radical, so MB removal could only be restrained to a certain extent.

$$\cdot OH + SO_4^{2-} \rightarrow SO_4^{\cdot -} + OH^- \tag{21}$$

Figure 6d reveals the effect of the addition of HA on the MB removal. As expected, the MB removal efficiency was obviously reduced with the increase in HA amount. The HA could rapidly react with ·OH and R-C·, with rate constants of 2.5×10^4 L mg^{-1} s^{-1} and 5.76×10^4 L mg^{-1} s^{-1}, respectively [36], which would consume reactive species in the

EC/Fe^{2+}/PAA. Moreover, the presence of HA would impede the Fe^{2+} catalytic activity by the complexation of HA and Fe^{2+} [45,61], thus affecting the PAA activation for removing the MB.

3.6. Electrical Energy Analysis

Table 1 illustrates the electrical energy consumption of the EC, EC/PAA, and EC/Fe^{2+}/PAA processes for removing MB, and the calculated EE/O values for the three systems were 5.601, 3.345, and 0.851 kWh/m^3, respectively. The EE/O of EC/Fe^{2+}/PAA was approximately 6.58 and 3.93 times lower than those of EC and EC/PAA, respectively, which embodied an obvious decline for the power consumption, while the PAA and Fe^{2+} corresponded with the EC process. It can be deduced that the synergistic effect could not only promote the dye removal, but also markedly improve the energy efficiency.

Table 1. Electrical energy consumption for different systems.

System	E (V)	I (A)	t (min)	V (L)	Log (C_0/C)	EE/O (kWh/m^3)
EC	7.4	0.06	60	0.1	4.585	5.601
EC/PAA	6.1	0.06	60	0.1	1.094	3.345
EC/Fe^{2+}/PAA	6.5	0.06	60	0.1	0.739	0.851

3.7. Proposed Mechanism

Based on the aforesaid outcomes and discussion, the proposed mechanism of PAA activation for the degradation of MB using electrochemical synergy with Fe^{2+} is summarized in Figure 7. Firstly, the PAA could be activated by electron transfer on the cathode of the EC system, generating ·OH, CH_3CO_2·, and CH_3CO_3·. Moreover, the ·OH generated from hydrolysis on the anode could also promote PAA activation. Secondly, the Fe^{2+} could activate PAA to produce ·OH, CH_3CO_2·, and CH_3CO_3·. Meanwhile, the Fe^{3+} could be persistently transformed into Fe^{2+} through the cathode reaction to encourage the circulation of Fe^{3+}/Fe^{2+}, which was conducive to the continuous activation of PAA. In addition, the cathode could donate electrons to stimulate O_2 molecules to produce $O_2^{\cdot-}$, and then generate $^1O_2$. On the other hand, the MB could lose electrons and be degraded through the anode oxidation reaction. Finally, under these radical and non-radical pathways, the dye was eliminated in the EC/Fe^{2+}/PAA synergistic system.

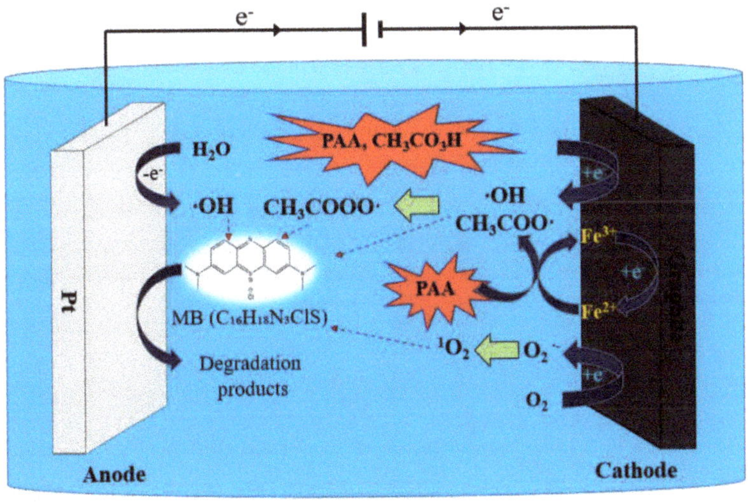

Figure 7. Possible mechanism for EC/Fe^{2+}/PAA system.

4. Conclusions

A remarkable cooperative effect was generated in the electron-Fenton process for PAA activation. The effects of the operational parameters on the MB removal were optimized, and the MB decolorization efficiency reached 98.97% after 30 min under 5.4 mmol L^{-1} PAA, 30 μmol L^{-1} Fe^{2+}, 15 mA cm^{-2} current intensity, and pH 2.9. The scavenging tests proved that ·OH, $CH_3CO_2·$, $CH_3CO_3·$, and $^1O_2$ were the main reactive species for MB removal in the EC/Fe^{2+}/PAA. The PAA decomposition was greatest in the EC/Fe^{2+}/PAA compared with that in the EC and EC/PAA, demonstrating the synergistic effect for PAA activation. The circulation of Fe^{3+}/Fe^{2+} can be maintained significantly with the assistance of the electric field in the EC/Fe^{2+}/PAA system. The calculated EE/O results indicated that the ternary process could reduce the electrical energy consumption considerably compared to that in the EC and EC/PAA. In addition, the EC/Fe^{2+}/PAA process was also used to test four other types of organic pollutants (MO, RhB, TCH, and MTZ), and they all achieved good removal effects. Therefore, the above results indicate that the EC/Fe^{2+}/PAA synergistic system was a potent and dependable AOT. Furthermore, this work provides a simple and practical collaborative process for improving electrochemical oxidation technology for organic wastewater treatment.

Supplementary Materials: The following supporting information can be downloaded at: https://www.mdpi.com/article/10.3390/coatings12040466/s1, Text S1: Materials and Reagents; Text S2: Equipment and Procedure; Figure S1: UV-Vis spectrum for MB removal in EC/Fe^{2+}/PAA system. Experimental conditions: $[MB]_0$ = 20 mg L^{-1}, $[Fe^{2+}]_0$ = 30 μmol L^{-1}, $[PAA]_0$ = 5.4 mmol L^{-1}, $[H_2O_2]_0$ = 11.25 mmol L^{-1}, $[NaNO_3]$ = 30 mmol L^{-1}, pH = 2.9, j = 15 mA cm^{-2}; Figure S2: TOC removal for MB in in EC/Fe^{2+}/PAA system. Experimental conditions: $[MB]_0$ = 20 mg L^{-1}, $[Fe^{2+}]_0$ = 30 μmol L^{-1}, $[PAA]_0$ = 5.4 mmol L^{-1}, $[H_2O_2]_0$ = 11.25 mmol L^{-1}, $[NaNO_3]$ = 30 mmol L^{-1}, pH = 2.9, j = 15 mA cm^{-2}; Figure S3: HPLC-MS analysis of MB before and after reaction.

Author Contributions: Writing—original draft, K.Y.; Conceptualization, D.Y.; Writing—review and editing, D.Y., L.X., Q.H., S.T.; Investigation, E.Z., X.L., M.S.; Methodology, M.S., Q.H.; Validation, L.X.; Formal analysis, Q.H.; Project administration, S.T. All authors have read and agreed to the published version of the manuscript.

Funding: The authors acknowledge the financial support of the National Natural Science Foundation of China (No. 51908485), the Natural Science Foundation of Hebei province (Nos. E2020203185, B2020203033), the Hebei Province Foundation for Returnees (No. C20210502), and the University Science and Technology Program Project of Hebei Provincial Department of Education (No. QN2020143).

Institutional Review Board Statement: Not applicable.

Informed Consent Statement: Not applicable.

Data Availability Statement: Not applicable.

Conflicts of Interest: The authors declare no conflict of interest.

References

1. Cai, Q.Q.; Wu, M.Y.; Hu, L.M.; Lee, B.C.Y.; Ong, S.L.; Wang, P.; Hu, J.Y. Organics removal and in-situ granule activated carbon regeneration in FBR-Fenton/GAC process for reverse osmosis concentrate treatment. *Water Res.* **2020**, *183*, 116119. [CrossRef] [PubMed]
2. Zeng, S.; Smirnov, A.Y.; Borisevich, V.D.; Sulaberidze, G.A. Comparison study of schemes for separation of intermediate isotope components. *Sep. Sci. Technol.* **2020**, *55*, 2092–2098. [CrossRef]
3. Liu, J.; Huang, S.; Wang, T.; Mei, M.; Chen, S.; Li, J. Peroxydisulfate activation by digestate-derived biochar for azo dye degradation: Mechanism and performance. *Sep. Purif. Technol.* **2021**, *279*, 119687. [CrossRef]
4. Ahmad, A.; Khan, M.; Khan, S.; Luque, R.; Abualnaja, K.M.; Alduaij, O.K.; Yousef, T.A. Bio-Construction of CuO Nanoparticles Using Texas Sage Plant Extract for catalytical degradation of Methylene blue Via Photocatalysis. *J. Mol. Struct.* **2022**, *1256*, 132522. [CrossRef]
5. Duan, J.; Pang, S.; Wang, Z.; Zhou, Y.; Gao, Y.; Li, J.; Guo, Q.; Jiang, J. Hydroxylamine driven advanced oxidation processes for water treatment: A review. *Chemosphere* **2021**, *262*, 128390. [CrossRef]

6. Ahamad, T.; Naushad, M.; Alshehri, S.M. Analysis of degradation pathways and intermediates products for ciprofloxacin using a highly porous photocatalyst. *Chem. Eng. J.* **2021**, *417*, 127969. [CrossRef]
7. Stoian, M.; Maurer, T.; Lamri, S.; Fechete, I. Techniques of Preparation of Thin Films: Catalytic Combustion. *Catalysts* **2021**, *11*, 1530. [CrossRef]
8. Yuan, D.; Zhang, C.; Tang, S.; Li, X.; Tang, J.; Rao, Y.; Wang, Z.; Zhang, Q. Enhancing CaO_2 fenton-like process by Fe(II)-oxalic acid complexation for organic wastewater treatment. *Water Res.* **2019**, *163*, 114861. [CrossRef]
9. Ahamad, T.; Naushad, M.; Al-Saeedi, S.I.; Almotairi, S.; Alshehri, S.M. Fabrication of MoS_2/ZnS embedded in N/S doped carbon for the photocatalytic degradation of pesticide. *Mater. Lett.* **2020**, *263*, 127271. [CrossRef]
10. Wang, J.; Zhuan, R. Degradation of antibiotics by advanced oxidation processes: An overview. *Sci. Total Environ.* **2020**, *701*, 135023. [CrossRef]
11. Kibbee, R.; Örmeci, B. Peracetic acid (PAA) and low-pressure ultraviolet (LP-UV) inactivation of Coxsackievirus B3 (CVB3) in municipal wastewater individually and concurrently. *Water Res.* **2020**, *183*, 116048. [CrossRef]
12. Du, P.; Liu, W.; Cao, H.; Zhao, H.; Huang, C. Oxidation of amino acids by peracetic acid: Reaction kinetics, pathways and theoretical calculations. *Water Res. X* **2018**, *1*, 100002. [CrossRef]
13. Zhang, K.; Zhou, X.; Du, P.; Zhang, T.; Cai, M.; Sun, P.; Huang, C. Oxidation of β-lactam antibiotics by peracetic acid: Reaction kinetics, product and pathway evaluation. *Water Res.* **2017**, *123*, 153–161. [CrossRef]
14. Ghanbari, F.; Giannakis, S.; Lin, K.A.; Wu, J.; Madihi-Bidgoli, S. Acetaminophen degradation by a synergistic peracetic acid/UVC-LED/Fe(II) advanced oxidation process: Kinetic assessment, process feasibility and mechanistic considerations. *Chemosphere* **2021**, *263*, 128119. [CrossRef]
15. Da Silva, W.P.; Carlos, T.D.; Cavallini, G.S.; Pereira, D.H. Peracetic acid: Structural elucidation for applications in wastewater treatment. *Water Res.* **2020**, *168*, 115143. [CrossRef]
16. Ao, X.; Eloranta, J.; Huang, C.; Santoro, D.; Sun, W.; Lu, Z.; Li, C. Peracetic acid-based advanced oxidation processes for decontamination and disinfection of water: A review. *Water Res.* **2021**, *188*, 116479. [CrossRef]
17. Wang, J.; Wang, Z.; Cheng, Y.; Cao, L.; Bai, F.; Yue, S.; Xie, P.; Ma, J. Molybdenum disulfide (MoS_2): A novel activator of peracetic acid for the degradation of sulfonamide antibiotics. *Water Res.* **2021**, *201*, 117291. [CrossRef]
18. Wang, J.; Wan, Y.; Ding, J.; Wang, Z.; Ma, J.; Xie, P.; Wiesner, M.R. Thermal Activation of Peracetic Acid in Aquatic Solution: The Mechanism and Application to Degrade Sulfamethoxazole. *Environ. Sci. Technol.* **2020**, *54*, 14635–14645. [CrossRef]
19. Cai, M.; Sun, P.; Zhang, L.; Huang, C. UV/Peracetic Acid for Degradation of Pharmaceuticals and Reactive Species Evaluation. *Environ. Sci. Technol.* **2017**, *51*, 14217–14224. [CrossRef]
20. Zhang, T.; Huang, C. Modeling the Kinetics of UV/Peracetic Acid Advanced Oxidation Process. *Environ. Sci. Technol.* **2020**, *54*, 7579–7590. [CrossRef]
21. Rokhina, E.V.; Makarova, K.; Lahtinen, M.; Golovina, E.A.; Van As, H.; Virkutyte, J. Ultrasound-assisted MnO_2 catalyzed homolysis of peracetic acid for phenol degradation: The assessment of process chemistry and kinetics. *Chem. Eng. J.* **2013**, *221*, 476–486. [CrossRef]
22. Wu, W.; Tian, D.; Liu, T.; Chen, J.; Huang, T.; Zhou, X.; Zhang, Y. Degradation of organic compounds by peracetic acid activated with Co_3O_4: A novel advanced oxidation process and organic radical contribution. *Chem. Eng. J.* **2020**, *394*, 124938. [CrossRef]
23. Wang, J.; Xiong, B.; Miao, L.; Wang, S.; Xie, P.; Wang, Z.; Ma, J. Applying a novel advanced oxidation process of activated peracetic acid by $CoFe_2O_4$ to efficiently degrade sulfamethoxazole. *Appl. Catal. B-Environ.* **2021**, *280*, 119422. [CrossRef]
24. Kim, J.; Zhang, T.; Liu, W.; Du, P.; Dobson, J.T.; Huang, C. Advanced Oxidation Process with Peracetic Acid and Fe(II) for Contaminant Degradation. *Environ. Sci. Technol.* **2019**, *53*, 13312–13322. [CrossRef]
25. Kim, J.; Du, P.; Liu, W.; Luo, C.; Zhao, H.; Huang, C. Cobalt/Peracetic Acid: Advanced Oxidation of Aromatic Organic Compounds by Acetylperoxyl Radicals. *Environ. Sci. Technol.* **2020**, *54*, 5268–5278. [CrossRef]
26. Carlos, T.D.; Bezerra, L.B.; Vieira, M.M.; Sarmento, R.A.; Pereira, D.H.; Cavallini, G.S. Fenton-type process using peracetic acid: Efficiency, reaction elucidations and ecotoxicity. *J. Hazard. Mater.* **2021**, *403*, 123949. [CrossRef]
27. Chen, H.; Lin, T.; Zhang, S.; Xu, H.; Tao, H.; Chen, W. Novel FeII/EDDS/UV/PAA advanced oxidation process: Mechanisms and applications for naproxen degradation at neutral pH and low FeII dosage. *Chem. Eng. J.* **2021**, *417*, 127896. [CrossRef]
28. Xiong, L.; Ren, W.; Lin, H.; Zhang, H. Efficient removal of bisphenol A with activation of peroxydisulfate via electrochemically assisted Fe(III)-nitrilotriacetic acid system under neutral condition. *J. Hazard. Mater.* **2021**, *403*, 123874. [CrossRef]
29. Midassi, S.; Bedoui, A.; Bensalah, N. Efficient degradation of chloroquine drug by electro-Fenton oxidation: Effects of operating conditions and degradation mechanism. *Chemosphere* **2020**, *260*, 127558. [CrossRef]
30. Long, Y.; Feng, Y.; Li, X.; Suo, N.; Chen, H.; Wang, Z.; Yu, Y. Removal of diclofenac by three-dimensional electro-Fenton-persulfate (3D electro-Fenton-PS). *Chemosphere* **2019**, *219*, 1024–1031. [CrossRef]
31. Cai, J.; Zhou, M.; Du, X.; Xu, X. Enhanced mechanism of 2,4-dichlorophenoxyacetic acid degradation by electrochemical activation of persulfate on Blue-TiO_2 nanotubes anode. *Sep. Purif. Technol.* **2021**, *254*, 117560. [CrossRef]
32. Sun, Z.; Li, S.; Ding, H.; Zhu, Y.; Wang, X.; Liu, H.; Zhang, Q.; Zhao, C. Electrochemical/Fe^{3+}/peroxymonosulfate system for the degradation of Acid Orange 7 adsorbed on activated carbon fiber cathode. *Chemosphere* **2020**, *241*, 125125. [CrossRef]
33. Liu, Z.; Ren, B.; Ding, H.; He, H.; Deng, H.; Zhao, C.; Wang, P.; Dionysiou, D.D. Simultaneous regeneration of cathodic activated carbon fiber and mineralization of desorbed contaminations by electro-peroxydisulfate process: Advantages and limitations. *Water Res.* **2020**, *171*, 115456. [CrossRef]

34. Yuan, D.; Yang, K.; Pan, S.; Xiang, Y.; Tang, S.; Huang, L.; Sun, M.; Zhang, X.; Jiao, T.; Zhang, Q.; et al. Peracetic acid enhanced electrochemical advanced oxidation for organic pollutant elimination. *Sep. Purif. Technol.* **2021**, *276*, 119317. [CrossRef]
35. Yang, Z.; Yu, A.; Shan, C.; Gao, G.; Pan, B. Enhanced Fe(III)-mediated Fenton oxidation of atrazine in the presence of functionalized multi-walled carbon nanotubes. *Water Res.* **2018**, *137*, 37–46. [CrossRef]
36. Chen, S.; Cai, M.; Liu, Y.; Zhang, L.; Feng, L. Effects of water matrices on the degradation of naproxen by reactive radicals in the UV/peracetic acid process. *Water Res.* **2019**, *150*, 153–161. [CrossRef]
37. Ahmad, A.; Davarpanah, A.; Thangavelu, L.; Bokov, D.O.; Alshgari, R.A.; Karami, A.M. Self-assembled pine-like CuCo/CP configuration as efficient electrocatalysts toward electrochemical water splitting. *J. Mol. Liq.* **2022**, *351*, 118635. [CrossRef]
38. Rokhina, E.V.; Makarova, K.; Golovina, E.A.; Van As, H.; Virkutyte, J. Free Radical Reaction Pathway, Thermochemistry of Peracetic Acid Homolysis, and Its Application for Phenol Degradation: Spectroscopic Study and Quantum Chemistry Calculations. *Environ. Sci. Technol.* **2010**, *44*, 6815–6821. [CrossRef]
39. Han, S.; Hassan, S.U.; Zhu, Y.; Zhang, S.; Liu, H.; Zhang, S.; Li, J.; Wang, Z.; Zhao, C. Significance of Activated Carbon Fiber as Cathode in Electro/Fe^{3+}/Peroxydisulfate Oxidation Process for Removing Carbamazepine in Aqueous Environment. *Ind. Eng. Chem. Res.* **2019**, *58*, 19709–19718. [CrossRef]
40. Pan, S.; Zhai, Z.; Yang, K.; Xiang, Y.; Tang, S.; Zhang, Y.; Jiao, T.; Zhang, Q.; Yuan, D. β-Lactoglobulin amyloid fibrils supported Fe(III) to activate peroxydisulfate for organic pollutants elimination. *Sep. Purif. Technol.* **2022**, *289*, 120806. [CrossRef]
41. Wang, Z.; Wang, J.; Xiong, B.; Bai, F.; Wang, S.; Wan, Y.; Zhang, L.; Xie, P.; Wiesner, M.R. Application of Cobalt/Peracetic Acid to Degrade Sulfamethoxazole at Neutral Condition: Efficiency and Mechanisms. *Environ. Sci. Technol.* **2020**, *54*, 464–475. [CrossRef]
42. Pueyo, N.; Ormad, M.P.; Miguel, N.; Kokkinos, P.; Ioannidi, A.; Mantzavinos, D.; Frontistis, Z. Electrochemical oxidation of butyl paraben on boron doped diamond in environmental matrices and comparison with sulfate radical-AOP. *J. Environ. Manag.* **2020**, *269*, 110783. [CrossRef]
43. Song, H.; Yan, L.; Wang, Y.; Jiang, J.; Ma, J.; Li, C.; Wang, G.; Gu, J.; Liu, P. Electrochemically activated PMS and PDS: Radical oxidation versus nonradical oxidation. *Chem. Eng. J.* **2020**, *391*, 123560. [CrossRef]
44. Liu, J.; Zhong, S.; Song, Y.; Wang, B.; Zhang, F. Degradation of tetracycline hydrochloride by electro-activated persulfate oxidation. *J. Electroanal. Chem.* **2018**, *809*, 74–79. [CrossRef]
45. Lin, J.; Hu, Y.; Xiao, J.; Huang, Y.; Wang, M.; Yang, H.; Zou, J.; Yuan, B.; Ma, J. Enhanced diclofenac elimination in Fe(II)/peracetic acid process by promoting Fe(III)/Fe(II) cycle with ABTS as electron shuttle. *Chem. Eng. J.* **2021**, *420*, 129692. [CrossRef]
46. Wang, S.; Wang, H.; Liu, Y.; Fu, Y. Effective degradation of sulfamethoxazole with Fe^{2+}-zeolite/peracetic acid. *Sep. Purif. Technol.* **2020**, *233*, 115973. [CrossRef]
47. Cheng, X.; Guo, H.; Zhang, Y.; Wu, X.; Liu, Y. Non-photochemical production of singlet oxygen via activation of persulfate by carbon nanotubes. *Water Res.* **2017**, *113*, 80–88. [CrossRef]
48. Hama Aziz, K.H.; Mahyar, A.; Miessner, H.; Mueller, S.; Kalass, D.; Moeller, D.; Khorshid, I.; Rashid, M.A.M. Application of a planar falling film reactor for decomposition and mineralization of methylene blue in the aqueous media via ozonation, Fenton, photocatalysis and non-thermal plasma: A comparative study. *Process Saf. Environ.* **2018**, *113*, 319–329. [CrossRef]
49. Wolski, L.; Ziolek, M. Insight into pathways of methylene blue degradation with H_2O_2 over mono and bimetallic Nb, Zn oxides. *Appl. Catal. B-Environ.* **2018**, *224*, 634–647. [CrossRef]
50. Wolski, L.; Sobańska, K.; Walkowiak, A.; Akhmetova, K.; Gryboś, J.; Frankowski, M.; Ziolek, M.; Pietrzyk, P. Enhanced adsorption and degradation of methylene blue over mixed niobium-cerium oxide—Unraveling the synergy between Nb and Ce in advanced oxidation processes. *J. Hazard. Mater.* **2021**, *415*, 125665. [CrossRef]
51. Yuan, D.; Sun, M.; Tang, S.; Zhang, Y.; Wang, Z.; Qi, J.; Rao, Y.; Zhang, Q. All-solid-state $BiVO_4/ZnIn_2S_4$ Z-scheme composite with efficient charge separations for improved visible light photocatalytic organics degradation. *Chin. Chem. Lett.* **2020**, *31*, 547–550. [CrossRef]
52. Yuan, D.; Tang, J.; Nie, Z.; Tang, S. Study on humic acid removal in water by ultraviolet activated sodium percarbonate. *J. Yanshan Univ.* **2021**, *45*, 220–226. [CrossRef]
53. Xiang, Y.; Yang, K.; Zhai, Z.; Zhao, T.; Yuan, D.; Jiao, T.; Zhang, Q.; Tang, S. Molybdenum co-catalytic promotion for Fe^{3+}/peroxydisulfate process: Performance, mechanism, and immobilization. *Chem. Eng. J.* **2022**, *438*, 135656. [CrossRef]
54. Yin, K.; Wu, J.; Deng, Q.; Wu, Z.; Wu, T.; Luo, Z.; Jiang, J.; Duan, J. Tailoring micro/nanostructured porous polytetrafluoroethylene surfaces for dual-reversible transition of wettability and transmittance. *Chem. Eng. J.* **2022**, *434*, 134756. [CrossRef]
55. Yang, B.; Zhou, P.; Cheng, X.; Li, H.; Huo, X.; Zhang, Y. Simultaneous removal of methylene blue and total dissolved copper in zero-valent iron/H_2O_2 Fenton system: Kinetics, mechanism and degradation pathway. *J. Colloid Interface Sci.* **2019**, *555*, 383–393. [CrossRef]
56. Yuan, D.; Zhang, C.; Tang, S.; Wang, Z.; Sun, Q.; Zhang, X.; Jiao, T.; Zhang, Q. Ferric ion-ascorbic acid complex catalyzed calcium peroxide for organic wastewater treatment: Optimized by response surface method. *Chin. Chem. Lett.* **2021**, *32*, 3387–3392. [CrossRef]
57. Li, D.; Li, S.; Zhang, S.; Sun, J.; Wang, L.; Wang, K. Aging state prediction for supercapacitors based on heuristic kalman filter optimization extreme learning machine. *Energy* **2022**, *250*, 123773. [CrossRef]
58. Wang, W.; Pang, J.; Su, J.; Li, F.; Li, Q.; Wang, X.; Wang, J.; Ibarlucea, B.; Liu, X.; Li, Y.; et al. Applications of nanogenerators for biomedical engineering and healthcare systems. *InfoMat* **2022**, *4*, e12262. [CrossRef]

59. Kim, J.; Huang, C. Reactivity of Peracetic Acid with Organic Compounds: A Critical Review. *ACS EST Water* **2021**, *1*, 15–33. [CrossRef]
60. Lee, Y.; Lee, G.; Zoh, K. Benzophenone-3 degradation via UV/H_2O_2 and UV/persulfate reactions. *J. Hazard. Mater.* **2021**, *403*, 123591. [CrossRef]
61. Liu, C.; Li, Q.; Wang, K. State-of-charge estimation and remaining useful life prediction of supercapacitors. *Renew. Sustain. Energy Rev.* **2021**, *150*, 111408. [CrossRef]

coatings

MDPI

Review

Recent Advances on Hydrogen Evolution and Oxygen Evolution Catalysts for Direct Seawater Splitting

Linzhou Zhuang, Shiyi Li, Jiankun Li, Keyu Wang, Zeyu Guan, Chen Liang and Zhi Xu *

State Key Laboratory of Chemical Engineering, School of Chemical Engineering, East China University of Science and Technology, Shanghai 200237, China; lzzhuang@ecust.edu.cn (L.Z.); y11200011@mail.ecust.edu.cn (S.L.); y20200023@mail.ecust.edu.cn (J.L.); y30191172@mail.ecust.edu.cn (K.W.); y30210038@mail.ecust.edu.cn (Z.G.); y20190082@mail.ecust.edu.cn (C.L.)

* Correspondence: zhixu@ecust.edu.cn

Abstract: Producing hydrogen via water electrolysis could be a favorable technique for energy conversion, but the freshwater shortage would inevitably limit the industrial application of the electrolyzers. Being an inexhaustible resource of water on our planet, seawater can be a promising alternative electrolyte for industrial hydrogen production. However, many challenges are hindering the actual application of seawater splitting, especially the competing reactions relating to chlorine at the anode that could severely corrode the catalysts. The execution of direct seawater electrolysis needs efficient and robust electrocatalysts that can prevent the interference of competing reactions and resist different impurities. In recent years, researchers have made great advances in developing high-efficiency electrocatalysts with improved activity and stability. This review will provide the macroscopic understanding of direct seawater splitting, the strategies for rational electrocatalyst design, and the development prospects of hydrogen production via seawater splitting. The non-precious metal-based electrocatalysts for stable seawater splitting and their catalytic mechanisms are emphasized to offer guidance for designing the efficient and robust electrocatalyst, so as to promote the production of green hydrogen via seawater splitting.

Keywords: seawater splitting; catalyst; oxygen evolution reaction; hydrogen evolution reaction; hypochlorite evolution reaction; anti-corrosion

Citation: Zhuang, L.; Li, S.; Li, J.; Wang, K.; Guan, Z.; Liang, C.; Xu, Z. Recent Advances on Hydrogen Evolution and Oxygen Evolution Catalysts for Direct Seawater Splitting. *Coatings* **2022**, *12*, 659. https://doi.org/10.3390/coatings12050659

Academic Editor: Marco Laurenti

Received: 28 March 2022
Accepted: 10 May 2022
Published: 12 May 2022

1. Introduction

With the increasing fossil energy depletion, environmental pollution, and global warming, the development of green and renewable resources, such as wind energy, hydro energy, solar energy, and ocean energy has become a topic that countries around the world attach great importance to. Most renewable energy has the inherent problem of being intermittent, random, and volatile, thus resulting in serious abandonment of wind, light, and water. As a storable energy source, hydrogen energy has continuously promoted the transformation of traditional fossil energy into green energy by virtue of its advantages, such as flexibility, high efficiency, being carbon-free, safe, renewable, and high energy density (140 MJ/kg), which is 3 times that of oil and 4.5 times that of coal. It is regarded as a secondary energy with the greatest development potential in the future energy transformation strategy [1]. At present, the hydrogen energy industry has achieved rapid development in China and many countries around the world. In 2020, China's central government and local governments have launched more than 40 hydrogen energy industry plans and special projects, such as "Made in China 2025". The relevant key industries are included in the "14th Five-Year Plan". At the same time, the United States, Japan, South Korea, the Netherlands, Germany, France, and other EU countries have successively issued various hydrogen energy development roadmaps and hydrogen energy strategic plans [2].

Hydrogen, as the core foundation of the upstream industry chain of hydrogen energy, has many preparation methods and a wide range of sources. Hydrogen is mainly produced by the reforming of natural gas, as well as gasification of coal and crude oil. Water splitting to produce hydrogen (and oxygen) has the advantage of producing high-purity hydrogen, but its applications are often constrained to a small scale. To date, scientists have developed various technologies to improve electrolyzers' performance and decrease their costs, including the proven alkaline electrolysis, proton exchange membrane (PEM) electrolysis, the newly developed anion exchange membrane (AEM) electrolysis, and the still immature high-temperature solid oxide electrolysis. These electrolyzers all possess cathode, which hydrogen evolution reaction (HER) processes, and anode, which oxygen evolution reaction (OER) causes. While readily available in the laboratory, a larger quantity of freshwater feeds could be an inevitable bottleneck for low-temperature water electrolyzer technologies if they would be applied widely in hot arid regions around the world, as the hot arid regions have poor access to freshwater. Unlike the scarce freshwater, seawater and the oceans represent about 96% of the total water reserves on our planet and thus are an almost unlimited resource. Therefore, water electrolyzing with eawater as direct feed could be a remarkable path for green hydrogen production [3]. Therefore, the development of high-efficiency seawater electrolysis hydrogen production technology can not only alleviate the shortage of freshwater resources but also promote the development of hydrogen energy economy and utilization of marine resources.

However, because the composition of seawater is much more complex, compared with the direct electrolysis of freshwater, seawater electrolysis faces more severe challenges. On the one hand, cations such as cobalt and magnesium in seawater are prone to generate insoluble substances during the electrolysis process on the surface of the catalyst [4], thereby reducing the reactivity of the catalyst [5,6]. On the other hand, seawater is rich in various electrochemically active anions, and these anions will directly interact with the anode during seawater electrolysis to compete with OER. Even the composition of seawater may vary greatly in different regions; the total ion concentration in seawater is always maintained at around 3.5% with pH of around 8 [7–10]. For the HER at the cathode, the lifetime of the catalyst is an unavoidable challenge in seawater electrolysis. This is due to the drastic increase in local pH near the electrode when HER occurs at the cathode. When the pH is greater than 9.5, calcium and magnesium ions in seawater will combine with hydroxide to form insoluble substances, such as calcium oxide and magnesium hydroxide. These impurities are easily attached to the surface of the catalyst, thereby hindering the direct contact between the electrolyte and the active site of the catalyst and thus severely reducing the service life of the catalyst in seawater electrolysis. Most of the reported HER catalysts would decay by more than 50% in seawater electrolysis within 24 h [11]. Besides, according to the standard redox potential of each ion in seawater, the oxidation of Cl^- or Br^- in seawater may compete with OER. However, since the content of Br^- ions in seawater is as low as 0.00087 mol L^{-1}, its effect can thus be almost negligible, while the concentration of Cl^- in seawater is as high as 0.5–0.6 mol L^{-1} [12]. Therefore, the oxidation reaction of Cl^- (chlorine evolution reaction, CER) during seawater electrolysis would easily compete with OER [13]. A comprehensive analysis of seawater electrolyzer chemistry at the anode has been presented by Dionigi et al. [14], and based on it, the chlorine evolution reaction (ClER) in an acidic solution and the hypochlorite formation (HCER) in an alkaline electrolyte could be the main competitive reactions of OER. The following two equations exhibit the chloride-related reactions in the electrolytes with low and high pH, respectively:

$$\text{ClER: } 2Cl^- \rightarrow Cl_2 + 2e^- \quad E^0 = 1.36 \text{ V vs. SHE, at pH} = 0 \tag{1}$$

$$\text{HCER: } Cl^- + 2OH^- \rightarrow ClO^- + H_2O + 2e^- \quad E^0 = 0.89 \text{ V vs. SHE, at pH} = 14 \tag{2}$$

During the CER process, the toxic chlorine gas or hydrochloric acid would be generated, which could endanger life safety and corrode reaction equipment, bringing about great production security risks. Therefore, the development of efficient catalysts for HER and OER in seawater is very important for reducing the power loss and improving the energy conversion efficiency in the hydrogen process.

In this review, we offer an overview concerning the development of different types of seawater electrocatalysts based on noble and nonprecious metals in the last few years. Moreover, we also highlight the applicable strategies to improve the activity, selectivity, and stability of the catalysts, as well as the major challenges and perspectives for the development of seawater electrolysis that could be practically applied.

2. HER Catalysts

Platinum (Pt) exhibits promising performance for HER in both alkaline and acidic seawater electrolyte. Many research groups have studied the HER performance of Pt in the electrolyte of the whole pH range. Other noble metals or alloys between noble metals and nonprecious metals have also attracted great interest owing to their promising binding strength of hydrogen and high stability in cruel conditions. Researchers have also developed nonprecious-metal-based catalysts to decrease the catalyst cost, reduce the energy barrier for initiating HER in tough conditions, and improve the ability of anti-fouling in seawater.

2.1. Precious Metal

Yang et al. strongly anchored a trace amount of precious metals on an anti-corrosion matrix (Figure 1a) [15]. For instance, the as-produced Pt/Ni-Mo needed an overpotential as low as 113 mV to obtain a high current density of 2000 mA cm^{-2} in the saline–alkaline electrolyte, revealing its excellent HER performance. It showed high activity and robust durability in the strongly alkaline seawater and at temperatures up to 80 °C. Moreover, it could be mass prepared at a low cost. Mu et al. designed an ultra-low Ru-incorporated cobalt-based oxide, denoted as Ru-Co_x/NF [16], effectively driving the electrolysis of water at high current densities in alkaline water and seawater. The electrolyzer apparatus assembled with Ru-Co_x/NF only required a low voltage of 2.2 and 2.62 V to achieve 1000 mA cm^{-2} in alkaline freshwater and seawater. Lee et al. proposed a catalyst of Co-CoO heterostructures incorporated with Rh atoms for efficient HER and OER catalysis in both seawater and freshwater [17]. Li et al. prepared a Pt-IrO_2/CC electrocatalyst modified by cyclic voltammetry, which reduced some metal cations in metal oxides (Figure 1b) [18]. This modification method resulted in the accumulation of negative charges at the active site of the metal, which significantly accelerated the HER. Notably, its precious metal load could be kept low at 36.6 μg $cm^{-2}{}_{(Ir+Pt)}$.

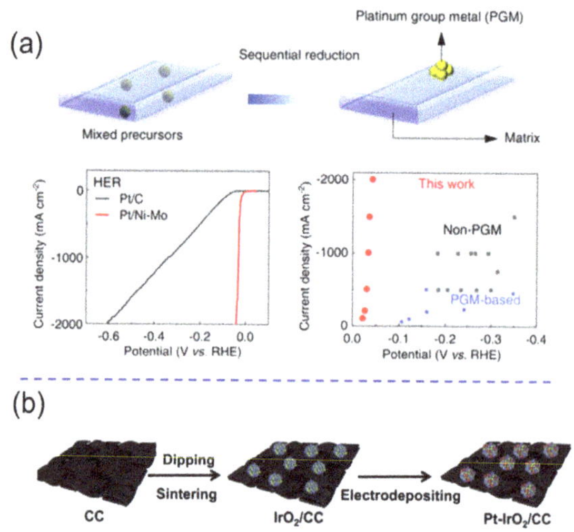

Figure 1. (**a**) The preparation schematic of Pt/Ni-Mo and its HER activity in alkaline seawater [15]; (**b**) illustration of the synthetic route of the Pt-IrO_2/CC [18].

2.2. Transition Metal Sulfide

Sun et al. designed the Cu_2S@Ni nanorod arrays for alkaline seawater, possessing a HER current density as large as ~500 mA cm^{-2} at the overpotential of <200 mV and a favorable stability [19]. The Ni-S interaction between the Ni surface and Cu_2S core could optimize the hydrogen adsorption energy, so as to improve the HER activity of Cu_2S@Ni. Zhao et al. prepared the Co-doped VS_2 nanosheets through a facile one-pot solvothermal method and annealing process (Figure 2). It was demonstrated that Co replacing V in VS_2 nanosheets could lead to abundant S defects and rich active edge sites, and thus it exhibited the Tafel slope of 214 mV dec^{-1} and stable operation duration of 12 h test in the alkaline seawater [20].

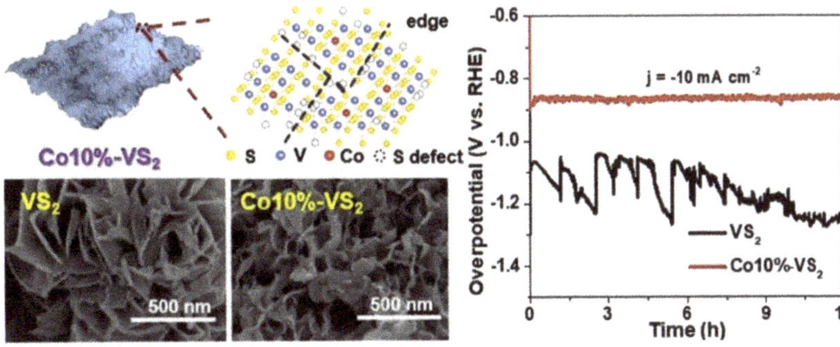

Figure 2. Preparation of Co-doped VS_2 nanosheets and its HER activity in alkaline seawater [20].

2.3. Transition Metal Selenide

Zhao et al. designed a cobalt selenide sample consisting of CoSe and Co_9Se_8 through calcining the Co foil with Se powder [21]. The Co charge state and HER activity of the catalysts could be manipulated by controlling the mass ratio of Co to Se. The authors found that a high Co charge state could favor the OER performance, while a low Co charge state could promote HER activity. Moreover, a current density of 10.3 mA cm^{-2} could be achieved at 1.8 V for overall seawater electrolysis [21].

2.4. Transition Metal Phosphide

Xu et al. synthesized the nanoporous C-Co_2P material on the basis of the Co-P-C precursor alloy by electrochemical etching (Figure 3a) [22]. The performance of C-Co_2P material showed the overpotential of 30 mV at 10 mA cm^{-2} in 1.0 M KOH. The authors concluded that doping C atoms could adjust the electronic structure of Co_2P to form C–H_{ad} intermediate that was conducive to hydrogen desorption [22]. Huang et al. prepared $Ni(OH)_2{\cdot}0.75H_2O$ as the precursor by hydrothermal method and then annealed $Ni(OH)_2{\cdot}0.75H_2O$ with $NaH_2PO_2H_2O$ in nitrogen atmosphere to form Ni_5P_4 and $Ni^{2+\delta}(OH)_{2-\delta}$, and finally, they synthesized Ni_5P_4@$Ni^{2+\delta}(OH)_{2-\delta}$ (NPNNS) (Figure 3b) [23]. Ni_5P_4 and $Ni^{2+\delta}(OH)_{2-\delta}$ can synergistically inhibit the properties of P-Hads bond, which impaired the HER activity, and thus, NPNNS achieved the overpotential of 87, 144, and 66 mV at 10 mA cm^{-2} in alkaline freshwater, alkaline seawater, and acidic electrolyte [23]. Wang et al. prepared a catalyst (Co-Fe_2P)-doped transition metal phosphide in Ni foam via a hydrothermal method, which showed promising activity for both HER and OER without being corroded by chloride ions (Figure 3c) [24]. In particular, Co-Fe_2P achieved overpotentials of 138 and 221 mV at 100 mA cm^{-2} in 1 M KOH and simulated seawater for HER [24].

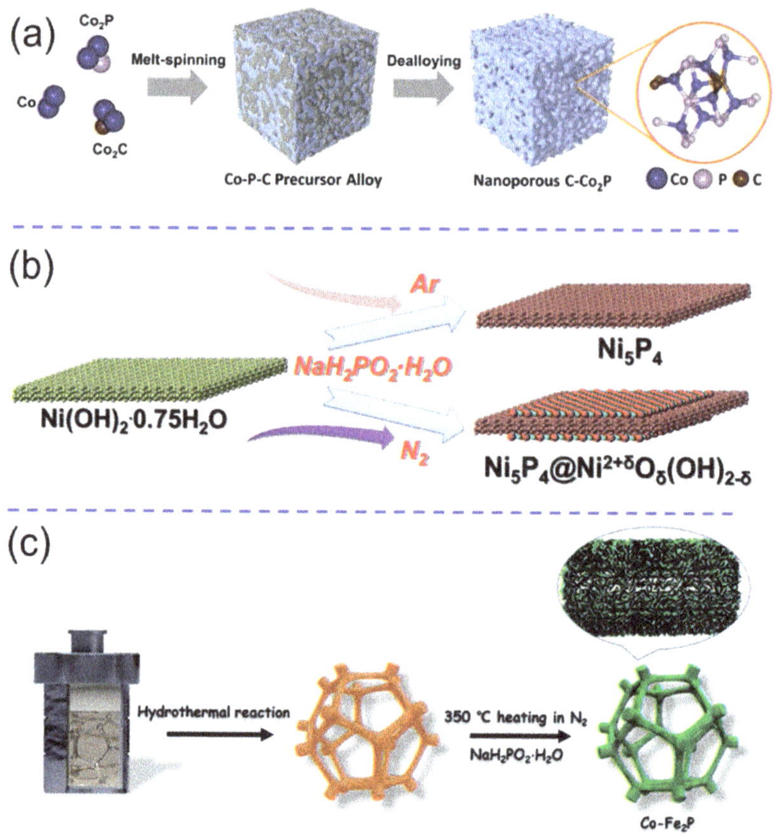

Figure 3. (**a**) The preparation procedure of C-Co_2P electrocatalysts [22]; (**b**) schematic illustration of NPNNS hybrid nanosheets [23]; (**c**) the synthesis process of the Co-Fe_2P electrocatalyst [24].

2.5. Transition Metal Carbide and Nitride

Yu et al. constructed a nanostructured NiCoN | Ni_xP | NiCoN catalyst HER catalysis [25], which exhibited a good HER activity and impressive stability owing to large surface area that exposed abundant active sites, high electrical conductivity, and improved intrinsic activity. It required a small overpotential of 165 mV to achieve 10 mA cm^{-2} in the seawater electrolyte. Wu et al. developed a multifunctional catalytic interface to propel HER in the electrolytes of various pH and seawater (Figure 4a) [26]. The catalytic interfaces among MXene, bimetallic carbide, and hybridized carbon endowed the prepared electrocatalysts with HER activity that was comparable to the commercial Pt/C in both 1.0 M KOH and 0.5 M H_2SO_4 and even outperformed it under pH 2.2–11.2. Zang et al. demonstrated that the Ni-N_3 catalyst with the atomically dispersed Ni in the triple nitrogen coordination could possess efficient HER activity in alkaline electrolyte (Figure 4b) [27]. It delivered a current density as large as 200 mA cm^{-2} at a lower overpotential than Pt/C, with no activity decay over 14 h. Qiu et al. reported a NiCo/Mxene-based catalyst for chlorine-free hydrogen production [28]. It required an overpotential of 49 and 235 mV to obtain 10 and 500 mA cm^{-2} in 1.0 M KOH. Moreover, NiCo@C/Mxene/CF could operate stably over 120 h for HER catalysis in seawater, demonstrating its robust stability in corrosive seawater.

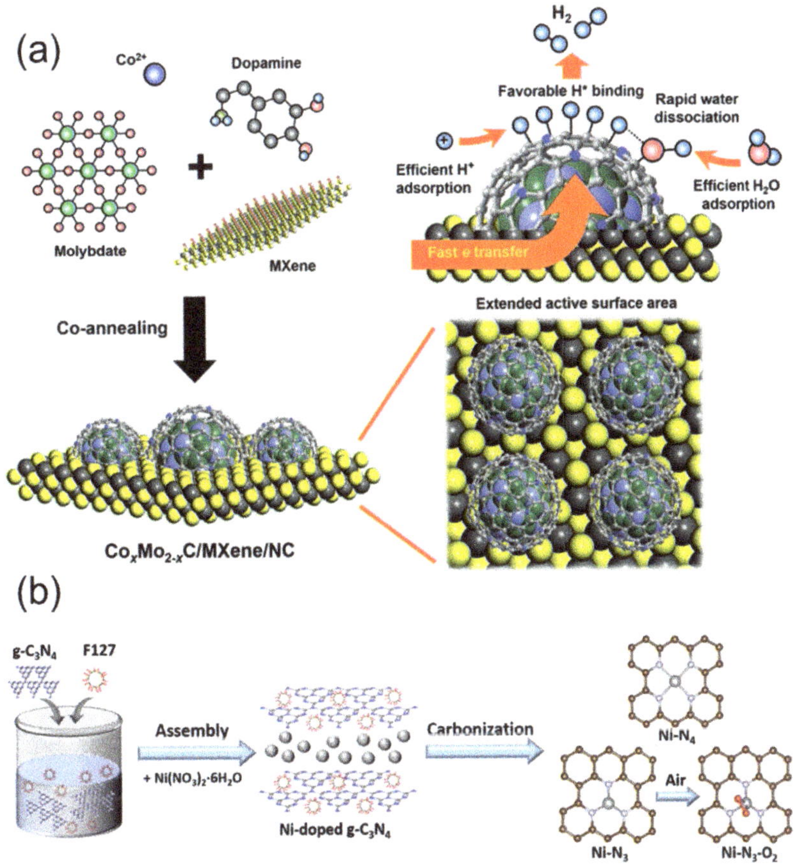

Figure 4. (**a**) The preparation strategy of the $Co_xMo_{2-x}C$/MXene/NC catalyst [26]; (**b**) preparation and analysis of Ni-SA/NC [27].

2.6. Transition Metal Alloys

Zhang et al. reported the anchoring of molybdenum-oxo functional groups on Cu substrates by in situ electrochemical reduction in CMO films [29]. The immobilized Cu plate achieved 150, 220, and 195 mA cm^{-2} at an overpotential of 270 mV in 0.5 M PBS, 1 M KOH, and 0.5 M H_2SO_4 for HER [29]. Lu et al. established a manganese-doped nickel/nickel oxide catalyst on Ni foam through pyrolyzing Mn-based MOF [11]. In the carbonization process of Mn-MOF, Ni atoms that form the Ni foam could interact with Mn to generate a Mn-NiO-Ni heteroatom structure, leading to a strong binding between the catalyst and substrate. The Mn-doped Ni/NiO catalyst exhibited a fantastic HER activity in natural seawater, close to a commercial Pt/C catalyst, with a low overpotential of −0.17 V at 10 mA cm^{-2}. Durability was tested under the current density of 66.7 mA cm^{-2}, and it could maintain the initial potential for nearly 14 h. In addition, after cleaning the deposition on the catalyst surface, it could recover high activity [11]. Yuan et al. fabricated the NiMo film catalyst on Ni foam through the electrodepositing method. The oxidation state of Ni and Mo could be modulated by tuning the composition of aqueous solution, which can affect the catalytic activity of the NiMo film [30]. The NiO species of NiMo film is conducive to the subsequent H adsorption/desorption during the HER process, so that a small overpotential of 31.8 mV could be attained in the 1 M KOH + 0.5 M NaCl solution. Furthermore, the NiMo film also exhibited excellent durability, with negligible increase in overpotential during the tests for 15 h [30]. Ros et al. loaded the Ni-Mo-Fe electrocatalyst on carbon to resist the corrosion of Cl^- in seawater. The Ni-Mo-Fe-based catalyst reconfigured into Ni-Mo-Fe alloy and Ni:$Fe(OH)_2$ redeposits in alkaline seawater under cathodic bias [31]. Additionally, high HER activity was achieved due to the increased hydrogen adsorption strength and the suitable electronic structure of Mo surface species. In addition, the Tafel slope was 256.1, 234.2, and 201.6 mV dec^{-1} in alkaline saline pure water, alkalinized seawater, and filtered seawater [31].

3. OER Catalysts

A large amount of Cl^- exists in seawater, and the competitive CER induced by it will reduce the Faradaic efficiency of the OER. In addition, calcium and magnesium ions abundant in seawater could be easily deposited on the cathode catalyst, resulting in catalyst deactivation. In this regard, researchers have developed indirect seawater electrolysis technology, that is, first purifying seawater through a reverse osmosis membrane and then electrolyzing the obtained freshwater to produce hydrogen and oxygen. However, the reverse osmosis treatment will produce a large amount of concentrated salt brine, and its discharge will seriously harm the marine ecological environment. At the same time, the reverse osmosis membrane requires regular maintenance during the operation process, which increases the operating cost. In contrast, the direct electrolysis of seawater for hydrogen production has the advantages of low investment cost, few system engineering problems, and small device footprint, but the key is to develop high-efficiency OER catalysts with high selectivity and high stability. Numerous catalysts based on precious metals or nonprecious metals have been reported.

3.1. Precious Metal

Du et al. prepared IrO_x-Cs@$BaCO_3$ [32] whose charge transfer interaction between IrO_x-Cs and $BaCO_3$ could force a more negative charge on $BaCO_3$, so as to repel Cl^- anions in seawater. The IrO_x-Cs@$BaCO_3$ with low content of Ir achieved a high mass activity of 1402 A g^{-1} in the real seawater, and its stability is close to that of IrO_2 [32]. Haq et al. embedded Au-modified Gd-Co_2B nanosheets into TiO_2 nanosheets grown on Ti chaff (Au-GdCo_2B@TiO_2) (Figure 5a) [33]. Because of the hydrophilic bimetallic boride structure, Au-GdCo_2B@TiO_2 possessed high density of active sites and thus achieved 500 and 1000 mA cm^{-2} at the overpotentials of 300 and 510 mV in alkaline seawater [33]. Ko et al. coordinated Ir atoms with heteroatoms, and even with a decreased loading of Ir to less than 6 wt% [34], the electrocatalyst could still show a low overpotential of 243 mV at 10 mA cm^{-2} in 0.1 M $HClO_4$ + 3.5 wt% NaCl electrolysis. Meanwhile, in neutral synthetic seawater, the OER showed a more favorable kinetics than CER, with nearly 560 mV overpotential [34]. Gayen et al. prepared the $Pb_2Ru_2O_{7-x}$ catalyst [35] and assembled it into a membrane electrode assembly as an anode catalyst with commercial Pt/C as a cathode catalyst. The electrolyzer obtained 275 mA cm^{-2} at a cell voltage of 1.80 V in the electrolyte of pH = 13 (Figure 5b). XPS confirmed that the Ru (V) oxidation state could stabilize OER intermediates through the formation of metal–oxygen bonds of large strength, while the quenching of oxygen vacancies existent in $Pb_2Ru_2O_{7-x}$ could facilitate water dissociation [35].

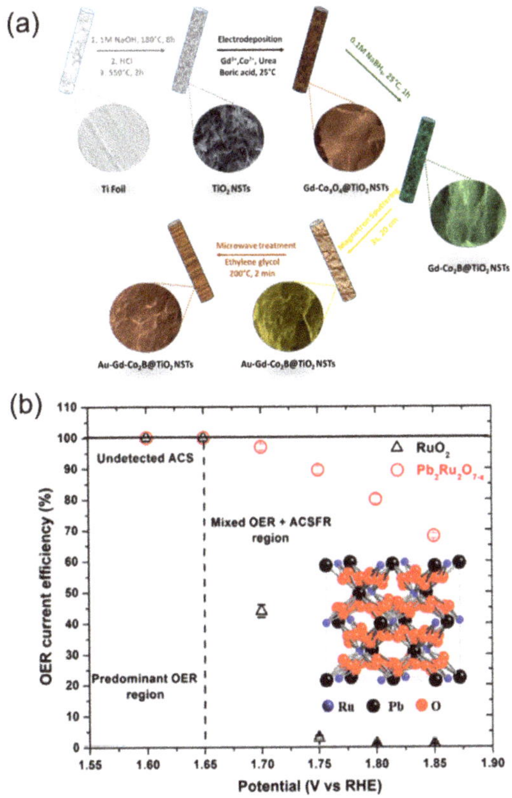

Figure 5. (**a**) Systematic illustration of the fabrication of Au-Gd-$Co_2B@TiO_2$ [33]; (**b**) the OER activity of $Pb_2Ru_2O_{7-x}$ in alkaline simulated seawater [35].

3.2. Transition Metal Oxide/Hydroxides

Ho et al. prepared the NiFe LDH nanosheet on a carbon cloth (CC) [24], and the NiFe-LDH/CC with a Ni/Fe ratio of 4:6 exhibited promising OER activity, reaching 10 mA cm^{-2} at the overpotential of 226 mV and 238 mV in freshwater and seawater, respectively. Meanwhile, the catalyst could stably operate at 100 mA cm^{-2} for over 450 h in freshwater and 165 h in seawater. Ren et al. developed an Fe^{2+}-driven spontaneous method of fabricating NiFe LDH [36], and the NiFe LDH could achieve 100 and 500 mA cm^{-2} at the low overpotentials of 247 and 296 mV, respectively [36]. Badreldin et al. reported the preparation of S, B-(CoFeCr) and S, B-(CoFeV) oxyhydroxides for OER in alkaline and neutral-pH saline water (Figure 6a) [37]. The stronger localized $Cr^{\delta+}(OOH)^{\delta-}$ bonds could form an electrostatic shielding layer to impede the anionic Cl^- attack. The S, B cooping could strongly enhance the hydrophilicity of catalysts, and S, B-(CoFeCr)OOH and S, B-(CoFeV)OOH could achieve 100 mA cm^{-2} at the overpotentials of 353 and 408 mV, respectively [37]. Furthermore, the catalysts loaded on Ni foam could be maintained at a fixed current density of 50 mA cm^{-2} in neutral pH saline environment for 50 h. Badreldin et al. also reported the fabrication of S, B-co-doped CoFe oxyhydroxide (Figure 6b) [38]. The as-prepared S,B-(CoFe)OOH-H showed a high OER selectivity of ~97% in alkaline and ~91% in neutral electrolyte at the current density of 10 mA cm^{-2}. Li et al. reported a simple dipping-and-heating method to synthesize FeNi oxide layer on Ni foam (FEN300) [39]. FEN300 could achieve 1000 mA cm^{-2} at the overpotential of 291 mV in the electrolyte of 1.0 M KOH + seawater. Meanwhile, its current density could be maintained with no decay for 50 h [39]. DFT calculation results indicated that modifying the electronic structure via formation of the interface between NiO and Fe_2O_3 could optimize the intermediate adsorption. The e^--e^- repulsion between Ni^{2+} and oxygen intermediates adjusted and optimized the interaction of Fe with oxygen intermediates. Haq et al. prepared the Gd-Mn_3O_4@CuO-$Cu(OH)_2$ catalyst to achieve the promising OER catalysis with freestanding amorphous nanostructure [40]. The surface oxygen vacancies could modulate the electronic structure of catalytic sites and optimize the reaction intermediates' adsorption energy. Meanwhile, the hierarchical surface structure possessed high conductivity, large surface-specific area, intrinsic activity, ionic mobility, and efficient charge transfer. Its potential to deliver 500 mA cm^{-2} was only 1.63 V (vs. RHE) in alkaline seawater, and its activity stability was maintained for over 75 h without any hypochlorite detected.

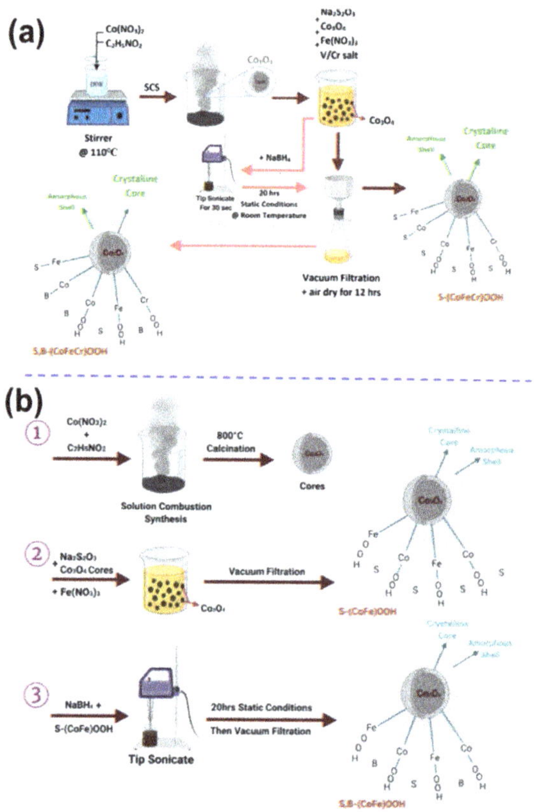

Figure 6. (**a**) The preparation of S, B-(CoFeCr)OOH and S, B-(CoFeV)OOH [37]; (**b**) preparation schematic of S,B-co-doped CoFe oxyhydroxide [38].

Abe et al. synthesized a manganese oxide film with rich oxygen vacancies for OER in neutral brine [41]. The catalyst was prepared by intercalating the layered manganese dioxide with Na^+ ions (Na | MnO_2) and then calcinating in the air at above 300 °C (Figure 7a). XPS results demonstrated that its oxygen vacancies began to form at 200 °C, while the valence state of Mn in the oxide clearly decreased. The Na | MnO_x catalyst treated at 400 °C displayed an OER selectivity of 87% at 10 mA cm^{-2}. Dresp et al. prepared the NiFe-LDH electrocatalysts by a microwave-assisted solvothermal route [42], which could maintain its activity for 100 h at 200 mA cm^{-2} in simulated seawater. The Faraday efficiency of the NiFe-LDH electrocatalysts was up to 88% [42]. Chen et al. took advantage of the rough surface and hydrophilic properties of wood aerogel and successfully attached the NiMoP alloys to wood aerogel (Figure 7b) [43]. NiMoP was then activated to form S,P-(Ni,Mo,Fe)OOH nanolayers. The overpotential of S,P-(Ni,Mo,Fe)OOH/NiMoP/wood aerogel was as low as 297 mV in the alkaline seawater to achieve 500 mA cm^{-2} [43]. Li et al. attached Fe(Cr)OOH and Fe_3O_4 onto Ni foam substrate to form the Fe(Cr)OOH/Fe_3O_4/NF electrocatalyst (Figure 7c). The synergistic effect among FeOOH, Fe_3O_4, and the doped Cr atoms endowed Fe(Cr)OOH/Fe_3O_4/NF with excellent OER activity, which achieved overpotentials of 198 and 241 mV at 10 and 500 mA cm^{-2} in 1 M KOH. Fe(Cr)OOH/Fe_3O_4/NF could work stably at 100 mA cm^{-2} for 100 h in alkaline seawater with no activity deterioration [44]. Wu et al. synthesized the CoP_x | | CoP_x@FeOOH electrocatalyst with core–shell structure for OER in seawater splitting [45], which possessed a high density of active centers and a stable structural strength. CoP_x | | CoP_x@FeOOH achieved the overpotentials of 480 and 637 mV at current densities of 100 and 500 mA cm^{-2} in 1 M KOH + simulated seawater and showed no current attenuation for 80 h of operation without being corroded by chloride ions [45]. Jung et al. took the porous carbon scaffold as the base and uniformly attached the NiFe-LDH nanosheets to the scaffold by chemical vapor deposition (CVD) and finally synthesized NiFe-LDH-S350 after sulfurization [46]. The complex two-dimensional nanosheet structure of NiFe-LDH-S350 could achieve the overpotential of 296 mV at 100 mA cm^{-2} in the electrolyte of 1 M KOH + 0.5 M NaCl for OER [46]. Gao et al. prepared the Ni/α-Ni$(OH)_2$ heterostructure featuring the karst landform on Ni foam substrate via a facile chemical and electrochemical corrosion method. The karst morphology expanded the substrate's surface area, so as to promote the exposure of active sites, while the catalysts attached better to the substrate to possess low charge resistance and impressive durability. Therefore, the Ni/α-Ni$(OH)_2$ catalyst showed high activity, with the overpotential of 560 mV, to achieve the OER current density of 10 mA cm^{-2} in natural seawater. Moreover, its cell voltage could remain steady for 24 h owing to the protection of polyatomic sulfate layers from Cl^- corrosion [47].

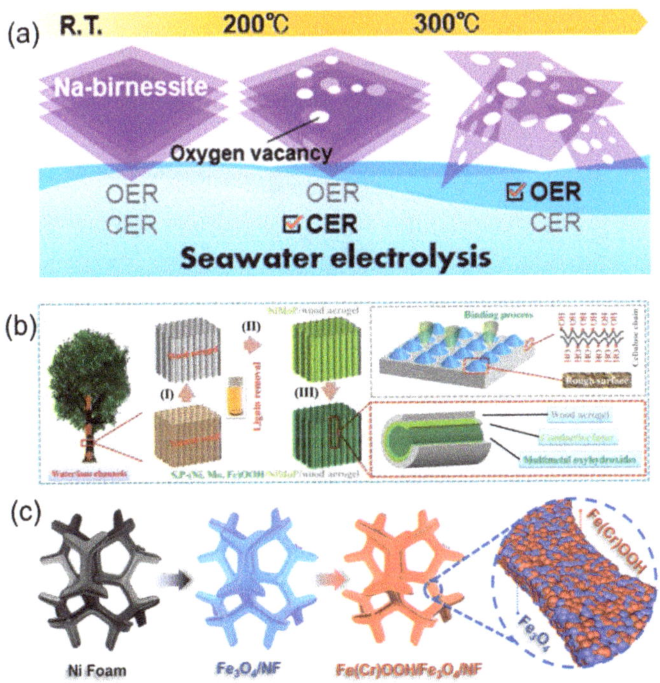

Figure 7. (**a**) The preparation and modulation of Na | MnO_2 followed by heat treatment [41]; (**b**) preparation procedure of the S,P-(Ni,Mo,Fe)OOH/NiMoP/wood [43]; (**c**) preparation of the Fe(Cr)OOH/Fe_3O_4/NF [44].

3.3. Transition Metal Phosphate

Song et al. used $CoCl_2$ and $NH_4H_2{\cdot}PO_4$ as a metal precursor and a phosphate precursor to synthesize the $NH_4CoPO_4{\cdot}H_2O$ nanosheet catalyst by chemical precipitation method (Figure 8a) [48]. In this process, Co (II) ions were activated to Co (III) species and became the active site of OER (Figure 8b). Due to its unique folding form, the two-dimensional nanosheet exhibited a large specific surface area, so as to load the high density of active centers. The overpotential of $NH_4CoPO_4{\cdot}H_2O$ was 252 mV and 268 mV at 1 M KOH + seawater to achieve 10 and 100 mA cm^{-2}, respectively [48]. Cong et al. accomplished $Co_{0.4}Ni_{1.6}P$ nanowire arrays derived from NiCo-MOF [49] and then deposited CeO_2 nanoparticles on its surface to construct a novel heterostructure (Figure 8c). In this process, rich oxygen vacancies were generated to promote the catalytic activity. When testing OER activity in the alkaline simulated seawater, an overpotential as low as 345 and 394 mV was needed to achieve 10 and 100 mA cm^{-2}, respectively. The catalysts also showed excellent durability because of the surface CeO_2 nanoparticles [49].

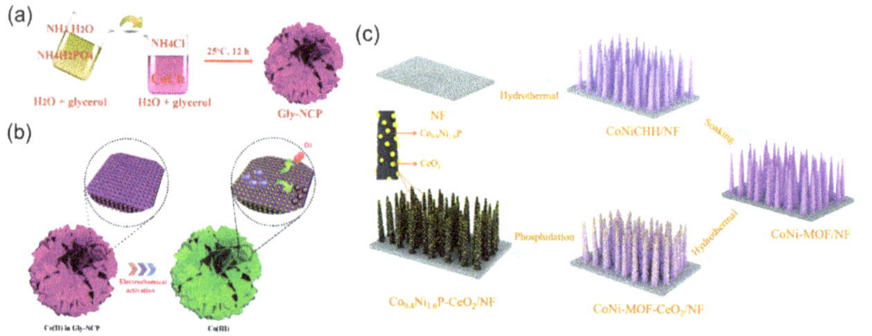

Figure 8. (**a**) The synthesis procedure of Gly-NCP nanosheets [48]; (**b**) the electrochemical activation process of Gly-NCP for improved OER activity [48]; (**c**) the schematic diagram for the preparation of $Co_{0.4}Ni_{1.6}P$-CeO_2/NF [49].

3.4. Transition Metal Chalcogenides

Song et al. synthesized a MoS_2-$(FeNi)_9S_8$ on Ni-Fe foam (MoS_2-$(FeNi)_9S_8$/NFF) for alkaline seawater oxidation (Figure 9a) [50]. In alkaline seawater, MoS_2-$(FeNi)_9S_8$/NFF could reach 100 and 500 mA cm^{-2} at the overpotential of 238 and 284 mV, respectively [50]. Moreover, it could operate stably at 50 mA cm^{-2} for over 70 h, demonstrating its promising stability. Its remarkable catalytic performance and stability should be attributed to the MoS_x layer with plenty of defects, which could optimize the adsorption strength of H*/OH* intermediates, as well as the strong interaction between S and Ni/Fe, which could enhance the corrosion resistance ability of the substrate [50]. Chang et al. reported the Fe,P-$NiSe_2$ NFs for high-efficiency direct seawater electrolysis (Figure 9b) [51]. The doped Fe atoms were identified as the HER active sites, while the adjacent Ni atoms as OER active sites. Consequently, a large current density of 0.8 A cm^{-2} could be achieved at 1.8 V, with the high OER selectivity and long-term stability for over 200 h. Hu et al. reported a Ni_3S_2 nanoarray-decorated Fe-$Ni(OH)_2$ for seawater oxidation [52]. The resulting Fe-$Ni(OH)_2$/Ni_3S_2 required a low overpotential of 269 mV to achieve 10 mA cm^{-2} and showed negligible activity deterioration at 10 mA cm^{-2} for 100 h. The DFT calculations indicated that the Fe sites were more favorable to OER than CER. Wang et al. uniformly grew NiCoS nanosheets onto Ni foam to form 3D self-supported catalyst (3D NiCoS NSAs) for alkaline seawater electrolysis [53]. Three-dimensional NiCoS NSAs had a high density of active sites and could effectively resist the corrosion by chloride ion. It achieved 500 and 1000 mA cm^{-2} at the overpotentials of 440 and 470 mV, respectively, in alkaline seawater. Three-dimensional NiCoS NSAs can maintain the current density of 800 mA cm^{-2} for 100 h in alkaline seawater with no current attenuation [53]. Yu et al. reported a facile method to synthesize S-(Ni,Fe)OOH on Ni foam within a few minutes [54]. It was demonstrated that the S atom was incorporating in the crystal lattice, and the ultra-fast etching process led to a porous structure with huge exposure area and abundant active sites. In addition, the catalyst showed promising activity, needing the overpotential of only 378 and 462 mV to reach 1000 mA cm^{-2} in 1 M KOH and 0.5 M NaCl + 1 M KOH, respectively. The S-(Ni,Fe)OOH catalyst exhibited almost no activity decay during the stability test, which could be related to the tight construction between S-(Ni,Fe)OOH and Ni foam [54].

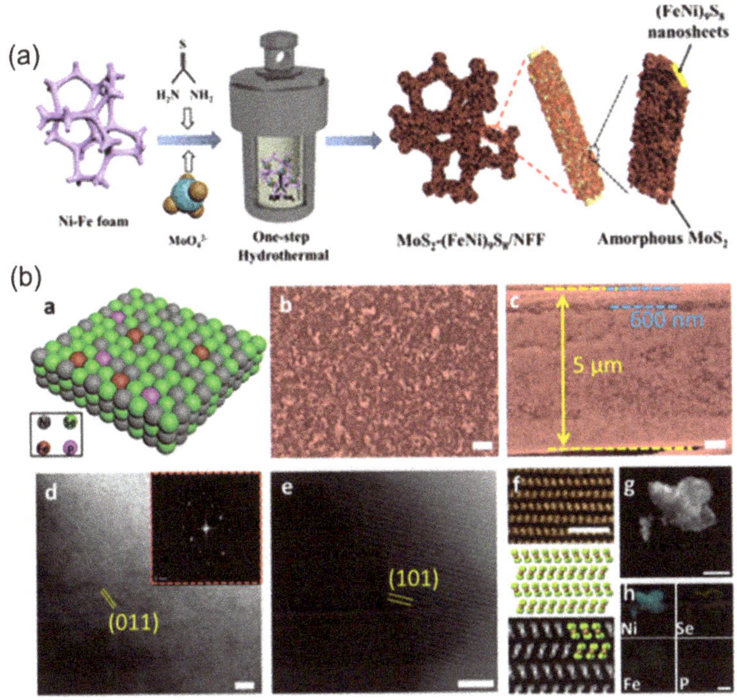

Figure 9. (**a**) Schematic diagram for the MoS_2-$(FeNi)_9S_8$/NFF preparation [50]; (**b**) chematic diagram for the Fe,P-$NiSe_2$ NFs preparation [51].

3.5. Transition Metal Phosphide

Wu et al. prepared the Ni_2P-Fe_2P nanosheet catalyst by soaking the Ni foam in an iron nitrate solution and hydrochloric acid, followed by phosphidation (Figure 10a) [55]. Owing to the abundant active sites, high intrinsic activity, and a superior transfer efficiency, this Ni_2P-Fe_2P showed a promising activity for overall water splitting, needing low voltages of 1.682 and 1.865 V to achieve 100 and 500 mA cm^{-2} in 1 M KOH, respectively [55]. It could achieve 100 and 500 mA cm^{-2} in 1 M KOH seawater at the voltages of 1.811 and 2.004 V, respectively [55]. Qi et al. used the Ni foam as the supporting base, wrapped the reduced graphene oxide (rGO) around it, and then attached Fe-Ni phosphides to rGO to synthesize the NiFeP/P-doped rGO/NF (NiFeP/P-rGO/NF) (Figure 10b) [56]. The overpotential of the as-obtained electrocatalyst was 290 mV and 340 mV in the electrolyte of 1 M KOH and 1 M NaCl, respectively, to achieve 100 and 400 mA cm^{-2}. Moreover, NiFeP/P-rGO/NF showed a high stability in seawater and could stably operate at 35 mA cm^{-2} for 450 h with no significant current attenuation [56]. Yan et al. synthesized the Cu-CoP NAs/CP that attached to carbon paper by the hydrothermal method and phosphorization process (Figure 10c) [57]. Cu-doped CoP could promote the adsorption and desorption of the intermediates, thus significantly improving the catalytic performance. Thus, Cu-CoP NAs/CP needed the overpotentials of 81 and 411 mV to achieve 10 mA cm^{-2} in the simulated seawater [57].

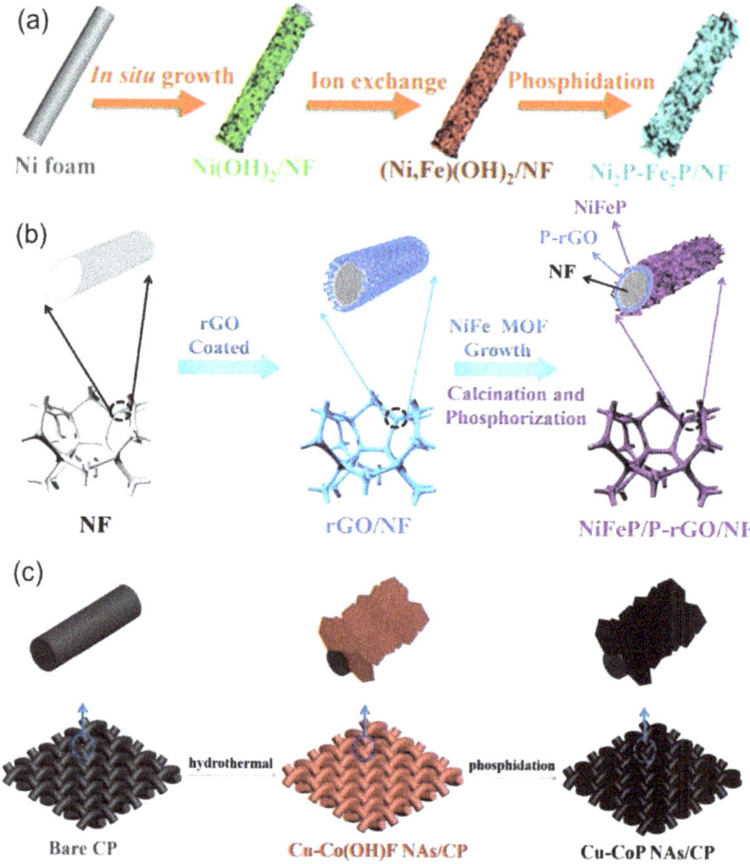

Figure 10. (**a**) The preparation process of Ni_2P-Fe_2P/NF [55]; (**b**) the fabrication process of NiFeP/P-rGO/NF [56]; (**c**) schematic illustration of the formation of Cu-CoP NAs/CP [57].

3.6. Transition Metal Nitrides

Ren et al. reported a Ni-foam-based three-dimensional NiMoN@NiFeN catalyst (Figure 11) [58]. It required the overpotentials as low as 277 and 337 mV to achieve 100 and 500 mA cm^{-2}, respectively, and could achieve 500 and 1000 mA cm^{-2} at the low voltages of 1.608 and 1.709 V when combined with NiMoN nanorods as HER catalyst at 60 °C.

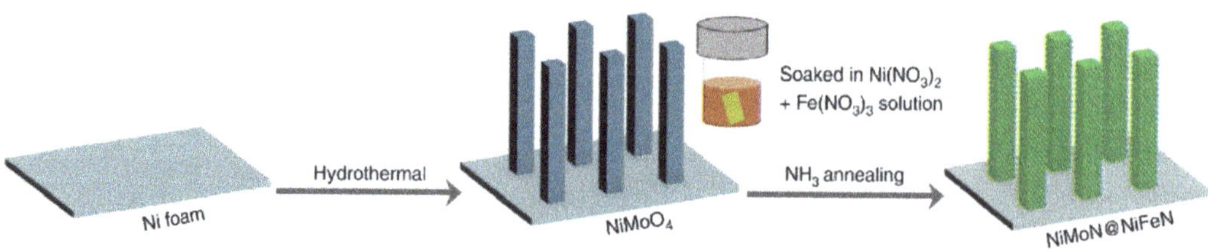

Figure 11. The schematic diagram of NiMoN@NiFeN preparation [58].

3.7. Micromolecule Additives to Form the Shielding Layer

Ma et al. reported that the service life of an anode catalyst can be significantly extended by adding SO_4^{2-} ions in alkaline seawater environment (Figure 12a) [59]. The presence of

SO_4^{2-} ions in the solution could preferentially adsorb on the anode surface, repel the Cl^- ions by electrostatic repulsion, so as to effectively alleviate the corrosion of Cl^- ions on the catalyst [59]. After adding the SO_4^{2-} ions, the NiFe-LDH nanoarrays/Ni foam anode can operate stably in seawater for 500–1000 h [59]. Zhuang et al. reported a novel structural buffer engineering strategy to endow the $Co_2(OH)_3Cl$ nanoplatelets with promising long-term operation stability and an OER selectivity as high as ~99.6% in seawater splitting (Figure 12b) [60]. The lattice Cl^- atoms of $Co_2(OH)_3Cl$ could play the role of structural buffers, and their continuous leaching during OER could leave vacancies for seawater Cl^- invasion, so that catalyst deactivation could be avoided. Consequently, $Co_2(OH)_3Cl$ could keep 99.9% of its initial current density after 60,000 s test, while that of $Co(OH)_2$ decayed quickly by 52.7% in 7000 s [60].

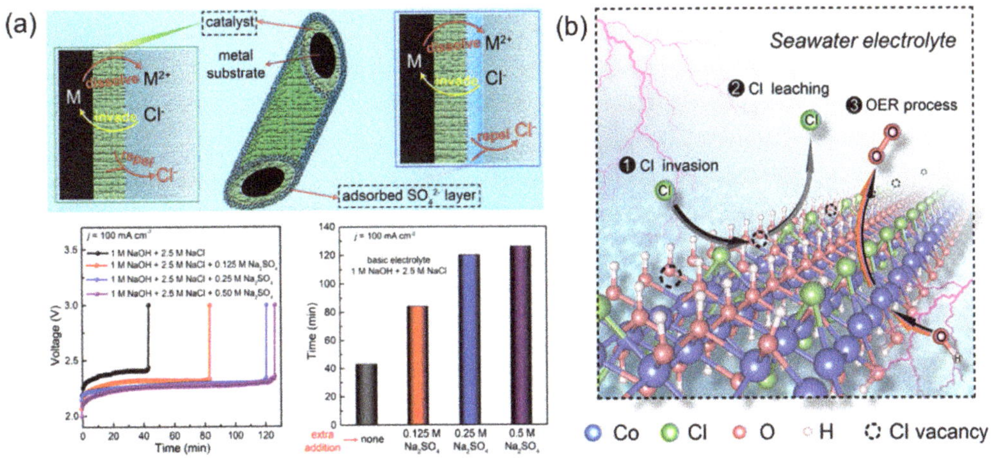

Figure 12. (**a**) Catalysts and electrolyte optimization to protect the metal substrate from Cl^- corrosion to improve the OER stability in seawater [59]; (**b**) Mechanism of the lattice Cl^- to protect the catalyst from Cl^- corrosion and deactivation [17].

3.8. Passivation Layer Covering

Li et al. reported that the design of NiFeB/$NiFeB_x$/NiFe alloy possessed a multi-layered structure (Figure 13a) [61]. The boron species were found to exist in the form of metaborate in the outermost oxidized $NiFeB_x$ layer, and their existence could promote the generation and stabilization of the catalytic active phase γ-(Ni,Fe)OOH [61]. Meanwhile, the $NiFeB_x$ interlayer could effectively prevent the anode material from excessive oxidative corrosion in the electrolyte containing chloride ions. Chen et al. discovered a PPy and tannic acid (TA) modified hollow MIL-88(FeCoNi) catalyst (Figure 13b) [62]. TA played an important role in etching MIL-88 to form a hollow structure, which exposed abundant active sites, while the released Fe^{3+} promoted the polymerization of pyrrole to enhance conductivity and stability. Owing to the coordination of PPy and TA, the HMIL-88@PPy-TA catalyst presented an extraordinary activity with a low overpotential of 370 mV at 1000 mA cm^{-2} in simulated seawater. In addition, under a constant current density of 100 mA cm^{-2}, it could maintain an initial overpotential for 100 h. It was proved that PPy protected the HMIL-88 from Cl^- corrosion, and the reconstruction of MIL-88(FeCoNi) to metal hydroxide during OER was beneficial for the maintenance of activity [62].

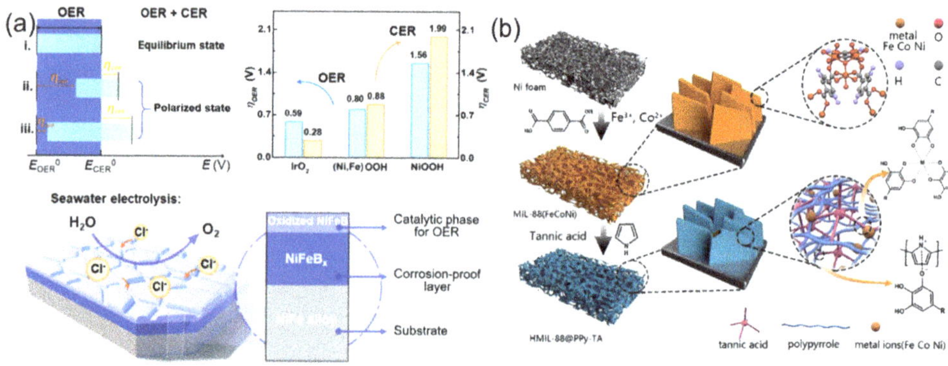

Figure 13. (**a**) Schematic diagram of the multilayer oxidized NiFeB/NiFeB$_x$/NiFe alloy [61]; (**b**) Preparation schematic of HMIL-88@PPy-TA catalyst [62].

4. Conclusions and Perspective

Nowadays, freshwater is becoming increasingly rarer, while the earth's oceans could provide an unlimited seawater resource. Against this global background, direct seawater electrolysis has become a promising electricity/hydrogen conversion and storage technology. It is expected to be effective in the regions with abundant renewable electricity and sufficient access to ocean seawater. However, because of the undesired electrochemical processes associated with the contaminants during seawater splitting, primarily the chloride chemistry, many research groups made intensive efforts in the exploration of robust and selective electrodes. Although researchers have had great achievements in seawater electrolysis in recent decades, various challenges should be overcome to fully promote seawater electrolysis for industrial H_2 production. The first challenge is to synthesize the highly active electrocatalysts, which could suppress Cl^- ions-associated electrochemical processes. They should exhibit high stability against corrosion by Cl_2, ClO^-, and so on. These strategies include doping various elements to modulate the electronic structure of active sites, tuning the binding strength of intermediates, synthesizing chlorine-blocking layers to promote OER, and constructing unique morphology. Second, the abundant ions in seawater could hinder the electrocatalytic reactions and result in unsatisfactory electrocatalytic performance. The promising electrode materials should effectively use the ions in seawater to improve the seawater electrolysis performance. Third, accurate identification and characterization of active sites should be brought to the forefront to understand the underlying mechanism and design the targeted electrocatalysts. Fourth, DFT calculations may play an important role in uncovering the reaction mechanisms and active sites of catalysts. The volcano plots, d-band center theory, and adsorption-free energy proposed by the computational simulations could benefit the study of intrinsic electrocatalytic properties. Meanwhile, the molecular dynamic simulation and finite element modeling may be applied to the effect of pore structure on the mass transfer and electrocatalytic activity of the catalysts. Lastly, the emerging technologies, such as machine learning, high-throughput theoretical computing, and artificial intelligence, may also be used to obtain the new high-level descriptors, so as to boost the development of catalyst design strategies. The construction of a clear structure–activity relationship could guide the rational design and fabrication of high-efficiency electrocatalysts for direct seawater splitting, so as to promote hydrogen production by seawater electrolysis.

Author Contributions: The manuscript was written through contributions of all authors. L.Z. and Z.X.: Conceptualization, Investigation, and Supervision. L.Z. and S.L.: Writing—original draft and image processing. J.L., K.W., Z.G. and C.L.: Validation, Resources, Investigation, Writing—review and editing. L.Z., S.L. and Z.X.: Visualization, Writing—review and editing. All authors have read and agreed to the published version of the manuscript.

Funding: This work was financially supported by the National Natural Science Foundations of China (Grant No. 21908054 and 22005098), and Central Government Funds for Guiding Local Science and Technology Development (Grant No. 2021Szvup040).

Institutional Review Board Statement: Not applicable.

Informed Consent Statement: Not applicable.

Data Availability Statement: Not applicable.

Conflicts of Interest: The authors declare no conflict of interest.

References

1. Cai, J.; Ding, J.; Wei, D.; Xie, X.; Li, B.; Lu, S.; Zhang, J.; Liu, Y.; Cai, Q.; Zang, S. Coupling of Ru and O-vacancy on 2D Mo-based electrocatalyst via a solid-phase interface reaction strategy for hydrogen evolution reaction. *Adv. Energy Mater.* **2021**, *11*, 2100141. [CrossRef]
2. Staffell, I.; Scamman, D.; Velazquez Abad, A.; Balcombe, P.; Dodds, P.E.; Ekins, P.; Shah, N.; Ward, K.R. The role of hydrogen and fuel cells in the global energy system. *Energy Environ. Sci.* **2019**, *12*, 463–491. [CrossRef]
3. Zhang, L.; Wang, Z.; Qiu, J. Energy-saving hydrogen production by seawater electrolysis coupling sulfion degradation. *Adv. Mater.* **2022**, *34*, 2109321. [CrossRef]
4. Oh, B.S.; Oh, S.G.; Hwang, Y.Y.; Yu, H.-W.; Kang, J.-W.; Kim, I.S. Formation of hazardous inorganic by-products during electrolysis of seawater as a disinfection process for desalination. *Sci. Total Environ.* **2010**, *408*, 5958–5965. [CrossRef]
5. Bennett, J.E. Electrodes for generation of hydrogen and oxygen from seawater. *Int. J. Hydrog. Energy* **1980**, *5*, 401–408. [CrossRef]
6. Kirk, D.W.; Ledas, A.E. Precipitate formation during sea water electrolysis. *Int. J. Hydrog. Energy* **1982**, *7*, 925–932. [CrossRef]
7. Cuartero, M.; Crespo, G.; Cherubini, T.; Pankratova, N.; Confalonieri, F.; Massa, F.; Tercier-Waeber, M.-L.; Abdou, M.; Schäfer, J.; Bakker, E. In Situ detection of macronutrients and chloride in seawater by submersible electrochemical sensors. *Anal. Chem.* **2018**, *90*, 4702–4710. [CrossRef]
8. Ji, S.M.; Jun, H.; Jang, J.S.; Son, H.C.; Borse, P.H.; Lee, J.S. Photocatalytic hydrogen production from natural seawater. *J. Photochem. Photobiol. A* **2007**, *189*, 141–144. [CrossRef]
9. Kester, D.R.; Duedall, I.W.; Connors, D.N.; Pytkowicz, R.M. Praparation of artifical seawater. *Limnol. Oceanogr.* **1967**, *12*, 176–179. [CrossRef]
10. Maeda, K.; Masuda, H.; Domen, K. Effect of electrolyte addition on activity of $(Ga_{1-x}Zn_x)(N_{1-x}O_x)$ photocatalyst for overall water splitting under visible light. *Catal. Today* **2009**, *147*, 173–178. [CrossRef]
11. Lu, X.; Pan, J.; Lovell, E.; Tan, T.H.; Ng, Y.H.; Amal, R. A sea-change: Manganese doped nickel/nickel oxide electrocatalysts for hydrogen generation from seawater. *Energy Environ. Sci.* **2018**, *11*, 1898–1910. [CrossRef]
12. Ha, H.; Jin, K.; Park, S.; Lee, K.-G.; Cho, K.H.; Seo, H.; Ahn, H.-Y.; Lee, Y.H.; Nam, K.T. Highly selective active chlorine generation electrocatalyzed by Co_3O_4 nanoparticles: Mechanistic investigation through in situ electrokinetic and spectroscopic analyses. *J. Phys. Chem. Lett.* **2019**, *10*, 1226–1233. [CrossRef] [PubMed]
13. Sun, H.; Hu, Z.; Xu, X.; He, J.; Dai, J.; Lin, H.-J.; Chan, T.-S.; Chen, C.-T.; Tjeng, L.H.; Zhou, W.; et al. Ternary phase diagram-facilitated rapid screening of double perovskites as electrocatalysts for the oxygen evolution reaction. *Chem. Mater.* **2019**, *31*, 5919–5926. [CrossRef]
14. Dionigi, F.; Reier, T.; Pawolek, Z.; Gliech, M.; Strasser, P. Design criteria, operating conditions, and nickel–iron hydroxide catalyst materials for selective seawater electrolysis. *ChemSusChem* **2016**, *9*, 962–972. [CrossRef] [PubMed]
15. Yang, F.; Luo, Y.; Yu, Q.; Zhang, Z.; Zhang, S.; Liu, Z.; Ren, W.; Cheng, H.-M.; Li, J.; Liu, B. A durable and efficient electrocatalyst for saline water splitting with current density exceeding 2000 mA cm^{-2}. *Adv. Funct. Mater.* **2021**, *31*, 2010367. [CrossRef]
16. Wu, D.; Chen, D.; Zhu, J.; Mu, S. Ultralow Ru incorporated amorphous cobalt-based oxides for high-current-density overall water splitting in alkaline and seawater media. *Small* **2021**, *17*, 2102777. [CrossRef]
17. Phan Khanh Linh, T.; Duy Thanh, T.; Malhotra, D.; Prabhakaran, S.; Kim, D.H.; Kim, N.H.; Lee, J.H. Highly effective freshwater and seawater electrolysis enabled by atomic Rh-modulated Co-CoO lateral heterostructures. *Small* **2021**, *17*, 2103826. [CrossRef]
18. Li, L.; Wang, B.; Zhang, G.; Yang, G.; Yang, T.; Yang, S.; Yang, S. Electrochemically modifying the electronic structure of IrO_2 nanoparticles for overall electrochemical water splitting with extensive adaptability. *Adv. Energy Mater.* **2020**, *10*, 2001600. [CrossRef]
19. Zhang, B.; Xu, W.; Liu, S.; Chen, X.; Ma, T.; Wang, G.; Lu, Z.; Sun, J. Enhanced interface interaction in Cu_2S@Ni core-shell nanorod arrays as hydrogen evolution reaction electrode for alkaline seawater electrolysis. *J. Power Sources* **2021**, *506*, 230235. [CrossRef]
20. Zhao, M.; Yang, M.; Huang, W.; Liao, W.; Bian, H.; Chen, D.; Wang, L.; Tang, J.; Liu, C. Synergism on electronic structures and active edges of metallic vanadium disulfide nanosheets via Co doping for efficient hydrogen evolution reaction in seawater. *ChemCatChem* **2021**, *13*, 2138–2144. [CrossRef]
21. Zhao, Y.; Jin, B.; Zheng, Y.; Jin, H.; Jiao, Y.; Qiao, S.-Z. Charge state manipulation of cobalt selenide catalyst for overall seawater electrolysis. *Adv. Energy Mater.* **2018**, *8*, 1801926. [CrossRef]
22. Xu, W.; Fan, G.; Zhu, S.; Liang, Y.; Cui, Z.; Li, Z.; Jiang, H.; Wu, S.; Cheng, F. Electronic structure modulation of nanoporous cobalt phosphide by carbon doping for alkaline hydrogen evolution reaction. *Adv. Funct. Mater.* **2021**, *31*, 2107333. [CrossRef]

23. Huang, Y.; Hu, L.; Liu, R.; Hu, Y.; Xiong, T.; Qiu, W.; Balogun, M.S.; Pan, A.; Tong, Y. Nitrogen treatment generates tunable nanohybridization of Ni_5P_4 nanosheets with nickel hydr(oxy)oxides for efficient hydrogen production in alkaline, seawater and acidic media. *Appl. Catal. B* **2019**, *251*, 181–194. [CrossRef]
24. Wang, S.; Yang, P.; Sun, X.; Xing, H.; Hu, J.; Chen, P.; Cui, Z.; Zhu, W.; Ma, Z. Synthesis of 3D heterostructure Co-doped Fe_2P electrocatalyst for overall seawater electrolysis. *Appl. Catal. B Environ.* **2021**, *297*, 120386. [CrossRef]
25. Yu, L.; Wu, L.; Song, S.; McElhenny, B.; Zhang, F.; Chen, S.; Ren, Z. Hydrogen generation from seawater electrolysis over a sandwich-like NiCoN | NixP | NiCoN microsheet array catalyst. *ACS Energy Lett.* **2020**, *5*, 2681–2689. [CrossRef]
26. Wu, X.; Zhou, S.; Wang, Z.; Liu, J.; Pei, W.; Yang, P.; Zhao, J.; Qiu, J. Engineering multifunctional collaborative catalytic interface enabling efficient hydrogen evolution in all pH range and seawater. *Adv. Energy Mater.* **2019**, *9*, 1901333. [CrossRef]
27. Zang, W.; Sun, T.; Yang, T.; Xi, S.; Waqar, M.; Kou, Z.; Lyu, Z.; Feng, Y.P.; Wang, J.; Pennycook, S.J. Efficient hydrogen evolution of oxidized $Ni\text{-}N_3$ defective sites for alkaline freshwater and seawater electrolysis. *Adv. Mater.* **2021**, *33*, 2003846. [CrossRef]
28. Sun, F.; Qin, J.; Wang, Z.; Yu, M.; Wu, X.; Sun, X.; Qiu, J. Energy-saving hydrogen production by chlorine-free hybrid seawater splitting coupling hydrazine degradation. *Nat. Commun.* **2021**, *12*, 4182. [CrossRef] [PubMed]
29. Zhang, X.; Zhang, Y.; Li, F.; Easton, C.D.; Bond, A.M.; Zhang, J. Oxomolybdate anchored on copper for electrocatalytic hydrogen production over the entire pH range. *Appl. Catal. B* **2019**, *249*, 227–234. [CrossRef]
30. Yuan, W.; Cui, Z.; Zhu, S.; Li, Z.; Wu, S.; Liang, Y. Structure engineering of electrodeposited NiMo films for highly efficient and durable seawater splitting. *Electrochim. Acta* **2021**, *365*, 137366. [CrossRef]
31. Ros, C.; Murcia-Lopez, S.; Garcia, X.; Rosado, M.; Arbiol, J.; Llorca, J.; Morante, J.R. Facing seawater splitting challenges by regeneration with Ni-Mo-Fe bifunctional electrocatalyst for hydrogen and oxygen evolution. *ChemSusChem* **2021**, *14*, 2872–2881. [CrossRef] [PubMed]
32. Du, G.; Sun, W.; Hu, Y.; Liao, J.; Tian, X.; Gao, H.; Ge, C. IrOx nanoclusters modified by $BaCO_3$ enable "Two Birds with One Stone" in solar-driven direct unbuffered seawater electrolysis. *ACS Appl. Mater. Interfaces* **2021**, *13*, 61088–61097. [CrossRef]
33. ul Haq, T.; Pasha, M.; Tong, Y.; Mansour, S.A.; Haik, Y. Au nanocluster coupling with $Gd\text{-}Co_2B$ nanoflakes embedded in reduced TiO_2 nanosheets: Seawater electrolysis at low cell voltage with high selectivity and corrosion resistance. *Appl. Catal. B* **2022**, *301*, 120836. [CrossRef]
34. Ko, J.S.; Johnson, J.K.; Johnson, P.I.; Xia, Z. Decoupling oxygen and chlorine evolution reactions in seawater using iridium-based electrocatalysts. *ChemCatChem* **2020**, *12*, 4526–4532. [CrossRef]
35. Gayen, P.; Saha, S.; Ramani, V. Selective seawater splitting using pyrochlore electrocatalyst. *ACS Appl. Energy Mater.* **2020**, *3*, 3978–3983. [CrossRef]
36. Ning, M.; Wu, L.; Zhang, F.; Wang, D.; Song, S.; Tong, T.; Bao, J.; Chen, S.; Yu, L.; Ren, Z. One-step spontaneous growth of NiFe layered double hydroxide at room temperature for seawater oxygen evolution. *Mater. Today Phys.* **2021**, *19*, 100419. [CrossRef]
37. Badreldin, A.; Nabeeh, A.; Youssef, E.; Mubarak, N.; ElSayed, H.; Mohsen, R.; Ahmed, F.; Wubulikasimu, Y.; Elsaid, K.; Abdel-Wahab, A. Adapting early transition metal and nonmetallic dopants on CoFe oxyhydroxides for enhanced alkaline and neutral pH saline water oxidation. *ACS Appl. Energy Mater.* **2021**, *4*, 6942–6956. [CrossRef]
38. Badreldin, A.; Youssef, K.; El Ghenymy, A.; Wubulikasimu, Y.; Ghouri, Z.K.; Elsaid, K.; Kumar, D.; Abdel-Wahab, A. Solution combustion synthesis of novel S,B-codoped CoFe oxyhydroxides for the oxygen evolution reaction in saline water. *ACS Omega* **2022**, *7*, 5521–5536. [CrossRef]
39. Li, L.; Zhang, G.; Wang, B.; Zhu, D.; Liu, D.; Liu, Y.; Yang, S. Fe_2O_3/NiO interface for the electrochemical oxygen evolution in seawater and domestic sewage. *ACS Appl. Mater. Interfaces* **2021**, *13*, 37152–37161. [CrossRef]
40. Ul Haq, T.; Mansour, S.; Haik, Y. Electronic and structural modification of Mn_3O_4 nanosheets for selective and sustained seawater oxidation. *ACS Appl. Mater. Interfaces* **2022**, *14*, 20443–20454. [CrossRef]
41. Abe, H.; Murakami, A.; Tsunekawa, S.; Okada, T.; Wakabayashi, T.; Yoshida, M.; Nakayama, M. Selective catalyst for oxygen evolution in neutral brine electrolysis: An oxygen-deficient manganese oxide film. *ACS Catal.* **2021**, *11*, 6390–6397. [CrossRef]
42. Dresp, S.; Dionigi, F.; Loos, S.; de Araujo, J.F.; Spoeri, C.; Gliech, M.; Dau, H.; Strasser, P. Direct Electrolytic splitting of seawater: Activity, selectivity, degradation, and recovery studied from the molecular catalyst structure to the electrolyzer cell level. *Adv. Energy Mater.* **2018**, *8*, 1800338. [CrossRef]
43. Chen, H.; Zou, Y.; Li, J.; Zhang, K.; Xia, Y.; Hui, B.; Yang, D. Wood aerogel-derived sandwich-like layered nanoelectrodes for alkaline overall seawater electrosplitting. *Appl. Catal. B* **2021**, *293*, 120215. [CrossRef]
44. Li, L.; Zhang, G.; Wang, B.; Yang, S. Constructing the Fe/Cr double (oxy)hydroxides on Fe_3O_4 for boosting the electrochemical oxygen evolution in alkaline seawater and domestic sewage. *Appl. Catal. B* **2022**, *302*, 120847. [CrossRef]
45. Wu, L.; Yu, L.; Zhu, Q.; McElhenny, B.; Zhang, F.; Wu, C.; Xing, X.; Bao, J.; Chen, S.; Ren, Z. Boron-modified cobalt iron layered double hydroxides for high efficiency seawater oxidation. *Nano Energy* **2021**, *83*, 105838. [CrossRef]
46. Jung, S.Y.; Kang, S.; Kim, K.M.; Mhin, S.; Kim, J.C.; Kim, S.J.; Enkhtuvshin, E.; Choi, S.; Han, H. Sulfur-incorporated nickel-iron layered double hydroxides for effective oxygen evolution reaction in seawater. *Appl. Surf. Sci.* **2021**, *568*, 150965. [CrossRef]
47. Gao, X.; Chen, Y.; Sun, T.; Huang, J.; Zhang, W.; Wang, Q.; Cao, R. Karst landform-featured monolithic electrode for water electrolysis in neutral media. *Energy Environ. Sci.* **2020**, *13*, 174–182. [CrossRef]
48. Song, Z.; Wang, K.; Sun, Q.; Zhang, L.; Li, J.; Li, D.; Sze, P.-W.; Liang, Y.; Sun, X.; Fu, X.Z.; et al. High-performance ammonium cobalt phosphate nanosheet electrocatalyst for alkaline saline water oxidation. *Adv. Sci.* **2021**, *8*, 2100498. [CrossRef]

49. Cong, Y.; Chen, X.; Mei, Y.; Ye, J.; Li, T.-T. CeO_2 decorated bimetallic phosphide nanowire arrays for enhanced oxygen evolution reaction electrocatalysis via interface engineering. *Dalton Trans.* **2022**, *51*, 2923–2931. [CrossRef]
50. Song, S.; Wang, Y.; Zhou, S.; Gao, H.; Tian, X.; Yuan, Y.; Li, W.; Zang, J.B. One-step synthesis of heterostructural MoS_2-$(FeNi)_9S_8$ on Ni-Fe foam synergistically boosting for efficient fresh/seawater electrolysis. *ACS Appl. Energy Mater.* **2022**, *5*, 1810–1821. [CrossRef]
51. Chang, J.; Wang, G.; Yang, Z.; Li, B.; Wang, Q.; Kuliiev, R.; Orlovskaya, N.; Gu, M.; Du, Y.; Wang, G.; et al. Dual-doping and synergism toward high-performance seawater electrolysis. *Adv. Mater.* **2021**, *33*, 2101425. [CrossRef]
52. Cui, B.; Hu, Z.; Liu, C.; Liu, S.; Chen, F.; Hu, S.; Zhang, J.; Zhou, W.; Deng, Y.; Qin, Z.; et al. Heterogeneous lamellar-edged Fe-$Ni(OH)_2$/Ni_3S_2 nanoarray for efficient and stable seawater oxidation. *Nano Res.* **2021**, *14*, 1149–1155. [CrossRef]
53. Wang, C.; Zhu, M.; Cao, Z.; Zhu, P.; Cao, Y.; Xu, X.; Xu, C.; Yin, Z. Heterogeneous bimetallic sulfides based seawater electrolysis towards stable industrial-level large current density. *Appl. Catal. B* **2021**, *291*, 120071. [CrossRef]
54. Yu, L.; Wu, L.; McElhenny, B.; Song, S.; Luo, D.; Zhang, F.; Yu, Y.; Chen, S.; Ren, Z. Ultrafast room-temperature synthesis of porous S-doped Ni/Fe (oxy)hydroxide electrodes for oxygen evolution catalysis in seawater splitting. *Energy Environ. Sci.* **2020**, *13*, 3439–3446. [CrossRef]
55. Wu, L.; Yu, L.; Zhang, F.; McElhenny, B.; Luo, D.; Karim, A.; Chen, S.; Ren, Z. Heterogeneous bimetallic phosphide Ni2P-Fe2P as an efficient bifunctional catalyst for water/seawater splitting. *Adv. Funct. Mater.* **2021**, *31*, 2006484. [CrossRef]
56. Qi, L.; Li, A.; Wang, M.; Zhang, Y.; Zhang, K.; Li, X. Stable and efficient oxygen evolution from seawater enabled by graphene-supported sub-nanometer arrays of transition metal phosphides. *Adv. Mater. Interfaces* **2022**, *9*, 2101720. [CrossRef]
57. Yan, L.; Zhang, B.; Zhu, J.; Li, Y.; Tsiakaras, P.; Shen, P.K. Electronic modulation of cobalt phosphide nanosheet arrays via copper doping for highly efficient neutral-pH overall water splitting. *Appl. Catal. B* **2020**, *265*, 118555. [CrossRef]
58. Yu, L.; Zhu, Q.; Song, S.; McElhenny, B.; Wang, D.; Wu, C.; Qin, Z.; Bao, J.; Yu, Y.; Chen, S.; et al. Non-noble metal-nitride based electrocatalysts for high-performance alkaline seawater electrolysis. *Nat. Commun.* **2019**, *10*, 5106. [CrossRef]
59. Ma, T.; Xu, W.; Li, B.; Chen, X.; Zhao, J.; Wan, S.; Jiang, K.; Zhang, S.; Wang, Z.; Tian, Z.; et al. The critical role of additive sulfate for stable alkaline seawater oxidation on nickel-based electrodes. *Angew. Chem. Int. Ed.* **2021**, *60*, 22740–22744. [CrossRef]
60. Zhuang, L.; Li, J.; Wang, K.; Li, Z.; Zhu, M.; Xu, Z. Structural buffer engineering on metal oxide for long-term stable seawater splitting. *Adv. Funct. Mater.* **2022**, 2201127. [CrossRef]
61. Li, J.; Liu, Y.; Chen, H.; Zhang, Z.; Zou, X. Design of a multilayered oxygen-evolution electrode with high catalytic activity and corrosion resistance for saline water splitting. *Adv. Funct. Mater.* **2021**, *31*, 2101820. [CrossRef]
62. Chen, Y.; Shen, L.; Wang, C.; Feng, S.; Zhang, N.; Zhang, K.; Yang, B. Utilizing tannic acid and polypyrrle to induce reconstruction to optimize the activity of MOF-derived electrocatalyst for water oxidation in seawater. *Chem. Eng. J.* **2022**, *430*, 132632. [CrossRef]

coatings

Article

Humic Acid Removal in Water via UV Activated Sodium Perborate Process

Deling Yuan [1,2], Zhihui Zhai [1], Eryu Zhu [1], Huilin Liu [1], Tifeng Jiao [1,2,*] and Shoufeng Tang [1,2,*]

1 Hebei Key Laboratory of Heavy Metal Deep-Remediation in Water and Resource Reuse, Hebei Key Laboratory of Applied Chemistry, School of Environmental and Chemical Engineering, Yanshan University, Qinhuangdao 066004, China; yuandl@ysu.edu.cn (D.Y.); zzhdyx06@163.com (Z.Z.); zhueryu2022@163.com (E.Z.); liuhl0094@163.com (H.L.)

2 State Key Laboratory of Metastable Materials Science and Technology, Yanshan University, Qinhuangdao 066004, China

* Correspondence: tfjiao@ysu.edu.cn (T.J.); tangshf@ysu.edu.cn (S.T.)

Abstract: Humic acid (HA) has complex molecular structure and is capable of adsorption, ion exchange, and chelation with organic and inorganic pollutants in water bodies, worsening water quality and jeopardizing human health and ecological environment. How to effectively remove HA from water is one of the research focuses of this paper. In this study, the UV-activated sodium perborate (SPB) synergistic system (UV/SPB) was established to eliminate HA in water. The effects of initial HA concentration, SPB dose, and initial pH value on the HA elimination were determined, and the main mechanisms of the synergy and HA degradation were explored. The outcomes show that the HA elimination ratio by the sole UV and only SPB system were only 0.5% and 1.5%, respectively. The HA removal of UV/SPB reached 88.8%, which can remove HA more effectively than other systems. Free radical masking experiment proved that hydroxyl radical produced by SPB activation is the main active substance for HA removal. The results of UV-vis absorption spectrum, absorbance ratio, specific UV absorbance, and excitation–emission matrix spectroscopy verified that the UV/SPB system can effectively decompose and mineralize HA.

Keywords: UV irradiation; sodium perborate; activation; humic acid removal; decomposition mechanism

Citation: Yuan, D.; Zhai, Z.; Zhu, E.; Liu, H.; Jiao, T.; Tang, S. Humic Acid Removal in Water via UV Activated Sodium Perborate Process. *Coatings* **2022**, *12*, 885. https://doi.org/10.3390/coatings12070885

Academic Editor: Ioannis V. Yentekakis

Received: 18 May 2022
Accepted: 21 June 2022
Published: 22 June 2022

1. Introduction

As a non-uniform macromolecular polymer, humic acid (HA) is the major ingredient of natural organic matter (NOM), which is the product of polymerization of diverse biologic remains for ages [1]. HA has a complex molecular structure and contains many organic functional groups, such as hydroxyl, carboxyl, carbonyl, methoxy, and quinone groups, which are capable of adsorption, ion exchange, and chelation with organic and inorganic pollutants in water bodies, worsening water quality and jeopardizing human health and ecological environment [2]. Finding a proper way to effectively remove HA from water is one of the focuses in environmental research.

The main methods for controlling HA in water are physical and chemical oxidation ways. Physical methods to remove HA include coagulation [3], flocculation [4], and adsorption [5], but these methods only transfer HA to solid phase, and subsequent solid waste treatment is still required. Chemical oxidation is of great interest because of its rapid decomposition and mineralization for HA [6]. Photocatalysis [7–9], Fenton oxidation [10], and electrochemical oxidation [11] are the common chemical oxidation methods for organic wastewater treatment; however, they universally have some shortcomings such as harsh reaction conditions and complex operation.

Sodium perborate (SPB, $NaBO_3$) is a widely used in situ oxidizing agent. Distinct from the sodium percarbonate, SPB is not an adduct from inorganic salt and H_2O_2 but contains cyclic perborate ions ($B_2O_8H_4^{2-}$) consisting of two peroxide chains without BO_3^-

ions [12]. SPB produces H_2O_2 stably after SPB is dissolved in water, so it is a good substitute for H_2O_2 [13]. Compared with liquid H_2O_2, the solid SPB is safe and easy to store and transport. Most importantly, the activation of SPB can form hydroxyl radicals (·OH) over a wide range. Ultraviolet (UV) radiation [14] and transition metal ions [15] are the main approaches for SPB activation. Researchers have used UV-activated perborate for organic pollutant removal [13]. Besides, the perborate has been employed as the oxidant in the homogeneous photo-Fenton and heterogeneous Fenton-like reactions for dye and phenol degradation [12,16], respectively. Among them, the UV activation method is easy operation and safe without secondary pollution, so it could efficaciously excite H_2O_2 to decompose organics from wastewater [17,18].

Although the UV-activated peroxide and even UV-activated SPB method have been applied for removing organics in water, the elimination of HA through the UV activated SPB is rarely reported. Therefore, the importance of this study was establishing a UV/SPB synergistic system (UV/SPB) for the reduction of HA in water. With the convenient and efficient spectrophotometric method [19], the effects of initial HA concentration, SPB dose, and initial pH value on HA decontamination were studied. The main active substances generated in the synergistic system for HA removal were determined by free radical masking experiment, and the degradation mechanism was systematically explored by UV spectra, total organic carbon, and three-dimensional excitation-emission matrix spectroscopy (3D-EEM).

2. Experimental

2.1. Chemicals

HA was purchased from Aladdin Industrial Co., Shanghai, China. The average molecular weight of HA was 2485 Da. Sodium perborate ($NaBO_3$, SPB) and sodium sulfate (Na_2SO_4) were also purchased from Aladdin Industrial Co., Shanghai, China. Sodium hydroxide (NaOH) and sulfuric acid (H_2SO_4) were purchased from Sinopharm Chemical Reagent Co., Shanghai, China. Sodium carbonate (Na_2CO_3) and sodium dihydrogen phosphate (NaH_2PO_4) were bought from Tianjin Kemiou Chemical Reagent Co., Ltd., Tianjin, China. Sodium nitrate ($NaNO_3$) was manufactured by Tianjin Hengxing Chemical Reagent Manufacturing Co., Ltd., Tianjin, China. Sodium bicarbonate ($NaHCO_3$) was supplied by Tianjin Kaitong Chemical Reagent Co., Ltd., Tianjin, China. Sodium chloride (NaCl) was obtained from Tianjin Fengchuan Chemical Reagent Co., Ltd., Tianjin, China. Tertiary butanol (TBA) was purchased from Tianjin Damao Chemical Reagent Factory, Tianjin, China. All chemicals used were of analytical grade without further purification.

2.2. Experimental Procedures

The HA removal experiments were conducted in 25 °C. A beaker served as the reactor was kept on the magnetic stirrer, and the UV lamp (power 16 W, wavelength 254 nm, Philips Co., Shanghai, China) was placed above the beaker with the distance 3.5 cm. The UV fluence in this study was computed to be 35.2 mJ cm^{-2} for 60 min. A certain volume of HA solution was diluted to 100 mL before the beginning of experiment. Then the SPB was introduced to the HA simulated wastewater, and the UV irradiation was begun to initiate the reaction. At specified time intervals, 2.5 mL aliquots were withdrawn for measuring the absorbance. All the experiments were performed at least two times. To confirm the generation of ·OH, TBA was used as the scavengers.

2.3. Method and Analyses

The absorbance of solution was detected through UV-vis spectrophotometer (SP-725 Shanghai Spectral Instrument Co., Ltd., Shanghai, China) with external standard method at a wavelength of 254 nm to determine the HA removal efficiency [20]. The equation was as follows:

$$\text{HA removal efficiency} = \frac{C_0 - C_t}{C_0} \times 100\% \quad (1)$$

where C_0 was the beginning HA amount, and C_t was the HA amount at the treatment time t.

The HA molecular structure variation was determined by a series of specific UV-vis adsorption spectra. The values of A_{253}, A_{203}, A_{254}, A_{436}, A_{250}, A_{365}, A_{465}, and A_{665} were determined by spectrophotometer at the wavelengths of 203 nm, 250 nm, 253 nm, 254 nm, 365 nm, 436 nm, 465 nm, and 665 nm, respectively [21]. In addition, continuous changes in absorbance of the solution at 200–800 nm were also measured to characterize the changes of HA molecular structure.

Total organic carbon (TOC) was measured by TOC analyzer (VCPH, Shimadzu, Japan). Specific UV absorbance ($SUVA_x$) was calculated with A_x and TOC [22].

$$SUVA_x = \frac{A_x}{TOC} \times 100\% \tag{2}$$

where A_x was the sample absorbance at x nm.

The 3D-EEM spectrum (FL4500, Hitachi, Tokyo, Japan) was used to explore the HA decomposition mechanism. The ranges emission and excitation wavelengths were from 280 to 550 nm and 200 to 400 nm, separately, and their corresponding slits were 10 and 5 nm, respectively.

3. Results and Discussion

3.1. Study on HA Removal by UV/SPB System

3.1.1. Comparison of HA Removal Performance during Different Systems

The removal of HA was firstly investigated in three processes, which were UV, SPB, and UV/SPB, as seen in Figure 1. The experimental conditions were as follows: HA concentration of 10 mg·L^{-1}, SPB dose of 1 mmol·L^{-1}, and initial pH 3. The HA was hardly removed, and the removal ratio was 0.5% after 60 min by the only UV process. The elimination of HA by the sole SPB process was also insignificant, and its decolorization ratio was 1.5% at 60 min. The decontamination efficiency by the UV/SPB process was 88.8%, which increased more significantly than the other two processes. Besides, with the same molecular weight of H_2O_2, the HA removal by the UV/H_2O_2 was conducted, and its elimination ratio only achieved 40.2% after 60 min. SPB is often used for in situ chemical oxidation, and SPB has the same function as H_2O_2 when SPB was dissolved in water (Equation (3)) [12]. In the sole SPB system, although H_2O_2 was generated in the system, it cannot be activated to produce ·OH, so HA was scarcely removed. In the UV/SPB system, H_2O_2 released from SPB can form ·OH after being irradiated by UV (Equation (4)) [14], which could oxidize and destroy functional groups in HA molecular structure.

$$NaBO_3 + H_2O \rightarrow NaBO_2 + H_2O_2 \tag{3}$$

$$H_2O_2 + hv \rightarrow 2 \cdot OH \tag{4}$$

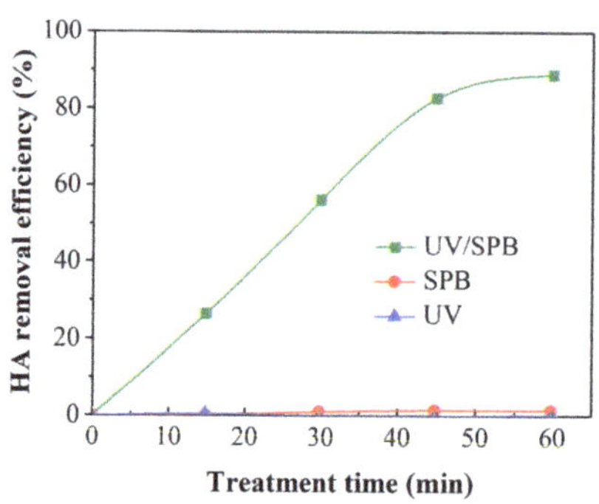

Figure 1. HA removal performance of different systems. $[HA]_0$ = 10 mg L^{-1}, $[SPB]_0$ = 1 mmol L^{-1}, pH = 3.

3.1.2. Effect of HA Concentration

The influence of HA concentration on the removal of HA by the UV/SPB system is shown in Figure 2a. The experimental conditions were SPB amount of 1 mmol L^{-1} and initial pH 3. The removal ratio decreased as the initial concentration of HA was increased. As the HA concentration was raised from 5 to 15 mg L^{-1}, the removal ration diminished from 89.8% to 70.8% after 60 min. It proved that the reactive substances formed during the UV/SPB process were persistently consumed because the active species generated in the system were insufficient to oxidize more and more contaminants in the solution, and the competition among HA molecules and reactive species became increasingly fierce with the gradual rise of HA concentration. Moreover, the HA amount enhancement would absorb more UV radiation [23], thereby inhibiting the activation of H_2O_2 and subsequently formation of ·OH radicals, which led to the decrease in HA removal.

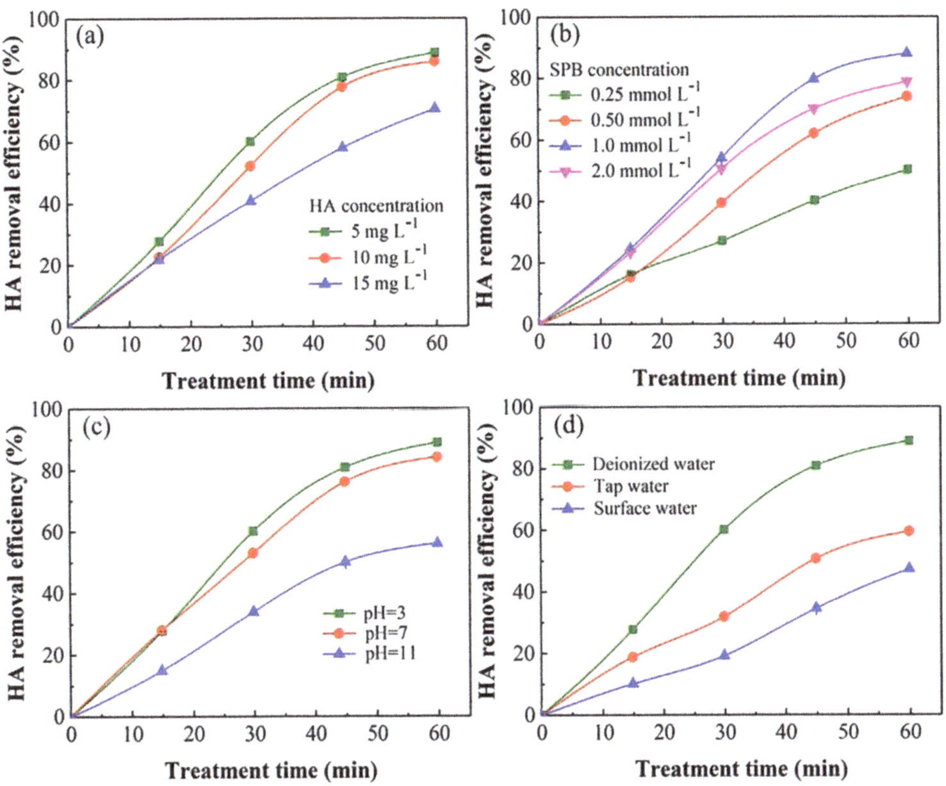

Figure 2. Influence of various parameters on HA decontamination during UV/SPB system: (**a**) HA concentration, (**b**) SPB concentration, (**c**) initial pH. ($[HA]_0$ = 5–15 mg L^{-1}, $[SPB]_0$ = 0.25–2 mmol L^{-1}, pH = 3–11). (**d**) HA removal in different waterbodies by UV/SPB. ($[HA]_0$ = 10 mg L^{-1}, $[SPB]_0$ = 1 mmol L^{-1}, pH = 3).

3.1.3. Influence of SPB Dose

The SPB is the source of active radicals which is significant for the synergetic process. The influence of SPB dosage on the HA elimination was investigated (Figure 2b). The experimental conditions were HA concentration of 10 mg L^{-1} and initial pH 3. The HA removal ratio enhanced from 53.0% to 88.8% as the SPB dosage was raised from 0.25 to 1.0 mmol L^{-1} after 60 min. With the increase of SPB concentration, the amount of active substances produced in the system was enhanced, which promoted the HA removal. However, excessive SPB would scavenge the ·OH and generate peroxy hydroxyl radical

$HO_2\cdot$ (Equation (5)) [24]. The redox potential of $HO_2\cdot$ is lower than that of $\cdot OH$. Therefore, the superfluous SPB (2.0 mmol L^{-1}) resulted in the decline of HA removal.

$$\cdot OH + H_2O_2 \rightarrow H_2O + HO_2^{\cdot} \quad (5)$$

3.1.4. Influence of Initial pH

Figure 2c shows the effect of different beginning pH on the HA removal during the UV/SPB treatment. The experimental conditions were HA concentration of 10 mg L^{-1} and SPB amount of 1 mmol L^{-1}. The HA removal decreased from 88.8% to 58.4% after 60 min while the pH value was elevated from 3 to 11. The HA molecule is neutral under strong acidic conditions, so its photochemical activity is stronger than those during neutral and basic circumstances. The redox potential $E_{\cdot OH,\, H2O}$ is also affected by pH. With the increase of pH from 3 to 11, the redox of $E_{\cdot OH,\, H2O}$ is declined from 2.62 V (pH = 3) to 2.15 V (pH = 11) [25]. Under the alkaline condition, the $\cdot OH$ would be transformed to $O\cdot^-$ (E = 1.78 V) via reaction (Equation (6)) [26], which had a lower oxidation capacity than that of $\cdot OH$. As the pH increased to 11, the predominant form of H_2O_2 is changed to HO_2^-, which has a higher rate constant with $\cdot OH$ (k = 7.5 × 10^9 M^{-1} s^{-1}, Equation (7)) than the reaction of H_2O_2 and $\cdot OH$ (k = 2.7 × 10^7 M^{-1} s^{-1}, Equation (8)) [27], leading to the consumption of $\cdot OH$.

$$OH^- + \cdot OH \rightarrow H_2O + O^- \quad (6)$$

$$HO_2^- + \cdot OH \rightarrow OH^- + HO_2^{\cdot}\ k = 7.5 \times 10^9\ M^{-1}\ s^{-1} \quad (7)$$

$$\cdot OH + H_2O_2 \rightarrow H_2O + HO_2^{\cdot}\ k = 2.7 \times 10^7\ M^{-1} \quad (8)$$

3.1.5. HA Removal in Different Water Bodies

Figure 2d indicates the HA removal in various water bodies by the UV/SPB system. After 60 min reaction, the removal ratios of HA during deionized water, tap water, and lake water were 88.8%, 59.6%, and 47.5%, separately. It showed that the lake water and tap water inhibited the removal of HA. The reasons were as follows. Firstly, the other NOM in lake would compete with $\cdot OH$ generated by UV/SPB. Secondly, the presence of various anions in the lake water and tap water that could inhibit the activity of the oxidative species, leading to a lower efficiency of HA elimination.

3.1.6. Effect of Common Anions in Water

Figure 3 demonstrates the effect of common anions such as CO_3^{2-}, HCO_3^-, Cl^-, NO_3^-, SO_4^{2-}, and $H_2PO_4^-$ for the HA elimination by the UV/SPB. In Figure 3a, when the CO_3^{2-} amount was increased from 1 to 10 mmol L^{-1}, the removal efficiency decreased from 63.7% to 44.9%. The reason was that $\cdot OH$ yielded in the system was consumed by CO_3^{2-} to produce $CO_3\cdot^-$ with weak oxidation capacity (Equation (9)) [28], leading to the decrease of HA removal.

$$OH\cdot + CO_3^{2-} \rightarrow CO_3^{\cdot -} + OH^- \qquad k = 4.2 \times 10^8\ M^{-1}\cdot s^{-1} \quad (9)$$

As shown in Figure 3b, with the increase of HCO_3^- concentration (1 to 10 mmol L^{-1}), the HA elimination efficiency decreased gradually (74.2% to 53.5%) at 60 min. HCO_3^- also consumed $\cdot OH$ produced in the system (Equation (10)). Besides, the solution pH would be increased after the HCO_3^- addition [29].

$$\cdot OH + HCO_3^- \rightarrow H_2O + CO_3^{\cdot -}\ k = 4.2 \times 10^8\ M^{-1}\cdot s^{-1} \quad (10)$$

In Figure 3c, as the Cl^- was enhanced from 1 to 30 mmol L^{-1}, the removal of HA decreased from 84.1% to 79.9%. Excessive Cl^- would consume $\cdot OH$ and form chlorine species (Equations (11) and (12)). The reduction in oxidation capacity led to a slight decrease in HA removal [30].

$$OH\cdot + Cl^- \rightarrow ClOH^{\cdot-} \quad k = 4.3 \times 10^9\ M^{-1}\cdot s^{-1} \tag{11}$$

$$Cl^- + Cl^- \rightarrow Cl_2^{\cdot-} \quad k = 8 \times 10^9\ M^{-1}\cdot s^{-1} \tag{12}$$

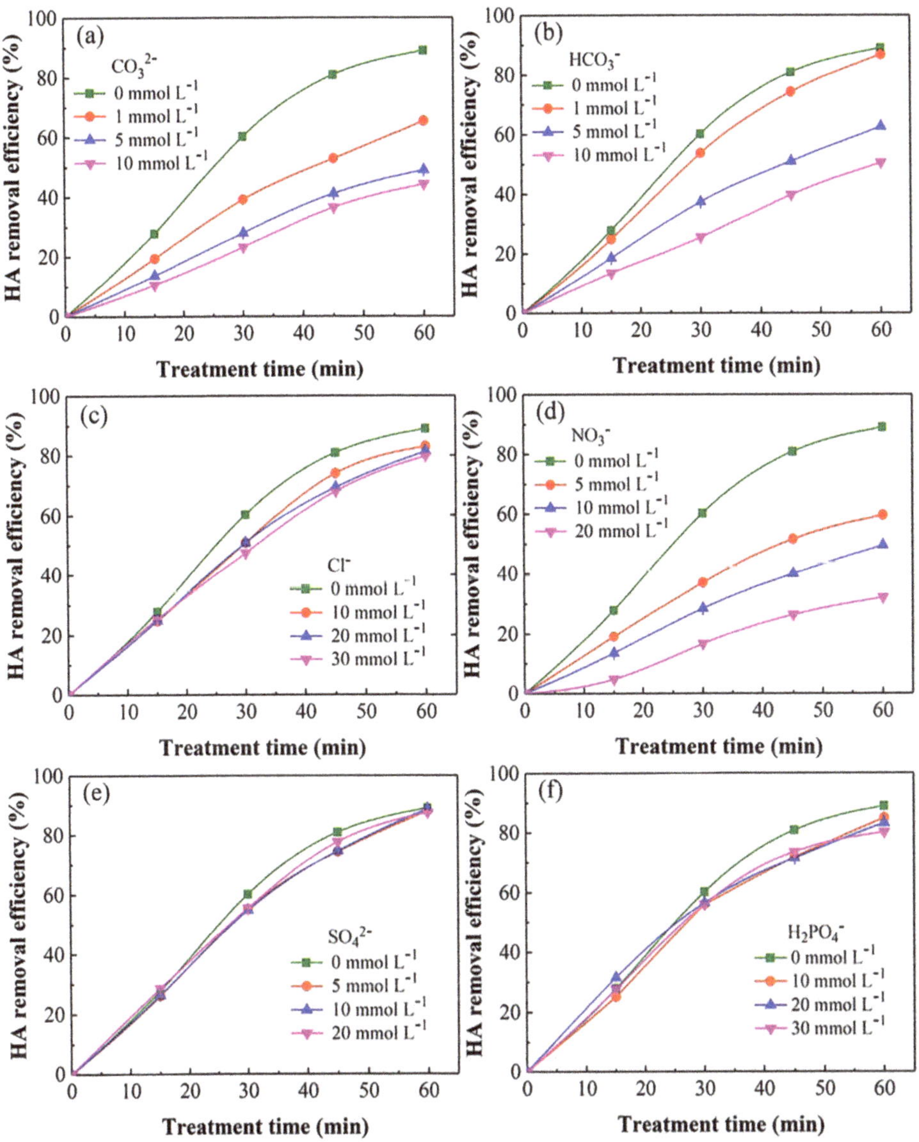

Figure 3. Effect of common anions on HA elimination in UV/SPB: (**a**) CO_3^{2-}, (**b**) HCO_3^-, (**c**) Cl^-, (**d**) NO_3^-, (**e**) SO_4^{2-}, (**f**) $H_2PO_4^-$. ($[HA]_0$ = 10 mg L^{-1}, $[SPB]_0$ = 1 mmol L^{-1}, pH = 3).

The more greater the addition of NO_3^-, the lower HA removal ratio was observed in Figure 3d. While 20 mmol L^{-1} NO_3^- was added, the HA removal was decreased to 33.4%. UV activated NO_3^- can yield free radical $NO_2\cdot$ (E_0 = 0.867 V), which has lower oxidation ability and would be annihilated during the UV/SPB (Equations (13)–(15)) [31]. In addition, NO_3^- could also directly consume ·OH (Equation (16)) [32].

$$NO_3^- + hv \rightarrow NO_2^{\cdot} + O^{\cdot-} \quad (13)$$

$$NO_3^- + hv \rightarrow NO_2^- + O \quad (14)$$

$$2NO_2^{\cdot} + H_2O \rightarrow NO_3^- + NO_2^- + 2H^+ \quad (15)$$

$$\cdot OH + NO_3^- \rightarrow NO_3^{\cdot} + OH^- \; k = 4.0 \times 10^5 \, M^{-1} \cdot s^{-1} \quad (16)$$

Figure 3e shows that the increase in SO_4^{2-} concentration did not affect the HA removal effect. When the SO_4^{2-} dosage enhanced to 20 mmol L^{-1}, the HA removal efficiency could reach 88.0%. According to the literature, SO_4^{2-} does not react with active species generated in the system [33,34], so it does not affect the HA removal remarkably.

As presented in Figure 3f, the HA elimination changed insignificantly with the addition of $H_2PO_4^-$. When the $H_2PO_4^-$ amount enhanced from 10 to 30 mmol L^{-1}, the HA elimination efficiencies were diminished from 86.3% to 82.4%. Although $H_2PO_4^-$ can react with ·OH· to form hydrogen phosphate radical (HPO_4^-) (Equation (17)) [35], the reaction rate is too low, which could not obviously impact the HA removal.

$$\cdot OH + H_2PO_4^- \rightarrow HPO_4^{\cdot -} + H_2O \; k = 2 \times 10^4 \, M^{-1} \cdot s^{-1} \quad (17)$$

3.2. Mechanism of HA Removal by UV/SPB

3.2.1. Scavenging Test

TBA could scavenge ·OH during oxidation process ($k_{TBA,\cdot OH}$ = 3.8–7.6 × 10^8 $M^{-1} \cdot s^{-1}$) [36]. Figure 4 demonstrates the influence of TBA addition on the HA removal in the UV/SPB. The experimental conditions were HA concentration of 10 mg L^{-1}, SPB dose of 1 mmol L^{-1}, and initial pH 3. As can be seen from the figure, adding TBA obviously restrained the HA elimination, which diminished from 16.5% to 11.5% with the addition of TBA enhanced from 0.05 to 0.5 mol L^{-1}. It testified that the main oxidizing species in the synergistic system could be ·OH. The elimination of HA was not completely inhibited after adding TBA, and this could be due to the fact that $HO_2\cdot$ produced by decomposition of H_2O_2 could also generate other reactive radicals, such as superoxide anion radicals ($O_2^{\cdot -}$) and single oxygen ($^1O_2$) (Equations (18)–(22)) [37,38], which also have a certain oxidation capacity and cannot be entirely suppressed by TBA.

$$H_2O_2 \rightarrow 2\cdot OH \quad (18)$$

$$\cdot OH + H_2O_2 \rightarrow HO_2^{\cdot} + H_2O \quad (19)$$

$$HO_2^{\cdot} \rightarrow O_2^{\cdot -} + H^+ \quad (20)$$

$$\cdot OH + O_2^{\cdot -} \rightarrow {}^1O_2 + OH^- \quad (21)$$

$$2O_2^{\cdot -} + 2H^+ \rightarrow {}^1O_2 + H_2O_2 \quad (22)$$

3.2.2. HA Decomposition Mechanism

The absorbance ratios (A_{253}/A_{203}, A_{250}/A_{365}, A_{254}/A_{436} and A_{465}/A_{665}) could reflect the transformation of HA molecular structure [39]. The evolution of these ratios in the UV/SPB system is shown in Figure 5a. With the increase in reaction time, the value of A_{253}/A_{203} decreased from 0.98 to 0.44, indicating that the stability of functional groups (such as carboxyl and carbonyl groups) in HA aromatic structure gradually diminished. The increase of A_{250}/A_{365} value from 2.42 to 3.20 suggested the decrease in the HA molecular weight. The A_{254}/A_{436} increased from 4.63 to 5.60, demonstrating that the HA chromophore was destroyed. The value of A_{465}/A_{665} decreased from 3.5 to 1.0 and proved the destruction of aromaticity in HA.

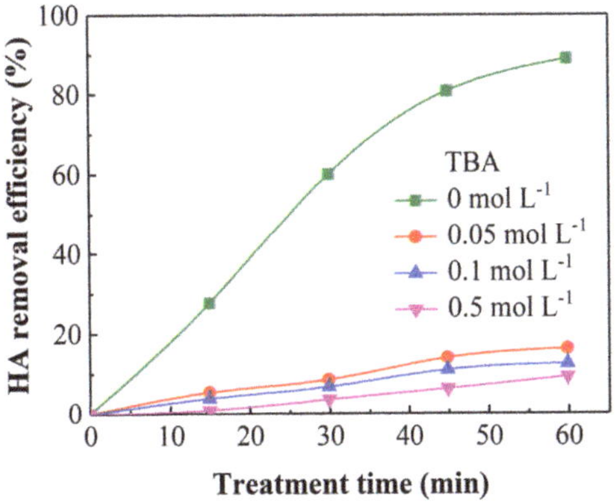

Figure 4. Influence of TBA addition on HA elimination. $[HA]_0 = 10$ mg L^{-1}, $[SPB]_0 = 1$ mmol L^{-1}, pH = 3.

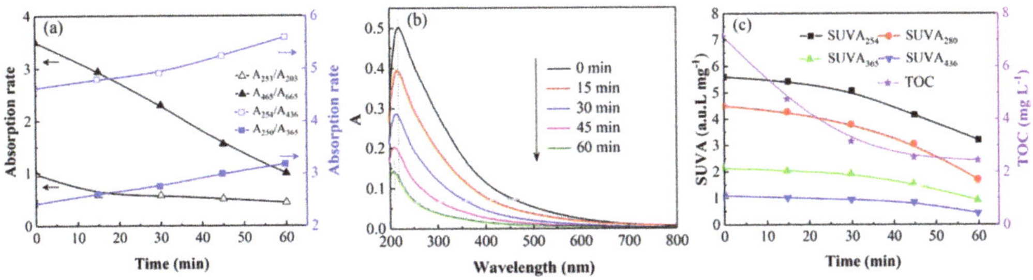

Figure 5. (**a**) UV absorption rate; (**b**) UV-vi spectrum; (**c**) $SUVA_x$ and TOC concentration for HA decomposition. $[HA]_0 = 10$ mg L^{-1}, $[SPB]_0 = 1$ mmol L^{-1}, pH = 3.

The UV-visible absorption spectrum can also manifest the changes of HA molecular structure. Figure 5b shows that the variation of the absorption spectrum of HA in the UV/SPB system over time. The absorption peak of HA at 200–250 nm weakened over time, indicating that ·OH generated in the UV/SPB system destroyed the chromophore groups and double bond structure of HA, and the unsaturated ketone was also oxidized. Moreover, the absorption peak moved to the direction of short wavelength, which was the phenomenon of blue shift. This demonstrated that the substitution reaction of a carbon atom occurred at the carbonylic group in the HA chromophore [40]. HA and fulvic acid are composed of linearly thickened aromatic rings and unsaturated carbons. When the carbon atom of chromophore, such as carbonyl, is substituted by some substituent groups, the absorption peak would shift to the short wavelength [41].

Specific ultraviolet absorbances ($SUVA_{254}$, $SUVA_{280}$, $SUVA_{365}$, and $SUVA_{436}$) were generally selected to characterize the degradation and mineralization for NOM. $SUVA_{254}$ reflects the molecular weight, $SUVA_{280}$ corresponds to the integrity of aromatic structure, $SUVA_{365}$ indicates the molecular volume, and $SUVA_{436}$ represents the chromophore situation in NOM [39]. It can be seen from Figure 5c that the decrease in the $SUVA_{254}$ and $SUVA_{280}$ values over time demonstrated that the molecular weight of organic compounds diminished, and the original aromatic structure was destructed after 60 min treatment in the UV/SPB. The decline of $SUVA_{365}$ indicated that the volume of organic molecules reduced as the reaction progressed. The decreasing value of $SUVA_{436}$ proved that the functional groups and chromophores were destructed by various reactive species. In addition, TOC in the system was lessened from 7.139 to 2.440 mg L^{-1}, and the mineralization efficiency reached 65.8%, demonstrating that most of HA had been mineralized into H_2O and CO_2. The results of UV spectrum and TOC proved that the UV/SPB synergistic treatment can effectively decompose the complex molecular structure of HA.

3D-EEM was used to further explore the degradation mechanism of HA in the UV/SPB system, and the results are shown in Figure 6. The scanning spectrum was divided into five regions because of the complexity of spectral response and scanning sample. The range of I and II can represent aromatic proteins in organic compounds which related to the structure of aromatic ring amino acids in NOM [42]. III region indicates fulminate-like substances related to hydroxyl and carboxyl groups in the humus structure. The range of region IV corresponds to the small molecular structure of organic matter [35]. The V region indicates to humic-like fluorescence [42,43]. It is clear that the fluorescence intensity of the five regions all diminished and gradually disappeared from 0 min (Figure 6a) to 15 min (Figure 6b) and 60 min (Figure 6c), which further proved that the HA molecular structure was decomposed and mineralized in this collaborative system.

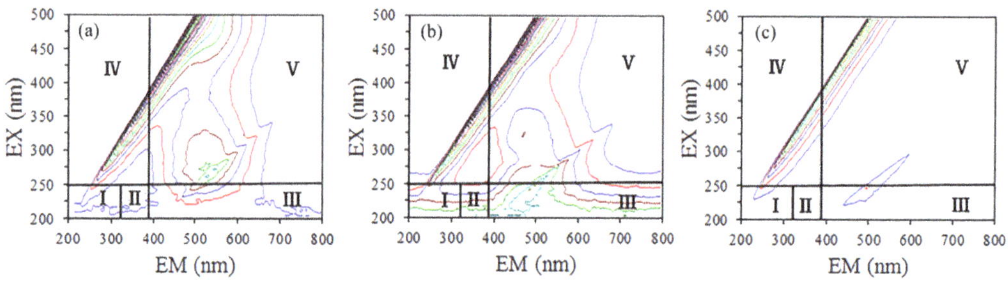

Figure 6. Time-varying 3D-EEM spectra of HA in UV/SPB system. (**a**) 0 min, (**b**) 15 min, (**c**) 60 min.

4. Conclusions

In this study, the UV/SPB synergistic system was established to remove HA from water, and the experimental results demonstrated that the UV/SPB process can effectively decompose HA. Under the experimental conditions of 10 mg L^{-1} HA concentration, 1 mmol L^{-1} SPB dosage, and beginning pH 3, the HA removal efficiency reached 88.8% after 60 min treatment. The HA removals in lake water and tap water were significantly lower than that in deionized water. The anion effect experiments confirmed that except for Cl^-, SO_4^{2-}, and $H_2PO_4^-$, the CO_3^{2-}, HCO_3^-, and NO_3^- inhibited the removal of HA to different degrees. The ·OH generated by SPB activation was suggested to be the main active substance for HA removal by masking experiments. The results of UV-vis spectrum, absorbance ratio, specific ultraviolet absorbance, and 3D-EEM jointly proved that the synergistic system could effectively degrade and mineralize HA in water.

Author Contributions: Conceptualization, D.Y. and S.T.; writing—review and editing, D.Y.; writing—original draft preparation, Z.Z.; investigation, E.Z. and H.L.; supervision, T.J.; project administration, S.T.; funding acquisition, D.Y. and S.T. All authors have read and agreed to the published version of the manuscript.

Funding: The authors acknowledge the Central Guidance on Local Science and Technology Development Fund of Hebei Province (No. 226Z3603G), the Natural Science Foundation of Hebei province (Nos. E2020203185, B2020203033), the Hebei Province Foundation for Returnees (No. C20210502), and the University Science and Technology Program Project of Hebei Provincial Department of Education (No. QN2020143).

Institutional Review Board Statement: Not applicable.

Informed Consent Statement: Not applicable.

Data Availability Statement: Not applicable.

Conflicts of Interest: The authors declare no conflict of interest.

References

1. Ryu, J.; Jung, J.; Park, K.; Song, W.; Choi, B.; Kweon, J. Humic acid removal and microbial community function in membrane bioreactor. *J. Hazard. Mater.* **2021**, *417*, 126088. [CrossRef] [PubMed]
2. Cui, L.; Zhang, Y.; He, K.; Sun, M.; Zhang, Z. Ti4O7 reactive electrochemical membrane for humic acid removal: Insights of electrosorption and electrooxidation. *Sep. Purif. Technol.* **2022**, *293*, 121112. [CrossRef]
3. Yue, Y.; An, G.; Lin, L.; Demissie, H.; Yang, X.; Jiao, R.; Wang, D. Design and coagulation mechanism of a new functional composite coagulant in removing humic acid. *Sep. Purif. Technol.* **2022**, *292*, 121016. [CrossRef]
4. Huang, X.; Wan, Y.; Shi, B.; Shi, J. Effects of powdered activated carbon on the coagulation-flocculation process in humic acid and humic acid-kaolin water treatment. *Chemosphere* **2019**, *238*, 124637. [CrossRef]
5. Chen, Y.; Qian, Y.; Ma, J.; Mao, M.; Qian, L.; An, D. New insights into the cooperative adsorption behavior of Cr(VI) and humic acid in water by powdered activated carbon. *Sci. Total Environ.* **2022**, *817*, 153081. [CrossRef] [PubMed]
6. Lee, S.; Roh, Y.; Koh, D.-C. Oxidation and reduction of redox-sensitive elements in the presence of humic substances in subsurface environments: A review. *Chemosphere* **2018**, *220*, 86–97. [CrossRef] [PubMed]
7. Oskoei, V.; Dehghani, M.H.; Nazmara, S.; Heibati, B.; Asif, M.; Tyagi, I.; Agarwal, S.; Gupta, V.K. Removal of humic acid from aqueous solution using UV/ZnO nano-photocatalysis and adsorption. *J. Mol. Liq.* **2015**, *213*, 374–380. [CrossRef]
8. Doustkhah, E.; Esmat, M.; Fukata, N.; Ide, Y.; Hanaor, D.A.; Assadi, M.H.N. MOF-derived nanocrystalline ZnO with controlled orientation and photocatalytic activity. *Chemosphere* **2022**, *303*, 124932. [CrossRef]
9. Hashemzadeh, B.; Alamgholiloo, H.; Pesyan, N.N.; Asgari, E.; Sheikhmohammadi, A.; Yeganeh, J.; Hashemzadeh, H. Degradation of ciprofloxacin using hematite/MOF nanocomposite as a heterogeneous Fenton-like catalyst: A comparison of composite and core−shell structures. *Chemosphere* **2021**, *281*, 130970. [CrossRef]
10. Huang, B.; Qi, C.; Yang, Z.; Guo, Q.; Chen, W.; Zeng, G.; Lei, C. Pd/Fe3O4 nanocatalysts for highly effective and simultaneous removal of humic acids and Cr(VI) by electro-Fenton with H2O2 in situ electro-generated on the catalyst surface. *J. Catal.* **2017**, *352*, 337–350. [CrossRef]
11. Maqbool, T.; Ly, Q.V.; He, K.; Cui, L.; Zhang, Y.; Sun, M.; Zhang, Z. Reactive electrochemical ceramic membrane for effective removal of high concentration humic acid: Insights of different performance and mechanisms. *J. Membr. Sci.* **2022**, *651*, 120460. [CrossRef]
12. Zhang, C.; Dong, Y.; Li, B.; Li, F. Comparative study of three solid oxidants as substitutes of H2O2 used in Fe (III)-oxalate complex mediated Fenton system for photocatalytic elimination of reactive azo dye. *J. Clean. Prod.* **2018**, *177*, 245–253. [CrossRef]
13. Sindelar, H.R.; Brown, M.T.; Boyer, T.H. Evaluating UV/H2O2, UV/percarbonate, and UV/perborate for natural organic matter reduction from alternative water sources. *Chemosphere* **2014**, *105*, 112–118. [CrossRef]
14. Gao, J.; Song, J.; Ye, J.; Duan, X.; Dionysiou, D.D.; Yadav, J.S.; Nadagouda, M.N.; Yang, L.; Luo, S. Comparative toxicity reduction potential of UV/sodium percarbonate and UV/hydrogen peroxide treatments for bisphenol A in water: An integrated analysis using chemical, computational, biological, and metabolomic approaches. *Water Res.* **2020**, *190*, 116755. [CrossRef]
15. Habibi, D.; Zolfigol, M.A.; Safaiee, M.; Shamsian, A.; Ghorbani-Choghamarani, A. Catalytic oxidation of sulfides to sulfoxides using sodium perborate and/or sodium percarbonate and silica sulfuric acid in the presence of KBr. *Catal. Commun.* **2009**, *10*, 1257–1260. [CrossRef]
16. LACSA, F. Oxidative Degradation of Phenol via Heterogeneous Fenton-Like Reaction over Fe-ZSM5 Catalyst Using Sodium Perborate and Sodium Percarbonate as Oxidants. Ph.D. Thesis, Ateneo de Manila University, Metro Manila, Philippines, 2017.
17. Chang, X.; Lin, T.; Mo, J.; Xu, H.; Tao, H.; Liu, W. Coagulation combined with ultraviolet irradiation activated sodium percarbonate as pretreatment prior to ultrafiltration: Analysis of free radical oxidation mechanism and membrane fouling control. *Chemosphere* **2022**, *287*, 132049. [CrossRef]
18. Gao, J.; Nunes, R.F.; O'Shea, K.; Saylor, G.L.; Bu, L.; Kang, Y.-G.; Duan, X.; Dionysiou, D.D.; Luo, S. UV/Sodium percarbonate for bisphenol A treatment in water: Impact of water quality parameters on the formation of reactive radicals. *Water Res.* **2022**, *219*, 118457. [CrossRef]
19. Alminshid, A.H.; Alalwan, H.A.; Abdulghani, H.A.; Mohammed, M.M. Spectrophotometric study of ephedrine hydrochloride in drug using molecular absorption UV–Visible. Spectrochim. *Acta Part A Mol. Biomol. Spectrosc.* **2022**, *270*, 120828. [CrossRef]
20. Zhou, X.-F.; Liang, J.-P.; Zhao, Z.-L.; Yuan, H.; Qiao, J.-J.; Xu, Q.-N.; Wang, H.-L.; Wang, W.-C.; Yang, D.-Z. Ultra-high synergetic intensity for humic acid removal by coupling bubble discharge with activated carbon. *J. Hazard. Mater.* **2021**, *403*, 123626. [CrossRef]
21. Cui, Y.; Yu, J.; Su, M.; Jia, Z.; Liu, T.; Oinuma, G.; Yamauchi, T. Humic acid removal by gas–liquid interface discharge plasma: Performance, mechanism and comparison to ozonation. *Environ. Sci. Water Res. Technol.* **2019**, *5*, 152–160. [CrossRef]
22. Yuan, D.; Tang, J.; Nie, Z.; Tang, S. Study on humic acid removal in water by ultraviolet activated sodium percarbonate. *J. Yanshan Univ.* **2021**, *45*, 220–226.
23. Li, X.; Wu, B.; Zhang, Q.; Xu, D.; Liu, Y.; Ma, F.; Gu, Q.; Li, F. Mechanisms on the impacts of humic acids on persulfate/Fe2+-based groundwater remediation. *Chem. Eng. J.* **2019**, *378*, 122142. [CrossRef]
24. Tan, C.; Gao, N.; Deng, Y.; Zhang, Y.; Sui, M.; Deng, J.; Zhou, S. Degradation of antipyrine by UV, UV/H2O2 and UV/PS. *J. Hazard. Mater.* **2013**, *260*, 1008–1016. [CrossRef] [PubMed]

25. Buxton, G.V.; Greenstock, C.L.; Helman, W.P.; Ross, A.B. Critical review of rate constants for reactions of hydrated electrons, hydrogen atoms and hydroxyl radicals (OH/O) in Aqueous Solution. *J. Phys. Chem. Ref. Data* **1988**, *17*, 513–886. [CrossRef]
26. Tang, S.; Yuan, D.; Rao, Y.; Li, M.; Shi, G.; Gu, J.; Zhang, T. Percarbonate promoted antibiotic decomposition in dielectric barrier discharge plasma. *J. Hazard. Mater.* **2019**, *366*, 669–676. [CrossRef]
27. Das, D.; Bordoloi, A.; Achary, M.P.; Caldwell, D.J.; Suri, R.P. Degradation and inactivation of chromosomal and plasmid encoded resistance genes/ARBs and the impact of different matrices on UV and UV/H2O2 based advanced oxidation process. *Sci. Total Environ.* **2022**, *833*, 155205. [CrossRef]
28. Yuan, D.; Zhang, C.; Tang, S.; Wang, Z.; Sun, Q.; Zhang, X.; Jiao, T.; Zhang, Q. Ferric ion-ascorbic acid complex catalyzed calcium peroxide for organic wastewater treatment: Optimized by response surface method. *Chin. Chem. Lett.* **2021**, *32*, 3387–3392. [CrossRef]
29. Fang, J.; Fu, Y.; Shang, C. The Roles of Reactive Species in Micropollutant Degradation in the UV/Free Chlorine System. *Environ. Sci. Technol.* **2014**, *48*, 1859–1868. [CrossRef]
30. Cai, L.; Li, L.; Yu, S.; Guo, J.; Kuppers, S.; Dong, L. Formation of odorous by-products during chlorination of major amino acids in East Taihu Lake: Impacts of UV, UV/PS and UV/H2O2 pre-treatments. *Water Res.* **2019**, *162*, 427–436. [CrossRef]
31. Ji, Y.; Zeng, C.; Ferronato, C.; Chovelon, J.-M.; Yang, X. Nitrate-induced photodegradation of atenolol in aqueous solution: Kinetics, toxicity and degradation pathways. *Chemosphere* **2012**, *88*, 644–649. [CrossRef]
32. Xu, Z.; Shan, C.; Xie, B.; Liu, Y.; Pan, B. Decomplexation of Cu(II)-EDTA by UV/persulfate and UV/H2O2: Efficiency and mechanism. *Appl. Catal. B Environ.* **2017**, *200*, 439–447. [CrossRef]
33. Tang, S.; Wang, Z.; Yuan, D.; Zhang, C.; Rao, Y.; Wang, Z.; Yin, K. Ferrous ion-tartaric acid chelation promoted calcium peroxide fenton-like reactions for simulated organic wastewater treatment. *J. Clean. Prod.* **2020**, *268*, 122253. [CrossRef]
34. Yuan, D.; Yang, K.; Zhu, E.; Li, X.; Sun, M.; Xiao, L.; Hari, Q.; Tang, S. Peracetic Acid Activated with Electro-Fe2+ Process for Dye Removal in Water. *Coatings* **2022**, *12*, 466. [CrossRef]
35. Tang, S.; Tang, J.; Yuan, D.; Wang, Z.; Zhang, Y.; Rao, Y. Elimination of humic acid in water: Comparison of UV/PDS and UV/PMS. *RSC Adv.* **2020**, *10*, 17627–17634. [CrossRef]
36. Zhao, M.; Xiang, Y.; Jiao, X.; Cao, B.; Tang, S.; Zheng, Z.; Zhang, X.; Jiao, T.; Yuan, D. MoS2 co-catalysis promoted CaO2 Fenton-like process: Performance and mechanism. *Sep. Purif. Technol.* **2021**, *276*, 119289. [CrossRef]
37. Cheng, M.; Zeng, G.; Huang, D.; Lai, C.; Xu, P.; Zhang, C.; Liu, Y. Hydroxyl radicals based advanced oxidation processes (AOPs) for remediation of soils contaminated with organic compounds: A review. *Chem. Eng. J.* **2016**, *284*, 582–598. [CrossRef]
38. Xiang, Y.; Liu, H.; Zhu, E.; Yang, K.; Yuan, D.; Jiao, T.; Zhang, Q.; Tang, S. Application of inorganic materials as heterogeneous cocatalyst in Fenton/Fenton-like processes for wastewater treatment. *Sep. Purif. Technol.* **2022**, *295*, 121293. [CrossRef]
39. Wang, T.; Qu, G.; Ren, J.; Yan, Q.; Sun, Q.; Liang, D.; Hu, S. Evaluation of the potentials of humic acid removal in water by gas phase surface discharge plasma. *Water Res.* **2016**, *89*, 28–38. [CrossRef]
40. Li, Y.; Qu, G.; Zhang, L.; Wang, T.; Sun, Q.; Liang, D.; Hu, S. Humic acid removal from micro-polluted source water using gas phase surface discharge plasma at different grounding modes. *Sep. Purif. Technol.* **2017**, *180*, 36–43. [CrossRef]
41. Wang, T.; Dongli, L.; Qu, G.; Sun, Q.; Liang, D.; Hu, S.; Zhu, L. Enhanced removal of humic acid from micro-polluted source water in a surface discharge plasma system coupled with activated carbon. *Environ. Sci. Pollut. Res.* **2017**, *24*, 21591–21600. [CrossRef]
42. Sarangapani, C.; Lu, P.; Behan, P.; Bourke, P.; Cullen, P. Humic acid and trihalomethane breakdown with potential by-product formations for atmospheric air plasma water treatment. *J. Ind. Eng. Chem.* **2018**, *59*, 350–361. [CrossRef]
43. Guo, H.; Yang, H.; Huang, J.; Tong, J.; Liu, X.; Wang, Y.; Qiao, W.; Han, J. Theoretical and experimental insight into plasma-catalytic degradation of aqueous p-nitrophenol with graphene-ZnO nanoparticles. *Sep. Purif. Technol.* **2022**, *295*, 121362. [CrossRef]

coatings

Article

Influence of Electrostatic Field on Mixed Aqueous Solution of Calcium and Ferrous Ions: Insights from Molecular Dynamics Simulations

Yong Han [1,2,*], Bingjia Wei [2], Xiaoqiang Guo [2] and Tifeng Jiao [1,*]

[1] State Key Laboratory of Metastable Materials Science and Technology, Yanshan University, Qinhuangdao 066004, China
[2] School of Electrical Engineering, Yanshan University, Qinhuangdao 066004, China
* Correspondence: hanyong@ysu.edu.cn (Y.H.); tfjiao@ysu.edu.cn (T.J.)

Abstract: In order to investigate the anti-scaling and anti-corrosion characteristics of an electrostatic anti-fouling system in the application process, the influence of an electrostatic field (EF) on the structure and dynamics of hydrated Ca^{2+} and hydrated Fe^{2+} in a mixed aqueous system was studied through the calculation and analysis of the radial distribution function (RDF), self-diffusion coefficients, viscosity, and hydrogen bond structure by using molecular dynamics simulation. The study results show that the EF can decrease the radius of the first water shell of hydrated Ca^{2+} but increase that of Fe^{2+}, which will reduce the possibility of forming calcite. The EF can make water molecules and Fe^{2+} more active, which can hinder iron release and thus decrease iron corrosion products. In addition, the EF can enhance the hydrogen structure of water molecules in the aqueous solution.

Keywords: electrostatic anti-fouling; calcium carbonate; ferrous ion; corrosion; molecular dynamics; electrostatic field

Citation: Han, Y.; Wei, B.; Guo, X.; Jiao, T. Influence of Electrostatic Field on Mixed Aqueous Solution of Calcium and Ferrous Ions: Insights from Molecular Dynamics Simulations. *Coatings* **2022**, *12*, 1165. https://doi.org/10.3390/coatings12081165

Received: 28 June 2022
Accepted: 10 August 2022
Published: 12 August 2022

1. Introduction

Fouling and iron corrosion are common problems in the water industry. Fouling results in degradation in the performance of heat exchangers. Iron corrosion can destroy the pipe and pollute the drinking water due to the suspension of iron particles. An electrostatic anti-fouling (EAF) system is one of the most common and effective ways to solve these problems. Water scale is almost entirely composed of calcium carbonate ($CaCO_3$), which naturally occurs in six various forms: calcite, aragonite, vaterite, two hydrate phases, and amorphous calcium carbonate (ACC). ACC is perhaps the most vital to growth processes since it can create the precursor to the crystallization of those phases and plays a critical role in the biomineralization process for this substance [1]. Thus, the electric field's impact on calcium carbonate has been the focus of the great majority of studies looking into the anti-fouling mechanism of EAF. It consists of two main aspects: how calcium carbonate crystals are formed and how the structure of scale crystals changes under electric fields. The impact of an electric field on calcium carbonate crystallization was studied by Qi et al. [2]. Their results show that the electric field applied to the $CaCO_3$ solution affects the lattice structure and crystal shape of $CaCO_3$ crystals. According to Li et al. [3], the scale sample produced by electrostatic water included more aragonite than before. Most equipment in industrial systems is made of carbon steel, and its main chemical composition is iron. Therefore, researchers who investigate the mechanism of metal corrosion tend to use iron as the study object. Lin et al. [4] found that the electric field accelerates steel corrosion, and the electric field favors the generation of plate-shaped γ-FeOOH and inhibits the formation of α-FeOOH. In addition, the scaling of calcium carbonate has a certain influence on iron corrosion. Sarin et al. [5] found that $CaCO_3$ plays a role in reducing iron release, which can alleviate water pollution problems, and Harris et al. found that the scaling tendency in water is the main factor determining the corrosion rate of iron [6]. The different contents of metal cations (including Fe^{2+}) in aqueous solutions also have a certain influence on the precipitation process of $CaCO_3$ [7]. Less attention has been paid to the impact of an electrostatic field (EF) on the microcosmic particle structure in aqueous systems.

Due to the relatively poor solubility of calcium carbonate, there are nearly zero $CO_3{}^{2-}$ ions in hard water, but $HCO_3{}^{-}$ ions are plentiful. As hydrated ions, Ca^{2+} and Fe^{2+} can be found in aqueous solutions. Additionally, these hydrated ions form hydrogen bonds with additional water molecules [8–10]. The formation, growth, and crystal structure of calcium carbonate precipitates are mainly determined by the structural and dynamic properties of Ca^{2+}. Fe^{2+} plays an important role in the metal corrosion process. The main reactions of iron corrosion are shown below.

The iron in the material loses two electrons and forms Fe^{2+}.

$$Fe \rightarrow Fe^{2+} + 2e^{-} \tag{1}$$

Electron acceptors such as oxygen or hydrogen ions in aqueous solutions serve to complete the reaction.

$$\frac{1}{2}O_2 + H_2O + 2e^{-} \leftrightarrow 2OH^{-} \tag{2}$$

$$2H^{+} + 2e^{-} \rightarrow H_2 \tag{3}$$

Ferrous solids such as ferrous hydroxide can form.

$$Fe^{2+} + 2OH^{-} \rightarrow Fe(OH)_2 \tag{4}$$

$Fe(OH)_2$ can be further oxidized to $Fe(OH)_3$ attached to the wall of the tube.

$$4Fe(OH)_2 + O_2 + 2H_2O \rightarrow 4Fe(OH)_3 \tag{5}$$

Therefore, by examining the impact of an EF on the structure and molecular dynamics of hydrated calcium ions and hydrated divalent iron ions, the mechanism of scale inhibition and corrosion under EAF in aqueous solutions may be more readily ascertained. This paper's study involves a blend of $CaCl_2$ and $FeCl_2$. The position of the first water shell of calcium ions and divalent iron ions, the framework of hydrated calcium ions and divalent iron ions, the self-diffusion coefficients of calcium ions, divalent iron ions, and water molecules, and hydrogen bonding in the liquid solution were all investigated using MD simulations to study the structural and kinetic properties of aqueous $CaCl_2$ solutions under an external EF. It should be noted that although the materials in this study and our previous work [11] are mixed aqueous solutions of two ions containing calcium ions, there are significant differences in the research details between the two. Our previous work mainly focused on the influence of magnesium ion impurities on the scaling process of calcium ions during the operation of the EAF system. However, in this study, we chose to pay attention to the influence of the EAF system on the corrosion process of iron water transmission pipes.

2. Model and Simulation Details

An equilibrium MD simulation was performed by using the Gromacs 5.0.2 software [12]. NPT ensembles were used for all simulations. The thermostat was adjusted to 300 K. The pressure was established at 1.01×10^5 Pa. As stated in Table 1, four sets of aqueous solution systems were constructed. The $CaCl_2$ aqueous solution had a concentration of 4 mol/kg, and there was only a 0.5 percent difference in concentration between models and tests [13]. In our study, the water molecule was represented by the SPCE model. In this simulation, ion–water interactions were simulated using the GROMOS force field [14–16]. A cutoff of 1.2 nm was used for Lennard-Jones interactions, and a real-space cutoff of 1.2 nm was used for electrostatic interactions when using the particle mesh Ewald (PME) [15] technique. Fourier spacing has a value of 0.12 nm. The Lennard-Jones potential parameters for the particles utilized in the GROMOS force field are shown in Table 2. The Lennard-Jones parameters of Fe^{2+} were taken from Fernanda et al. [17]. All of the simulations employed a 2.0 fs time step, and a 2.5×10^4 ps simulation time was used to confirm the stability of the system setup. Following a 1×10^4 ps period of system equilibration, the structures and characteristics of the sample (apart from the self-diffusion coefficient data) were determined throughout a 1.5×10^4 ps period (the final 1.5×10^4 ps of each simulation). By computing the average result of 1.5×10^4 ps, which was separated into ten segments, the average values of the parameters were acquired. The leapfrog method is a type of integration algorithm used for motion equations. The simulation box measures 4 nm, 4 nm, and 3 nm in length. The EF was 1 V/μm in intensity. The EF was used in a certain way (corresponding to the y-axis). The simulations were run as described in the step above. Previous research has confirmed the accuracy of our MD simulations and error analyses [18].

Table 1. Composition of the $CaCl_2$ solution studied by Gromacs.

Sample	[H_2O]	[Fe^{2+}]	[Ca^{2+}]	[Cl^-]	[Total]
0F	1110	0	80	160	1350
10F	1110	10	80	180	1380
20F	1110	20	80	200	1410
30F	1110	30	80	220	1440

NOTE: "0F" indicates no Fe^{2+} ions were added, "10F" indicates 10 Fe^{2+} were added, "20F" indicates 20 Fe^{2+} were added, and "30F" indicates 30 Fe^{2+} were added.

Table 2. The MD simulations employed the Lennard-Jones potential parameters according to the combining rules of Lorentz and Berthelot, i.e., $\varepsilon_{ij} = (\varepsilon_{ii}\varepsilon_{jj})^{1/2}$, $\sigma_{ij} = (\sigma_{ii} + \sigma_{jj})/2$.

	ε_{ii}/(kj/mol)	σ_{ii}/nm
Ca^{2+}	0.5069	0.2813
Fe^{2+}	2.135×10^{-6}	1.9129
Cl^-	1.2889	0.3470
O	1.7250	0.2626

3. Results and Discussion

3.1. Structural Parameters

The hydration shell regulates the mobility of ion pairs in the solution and the interaction between ion pairs [19]. A change in the hydration shell can affect calcium carbonate precipitation [20], and the layers of hydration around these ions define the route(s) for crystal formation in aquatic settings [21]. In this segment, the average radii of the first water shells of Ca^{2+} ($R_{Ca\text{-}O}$), Fe^{2+} ($R_{Fe\text{-}O}$), Ca^{2+} ($n_{Ca\text{-}O}$), Fe^{2+} ($n_{Fe\text{-}O}$), Ca^{2+} ($n_{Ca\text{-}O}$), and Fe^{2+} ($n_{Fe\text{-}O}$) under an external EF and under no EF are provided.

It is possible to determine $R_{Ca\text{-}O}$ and $R_{Fe\text{-}O}$ from the RDF [g(r)]; [g(r)] [22] is calculated as shown in Equation (6):

$$g_{AB}(r) = \frac{\langle \rho_B(r) \rangle}{\langle \rho_B \rangle_{local}} = \frac{1}{\langle \rho_B \rangle_{local}} \frac{1}{N_A} \sum_{i \in A}^{N_A} \sum_{j \in B}^{N_B} \frac{\delta(r_{ij} - r)}{4\pi r^2} \tag{6}$$

where $\langle \rho_B(r) \rangle$ is the particle density of particle B on all spheres surrounding particle A with radius r_{max}, $\langle \rho_B \rangle_{local}$ is the particle density of particle B at distance r around particle A, N_A denotes the number of particles A, and N_B denotes the number of particles B. Typically, r_{max} has a value equal to half of the box's minimum length.

The values of $n_{Ca\text{-}O}$ and $n_{Fe\text{-}O}$ are calculated using Equation (7) [14]:

$$n_{Ca}^{O}(r_s) = 4\pi\rho_0 \int_0^{r_s} g_{CaO}(r) r^2 dr \tag{7}$$

For $n_{Ca\text{-}O}$, r_s is the location of the first local minimum of $g_{CaO}(r)$, which is also the location of $R_{Ca\text{-}O}$; for $n_{Fe\text{-}O}$, r_s is the location of the first local minimum of $g_{FeO}(r)$, which is also the location of $R_{Fe\text{-}O}$.

Figure 1 shows RDFs and the corresponding running integration number of the $CaCl_2$ solution containing different numbers of Fe^{2+}. The coordinates of the first peak in $g_{CaO}(r)$, the coordinates of the first peak in $g_{FeO}(r)$, $R_{Ca\text{-}O}$, $R_{Fe\text{-}O}$, $n_{Fe\text{-}O}$ and $n_{Ca\text{-}O}$ are presented.

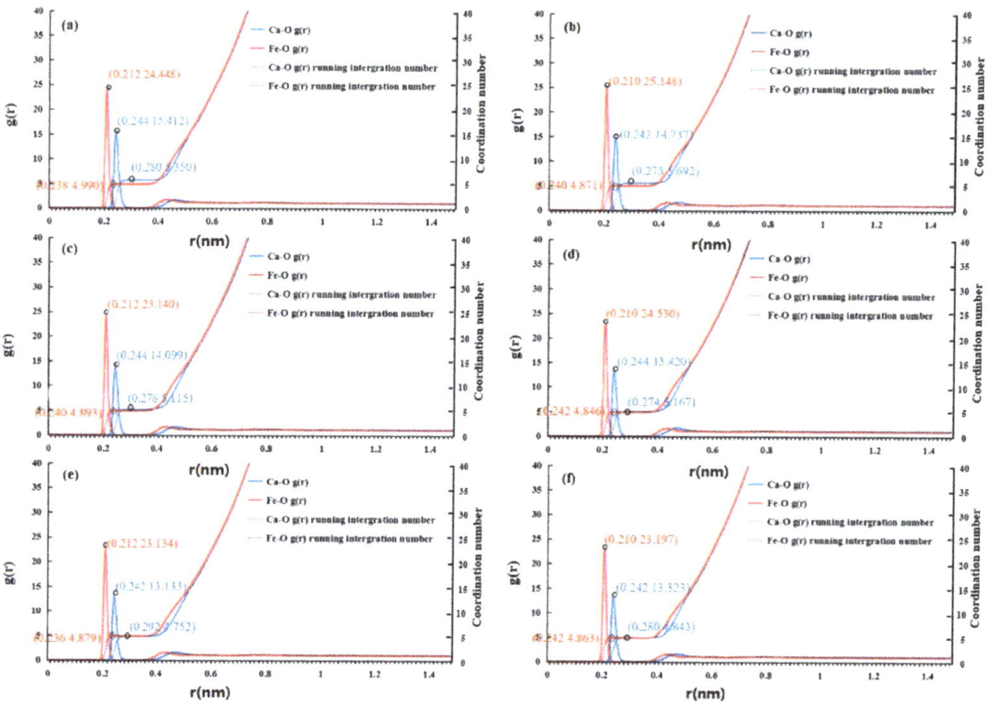

Figure 1. RDFs (curves with one obvious peak) and corresponding running integration numbers (single increment curves) of $CaCl_2$ solution containing different numbers of Fe^{2+} with an electric field applied (**b**,**d**,**f**) or without an electric field (**a**,**c**,**e**): (**a**,**b**) 10 Fe^{2+} added, (**c**,**d**) 20 Fe^{2+} added, and (**e**,**f**) 30 Fe^{2+} added.

According to Figure 1a,c,e, it can be seen that the height of the first peak in Fe^{2+}-O g(r) is higher than its Ca^{2+}-O g(r) counterpart. The first peak position of Fe^{2+}-O g(r) is less than the first peak position of Ca^{2+}-O g(r), which implies that Fe^{2+} has a smaller size than Ca^{2+}, and the hydrated structure of Fe^{2+} is more stable than Ca^{2+} [23]. Furthermore, the higher the content of Fe^{2+} ions, the lower the heights of the first peak in Ca^{2+}-O g(r). Fe^{2+} was added, but it had no discernible impact on $R_{Ca\text{-}O}$. With an increase in the amount of Fe^{2+} supplied, the $n_{Ca\text{-}O}$ drops. The decrease in $n_{Ca\text{-}O}$ could be due to the more stable hydrated structure of Fe^{2+} compared to Ca^{2+}, which means that fewer water molecules remain in the water shell of Ca^{2+}. In the $CaCl_2$ solution containing Fe^{2+} under an EF, Figure 1b,d,f demonstrate that the values of $R_{Ca\text{-}O}$ decrease while the values of $n_{Ca\text{-}O}$ increase, which is consistent with our earlier research on "pure" $CaCl_2$ aqueous solutions [24]. This is because the average binding ability of water molecules increases monotonically as the hydration shell becomes smaller [25,26]. In addition, the larger height and sharpness of the ferrous peak suggest a more stable structure under the EF, which means that the EF can make the Fe^{2+} radius smaller. The enthalpy of formation of ACC clusters decreases as the coordination number increases in the first hydration shell of Ca^{2+}, and the stability of

ACC correlates with water content. Experimentally, it has been proposed that "transient" ACC is largely anhydrous, while "stable" ACC contains some water [20]. Therefore, the EF increases the coordination number in the first hydration shell of Ca^{2+} and decreases the enthalpy of ACC, which can enhance the stability of ACC. To a certain extent, the EF reduces the possibility of ACC converting to calcite by changing the structure of the first hydration shell of Ca^{2+}, which will play a role in scale inhibition (according to our previous study [27]). The creation of carbonate ions close to Ca^{2+} is made easier by the reduced radius of the initial water shell of Ca^{2+}, which promotes the development of calcium carbonate crystals [24].

Figure 1 shows that the values of $R_{Fe\text{-}O}$ range between 0.230 and 0.240 nm, and the values of $n_{Fe\text{-}O}$ are distributed between 4.800 and 5.000, which is consistent with the results reported in Reference [28]. In $CaCl_2$ solution under an EF, the values of $R_{Fe\text{-}O}$ are larger, and the values of $n_{Fe\text{-}O}$ are lower. According to experimental findings [29], this suggests that the EF decreases hydration between Fe^{2+} and water molecules.

In addition, we also calculated the RDF of chloride ions around Ca^{2+} and Fe^{2+} under an EF (not shown here). The calculation results show that although the influence of the EF on the distribution of chloride ions around cations is small, it enhances the interaction between cations and anions (the number of anions around cations increases slightly). This is consistent with the conclusion of our previous paper. Therefore, the distribution of chloride ions is not analyzed in detail in this section.

3.2. Self-Diffusion Coefficients of Ca^{2+} and Fe^{2+} and Water Molecule

This section examines Ca^{2+}, Fe^{2+}, and water molecule self-diffusion coefficients in the presence of an exogenous EF in the $CaCl_2$ aqueous solution system. An important factor that may affect the course of chemical reactions in aqueous systems is the self-diffusion coefficient of ions. In addition, the self-diffusion coefficient may change significantly when the structure of the cation or anion is changed or another ion is involved [30]. It is worth noting that, in the simulation model in this paper, the solution seems to be strongly concentrated since there is 4 mol/kg $CaCl_2$, which means 4 mol of Ca^{2+} and 8 mol of Cl^- dissolved in 55.6 mol of water that solvates them (see Table 1). It may be considered that there are no free water molecules, because all water molecules seem to become ionic coordination water. This may make water molecules lack independent diffusivity. In fact, as the ions move in the water, the coordination water around the ions is not fixed, and they will be constantly replaced by the surrounding water molecules [13]. Therefore, the water molecules in this paper can be considered to have independent diffusivity, but this diffusivity will be affected by ions.

In this research, the "Einstein relation" [31] was used to calculate the self-diffusion coefficients of Ca^{2+}(D_{ca}) and water molecules (D_o):

$$D = \lim_{t\to\infty} \frac{1}{6t} \langle [r_i(t) - r_i(0)]^2 \rangle \quad (8)$$

where $r_i(0)$ is the starting position, the parentheses represent the set mean, and $r_i(t)$ is the numerator position i at time t. Fitting the slope of the linear component of the mean square displacement (MSD) and dividing by six are the particular steps in computing the self-diffusion coefficient of a particle. The RANSAC algorithm determines the linear component of the MSD. RANSAC is a technique for robust model fitting in the presence of outliers in the data [32].

The impact of Fe^{2+} on the values of D_{ca} and D_o with and without an EF is shown in Table 3. With regard to the self-diffusion coefficient, the standard deviation of the simulation results presented in Table 3 is conservatively estimated to be in the region of 0.41%–5.11%. Table 3 shows that whether or not there is an EF, the values of D_{ca} and D_o both drop as the amount of Fe^{2+} increases. Moreover, the higher the Fe^{2+} ion content, the lower the values of D_{ca} and D_o. In addition, Table 3 shows that the external EF can increase the values of D_{ca} and D_o no matter how much Fe^{2+} exists in the $CaCl_2$ aqueous solution during all simulations.

Table 3. The 10^5 self-diffusion coefficients of Ca^{2+} (D_{ca}) and water molecules (D_o) with no Fe^{2+} ions added (*0F*) and ten Fe^{2+} added (*10F*), twenty Fe^{2+} added (*20F*), and thirty Fe^{2+} added (*30F*) with and without EF.

Sample	Average 10^{-5} $cm^2 \cdot s^{-1}$	St.dev 10^{-5} $cm^2 \cdot s^{-1}$	Average 10^{-5} $cm^2 \cdot s^{-1}$	St.dev 10^{-5} $cm^2 \cdot s^{-1}$
	Without Electrostatic Field		**Under Electrostatic Field**	
D_{ca}				
0F	0.0608	0.0003	0.0707	0.0001
10F	0.0185	0.0005	0.0239	0.0001
20F	0.0124	0.0003	0.0137	0.0007
30F	0.0108	0.0005	0.0128	0.0001
D_o				
0F	0.6626	0.0004	0.7276	0.0003
10F	0.1837	0.0004	0.1870	0.0006
20F	0.1573	0.0005	0.1596	0.0001
30F	0.1283	0.0005	0.1343	0.0003
D_{Fe}				
10F	0.0188	0.0009	0.0195	0.0004
20F	0.0158	0.0004	0.0180	0.0002
30F	0.0142	0.0002	0.0172	0.0004

The valence and structural order of cations in aquatic environments impact the rate at which water molecules diffuse [33]. According to Table 3, the values of D_o are 0.6626×10^{-5} $cm^2 \cdot s^{-1}$ and 0.7276×10^{-5} $cm^2 \cdot s^{-1}$ without and with an EF, respectively, when Fe^{2+} is not in the solutions, in contrast to the values of D_o in the range of 0.12×10^{-5} $cm^2 \cdot s^{-1}$–0.18×10^{-5} $cm^2 \cdot s^{-1}$ when different numbers of Fe^{2+} ions are present. It is worth noting that when iron ions exist, the existence of an electric field can only slightly increase the diffusion coefficient of water molecules (with an increase of about 5%), which may be due to the binding of iron ions to water molecules, weakening the influence of the electric field. Water molecules that are coupled to metal ions are successfully immobilized in relation to the bulk water by being contained in a hydration layer [21]. As a result, when Fe^{2+} is available in the $CaCl_2$ solution, its significant capacity to bind to water molecules restricts the activity of the water molecules. The D_o values decline. The presence of Fe^{2+} is responsible for the decrease in the self-diffusion coefficient of Ca^{2+}. The $CaCl_2$ aqueous solution becomes more compact due to the electrostatic contact between Fe^{2+} and the water molecule. Ca^{2+} activity is constrained, and its self-diffusion coefficient is decreased.

The EF lowers the water molecule's self-diffusion coefficient while increasing the hydrogen bond network's binding capacity in pure water solutions [34]. However, the findings in Table 3 demonstrate that for the four example $CaCl_2$ aqueous solutions, the EF may increase the water molecule's self-diffusion coefficient. Therefore, the rise in the diffusion coefficient of water molecules in aqueous $CaCl_2$ solutions can be attributed to the activity of metal ions in the solution. When Ca^{2+} and Fe^{2+} diffuse, they either displace their solvated shells along their trajectories or break the strong connections with the initial water shell [35]. Thus, the EF increases the activity of Ca^{2+} and Fe^{2+} and influences the relatively stable water molecule hydrogen bond network.

As can be seen in Table 3, the self-diffusion coefficients of water molecules and Fe^{2+} in aqueous $CaCl_2$ solutions grow in the presence of EF. The speed at which oxidants are transported to the scale's surface may be increased by increasing the activity of water molecules, which can reduce iron release and the amounts of iron corrosion products that dissolve in water [36,37]. Water contamination issues may be resolved. Additionally, the rise in D_o and D_{Fe} may indicate that the EF has the ability to accelerate the oxidation of Fe^{2+}.

3.3. Viscosity of the $CaCl_2$ Aqueous Solution

Generally, the formation of calcium carbonate precipitation involves three stages: nucleation, grain growth, and aggregation of $CaCO_3$ crystals [38]. The viscosity of the solution affects the precipitation of calcium carbonate and the corrosion behavior of iron in the solution [39–43]. According to experimental findings, the crystallinity of scale-forming ions mixed with water is inversely related to the viscosity of a liquid [44]. A liquid's dynamic viscosity is another kinetic characteristic that affects how quickly solutes in the liquid alter their conformation. It is susceptible to an EF [44].

The following equation can be used to determine the viscosity (η):

$$\eta = \frac{A}{V}\frac{\rho}{k^2} \tag{9}$$

where ρ and V stand for density and velocity, respectively; A is a constant [44], and we can deduce the following from a cosine that meets both requirements:

$$a_x = A\cos(kz), k = 2\pi/l_z \tag{10}$$

where l_z is the box's height. In the simulation, the instantaneous V is defined as a Fourier coefficient:

$$V(t) = \frac{2\sum_{i=1}^{N} m_i v_{i,x}(t)\cos[kr_{i,z}(t)]}{\sum_{i=1}^{N} m_i} \tag{11}$$

where m_i is the mass of an atom, $r_{i,z}$ is the z coordinate, and $v_{i,x}$ is the x component of the velocity. The amplitude of the velocity profile can be completely formed before measuring the average for V.

Four sample $CaCl_2$ aqueous solutions' viscosities under an EF are shown in Figure 2. For comparison, the values in the absence of the EF are shown. As shown in Figure 2, the presence of Fe^{2+} increases the viscosity values in the $CaCl_2$ system, and four typical aqueous $CaCl_2$ solutions become viscous in the presence of Fe^{2+}. The EF also reduces the viscosity of aqueous $CaCl_2$ solutions.

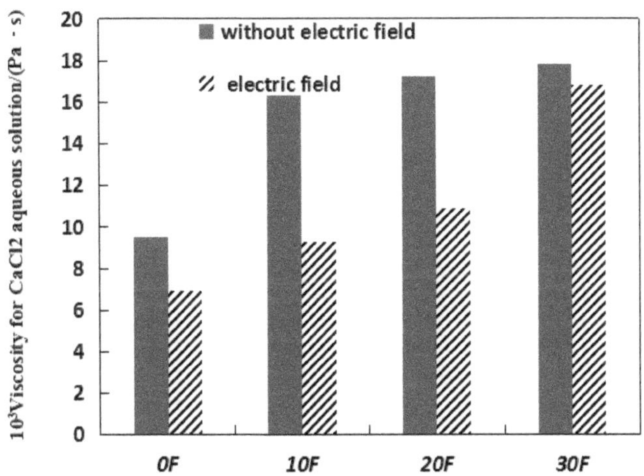

Figure 2. The 10^3 average viscosity with no Fe^{2+} ions added (*0F*) and 10 Fe^{2+} added (*10F*), 20 Fe^{2+} added (*20F*), and 30 Fe^{2+} added (*30F*) with and without EF and EF.

The molecular weight of the particles inside the system is inversely proportional to the dynamic viscosity, which is a reflection of a sort of friction brought on by fluid motion in the solution system. Bakker [45] found that viscosity is a distinct feature that describes the typical behavior of a large number of water molecules in aqueous solutions and that specific ions have a significant effect on the viscosity of water because of the extended lifetime of their initial hydration shells. Viscosity is substantially influenced by the water molecules' extended residence periods in ion hydration shells. The aqueous solution of $CaCl_2$ also has no carbon–hydrogen bonds, making it an inorganic salt. Therefore, changes in the viscosity of $CaCl_2$ solutions can be attributed to changes in the structural and kinetic properties of the water molecules.

Figure 2 demonstrates that without the EF, the viscosities of the four example $CaCl_2$ aqueous solutions are 9.49×10^{-3} Pa·s, 16.31×10^{-3} Pa·s, 17.24×10^{-3} Pa·s, and 17.81×10^{-3} Pa·s. This means that the addition of Fe^{2+} can cause a relatively high viscosity in $CaCl_2$ aqueous solutions. Dian et al. [46] studied Ca^{2+} solvation using polarizable atomic multipole potential, and their results show that the lifetime of water molecules in the first solvation shell of Ca^{2+} is 18 picoseconds (ps) when the temperature is 298 K. The experimental [47] values of the lifetime of water molecules in the first solvation shell of Fe^{2+} were determined in the range of 10^5–10^6 ps when the temperature is 298.15 K, in contrast to only a few picoseconds for Ca^{2+}. Therefore, the addition of Fe^{2+} forms a water shell, which can lead to a longer residence time for water molecules in Fe^{2+} hydration shells compared to Ca^{2+}. Therefore, the viscosity of the solution is greatly increased by Fe^{2+}. Furthermore, the mobility of water molecules is related to the viscosity of the solution. According to Table 3, the activity of water molecules is more limited when the Fe^{2+} ion content increases, and this may result in a rise in the solution's viscosity.

Figure 2 demonstrates that when Fe^{2+} ions are available in the solution, the EF can reduce the viscosity of $CaCl_2$ aqueous solutions. This is due to the fact that the EF reduces the mean dynamic residence time of water at Ca^{2+} and increases the activity of water molecules, both of which lead to a drop in dynamic viscosity. The viscosity of the salt solution can be reduced by the EF [48]. Electrostatic interaction between the water molecule and Fe^{2+} might raise the viscosity of the solution. This indicates that the dynamic viscosity of the $CaCl_2$ solution can be reduced by the presence of Fe^{2+} and an external EF. In conclusion, the aqueous $CaCl_2$ solution can become more viscous when Fe^{2+} ions are present, and the viscosity can be reduced when an EF is applied to an aqueous $CaCl_2$ solution that contains Fe^{2+}. The decrease in viscosity of $CaCl_2$ aqueous solutions, on the one hand, means that the scale-forming ions in the solution are reduced [40] and the fluidity of the solution is enhanced [49]. On the other hand, the decrease in the viscosity of $CaCl_2$ aqueous solutions reduces the aggregation of calcium carbonate precipitates [41–43], which plays an effective role in anti-fouling. The decrease in solution viscosity reduces the adhesion of the electrode surface to ions and increases the rate of transport of oxidants, which can influence the capacitance values of the double layer and reflect changes in the surface state of the corrosion metal [36,37].

3.4. Effect of Fe^{2+} and Electric Field on Hydrogen Bonds

The hydrogen bond network that is generated by water molecules has a significant impact on the precipitation of calcium carbonate [20,21] and is strongly connected to the chemical and physical characteristics of aqueous solutions [50,51]. The dissolution of calcium carbonate and corrosion products is influenced by the nature of the hydrogen bond network. According to Gromacs software, a hydrogen bond occurs when a donor and acceptor are separated by a van der Waals (VDW) length less than 3.5Å(r) and when the hydrogen donor–acceptor angle is less than 30° [52]. The geometrical hydrogen bond requirement is displayed in Figure 3. The oxygen atoms in water molecules are the study's donors and acceptors. This work estimated the hydrogen bond angle as $\angle O_D$-H_D- $O_A(\theta)$, which is a suitable angle that is frequently used in the literature and offers helpful insight into the flexibility of donor hydrogen bonds [53]. Calculating the number of hydrogen bonds allows one to assess the overall importance of a hydrogen bond. The number of hydrogen bonds and the strength of the hydrogen bonds are positively correlated [52].

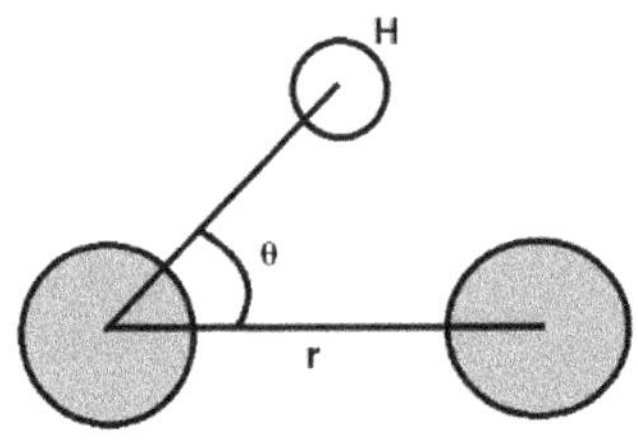

Figure 3. Geometrical hydrogen bond criterion.

The current study looked at both the number of hydrogen bonds (n_{HB}) and the percentage of water molecules with n or more hydrogen bonds (f_n). Without the addition of Fe^{2+} (*0F*) and with the addition of 10 Fe^{2+} (*10F*), 20 Fe^{2+} (*20F*), and 30 Fe^{2+} (*30F*), the values of f_n and n_{HB} in $CaCl_2$ aqueous solutions are shown in Table 4. (*30F*). Additionally, the effect of the EF is demonstrated.

Table 4. The proportion of water molecules that have n hydrogen bonds (f_n), the amount of Fe^{2+} that was added $F(n^*F)$, and the average number of hydrogen bonds per water molecule (n_{HB}).

	Without Electrostatic Field				Under Electrostatic Field			
	0F	*10F*	*20F*	*30F*	*0F*	*10F*	*20F*	*30F*
f_0(%)	58.09	61.54	62.20	65.02	57.99	60.93	61.10	64.08
f_1(%)	31.17	29.75	29.16	28.39	31.08	29.88	29.80	29.36
f_2(%)	8.23	7.37	7.30	6.59	8.34	7.80	7.70	7.04
f_3(%)	1.56	1.29	1.28	1.18	1.63	1.34	1.36	1.27
n_{HB}	580.67	536.78	528.55	512.19	584.43	549.33	546.80	524.60

Table 4 demonstrates how the EF may alter the percentage of molecules without hydrogen bonds while increasing the percentage of molecules with one to three hydrogen bonds and the average number of hydrogen bonds in a water molecule, regardless of the presence or absence of Fe^{2+}. The proportion of water molecules with *n* or more hydrogen bonds might be used to define the dimension of the water cluster. For instance, the size of the water cluster increases with the proportion of molecules with two or three hydrogen bonds. The decrease in f_0 and the rise in n_{HB}, f_1, f_2, and f_3 indicate that water molecule activity is constrained and that more water molecules are shifting from the free state to the hydrogen-bonded state. This indicates that the EF improves the hydrogen structure of the water molecules to some extent. The findings in Table 4 further demonstrate that, regardless of whether the EF is administered or not, Fe^{2+} can enhance f_0 and reduce n_{HB}, f_1, f_2, and f_3. Additionally, in the $CaCl_2$ systems, n_{HB}, f_1, f_2, and f_3 decrease as the amount of Fe^{2+} rises, whereas f_0 increases. This is the result of the Fe^{2+} ion's ability to attract ion–dipole interactions, which retains water molecules in the ion hydration shell and reduces the number of water molecules bound by hydrogen bonds [54]. This results in fewer hydrogen bonds being created, an increase in the percentage of molecules without hydrogen bonds, a drop in the percentage of molecules with one to three hydrogen bonds, and a rise in the mean number of hydrogen bonds per water molecule.

4. Conclusions

A methodical long-time balanced molecular dynamics simulation was used to examine the structural and dynamic characteristics of mixed solutions of $CaCl_2$ and $FeCl_2$ under an EF. Some significant findings were made.

(1) Compared to Ca^{2+}, the hydrated structure of Fe^{2+} is more stable. The EF may raise the first water coordination number for Ca^{2+} and reduce the radius of hydrated Ca^{2+}. The EF has the ability to reduce the first water coordination number for hydrated Fe^{2+}

and expand its radius. The presence of Fe^{2+} has no discernible impact on the radius of hydrated Ca^{2+}; however, it can lower the first water coordination number of Ca^{2+}.

(2) The viscosity of $CaCl_2$ in an aqueous solution can be efficiently raised by Fe^{2+}. The presence of an EF can make the mixed aqueous solution less viscous.

(3) The self-diffusion coefficient of Ca^{2+} can be lowered by Fe^{2+}. Ca^{2+}, Fe^{2+}, and water molecules' self-diffusion coefficients can all be increased by an EF.

(4) An EF can enhance the hydrogen structure of the water molecules in the aqueous solution.

To sum up, the EF generated by an EAF system can change the structure of the solvent layer of Ca^{2+}, resulting in a reduction in the possibility of Ca^{2+} forming high-density calcite scale. In addition, the existence of an EF can also increase the activity of water molecules and reduce the viscosity of the solution, which will effectively hinder the release of Fe^{2+} in water and the formation of iron corrosion products.

Author Contributions: Conceptualization, methodology, software, and validation, Y.H.; data curation and writing—original draft preparation, B.W.; visualization and investigation, X.G.; supervision and writing—reviewing and editing, T.J. All authors have read and agreed to the published version of the manuscript.

Funding: This work was financially supported by the National Natural Science Foundation of China (No. 51408525), University Science and Technology Research Projects of Hebei Province (No. ZD2020324), The Central Guidance on Local Science and Technology Development Fund of Hebei Province (No. 216Z3601G), and the Natural Science Foundation of Hebei Province (No. E2022203061).

Institutional Review Board Statement: Not applicable.

Informed Consent Statement: Not applicable.

Data Availability Statement: Data sharing is not applicable to this article.

Conflicts of Interest: The authors declare no conflict of interest.

References

1. Weiner, S.; Sagi, I.; Addadi, L. Choosing the crystallization path less traveled. *Science* **2005**, *309*, 1027–1028. [CrossRef] [PubMed]
2. Qi, J.Q.; Guo, R.; Wang, Y.; Liu, X.W.; Chan, H.L.W. Electric field-controlled crystallizing $CaCO_3$ nanostructures from solution. *Nanoscale Res. Lett.* **2016**, *11*, 120. [CrossRef] [PubMed]
3. Li, H.H.; Liu, Z.F.; Gao, Y.H.; Zhang, L.H.; Li, X.H. Influence of electrostatic water treatment on crystallization behavior of $CaCO_3$ and synergistic scale inhibition with a green scale inhibitor. *CIESC J. Chem. Ind. Eng.* **2013**, *64*, 1736–1742. [CrossRef]
4. Lin, D.; Dai, N.; Chen, Y.; Ni, Q.; Zhang, X.; Zhang, J. Effect of direct current electric field on initial corrosion of steel exposed to simulated marine environment. *Corros. Sci. Protetion Technol.* **2017**, *29*, 63–67. [CrossRef]
5. Sarin, P.; Snoeyink, V.L.; Lytle, D.A.; Kriven, W.M. Iron corrosion scales: Model for scale growth, iron release, and colored water formation. *J. Environ. Eng.* **2004**, *130*, 364–373. [CrossRef]
6. Finan, M.A.; Harris, A.; Marshall, A. Ecologically acceptable method of preventing corrosion/scale problems by control of water conditions. *Natl. Assoc. Corros. Eng.* **1980**, *204*, 24–29.
7. de Leeuw, N.H. Molecular dynamics simulations of the growth inhibiting effect of Fe^{2+}, Mg^{2+}, Cd^{2+}, and Sr^{2+} on calcite crystal growth. *J. Phys. Chem. B* **2002**, *106*, 5241–5249. [CrossRef]
8. Zhao, G.-J.; Liu, J.-Y.; Zhou, L.-C.; Han, K.-L. Site-selective photoinduced electron transfer from alcoholic solvents to the chromophore facilitated by hydrogen bonding: A new fluorescence quenching mechanism. *J. Phys. Chem. B* **2007**, *111*, 8940–8945. [CrossRef]
9. Zhao, G.J.; Han, K.L. Hydrogen bonding in the electronic excited state. *Accounts. Chem. Res.* **2012**, *45*, 404. [CrossRef]
10. Chen, J.S.; Yuan, M.H.; Wang, J.P.; Yang, Y.; Chu, T.S. Sensing mechanism for biothiols chemosensor DCO: Roles of excited-state hydrogen-bonding and intramolecular charge transfer. *J. Phys. Chem. A* **2014**, *118*, 8986–8995. [CrossRef]
11. Zhu, L.; Han, Y.; Zhang, C.; Zhao, R.; Tang, S. Molecular dynamics simulation for the impact of an electrostatic field and impurity Mg^{2+} ions on hard water. *RSC Adv.* **2017**, *7*, 47583. [CrossRef]
12. Nedyalkova, M.; Madurga, S.; Pisov, S.; Pastor, I.; Vilaseca, E.; Mas, F. Molecular dynamics simulation of the spherical electrical double layer of a soft nanoparticle: Effect of the surface charge and counterion valence. *J. Chem. Phys.* **2012**, *137*, 174701. [CrossRef] [PubMed]
13. Li, M.; Duan, Z.; Zhang, Z.; Zhang, C.; Weare, J. The structure, dynamics and solvation mechanisms of ions in water from long time molecular dynamics simulations: A case study of $CaCl_2$ (Aq) aqueous solutions. *Mol. Phys.* **2008**, *106*, 2685–2697. [CrossRef]

14. Chialvo, A.A.; Simonson, J.M. The structure of $CaCl_2$ aqueous solutions over a wide range of concentration. Interpretation of diffraction experiments via molecular simulation. *J. Chem. Phys.* **2003**, *119*, 8052–8061. [CrossRef]
15. Isele-Holder, R.E.; Mitchell, W.; Ismail, A.E. Development and application of a particle-particle particle-mesh ewald method for dispersion interactions. *J. Chem. Phys.* **2012**, *137*, 174107. [CrossRef] [PubMed]
16. Gunsteren, W.F.V.; Billeter, S.R.; Eising, A.A.; Hünenberger, P.H.; Krüger, P.K.H.C.; Mark, A.E.; Scott, W.R.P.; Tironi, I.G. *Biomolecular Simulation: The GROMOS96 Manual and User Guide*; Vdf Hochschulverlag AG an der ETH Zürich: Zürich, Switzerland, 1996; Volume 86, pp. 1–1044.
17. Duarte, F.; Bauer, P.; Barrozo, A.; Amrein, B.A.; Purg, M.; Åqvist, J.; Kamerlin, S.C.L. Force field independent metal parameters using a nonbonded dummy model. *J. Phys. Chem. B* **2014**, *118*, 4351–4362. [CrossRef] [PubMed]
18. Han, Y.; Zhao, Y. Influence to the structure and dynamics of hydrated Ca^{2+} and water molecules in $CaCl_2$ aqueous solution caused by external AC electric field: A mechanism study by molecular dynamics simulation. *Int. J. Electrochem. Sci.* **2012**, *7*, 10008–10026.
19. Lopez-Berganza, J.A.; Diao, Y.; Pamidighantam, S.; Espinosa-Marzal, R.M. Ab initio studies of calcium carbonate hydration. *J. Phys. Chem. A* **2015**, *119*, 11591–11600. [CrossRef]
20. Raiteri, P.; Gale, J.D. Water is the key to nonclassical nucleation of amorphous calcium carbonate. *J. Am. Chem. Soc.* **2010**, *132*, 17623–17634. [CrossRef]
21. Dorvee, J.R.; Veis, A. Water in the formation of biogenic minerals: Peeling away the hydration layers. *J. Struct. Biol.* **2013**, *183*, 278–303. [CrossRef]
22. York, D.M.; Darden, T.A.; Pedersen, L.G.; Anderson, M.W. Molecular dynamics simulation of HIV-1 protease in a crystalline environment and in solution. *Biochemistry* **1993**, *32*, 3196. [CrossRef]
23. Ye, Q.; Chen, X.; Zhou., J.; Zhao, H.F.; Chu, W.S.; Zheng, X.S.; Marcelli, A.; Wu, Z.Y. Local hydrated structure of an Fe^{2+}/Fe^{3+} aqueous solution: An investigation using a combination of molecular dynamics and X-ray absorption fine structure methods. *Chin. Phys. C* **2013**, *37*, 118–123. [CrossRef]
24. Han, Y.; Zhu, L.; Zhang, Y. Molecular dynamics simulation for the impact of external electric fields on $CaCl_2$ aqueous solution. *Chem. Res. Chin. Univ.* **2016**, *32*, 641–646. [CrossRef]
25. Cunningham, A.B.; Phillips, A.J.; Troyer, E.; Lauchnor, E.; Hiebert, R.; Gerlach, R.; Spangler, L. Wellbore leakage mitigation using engineered biomineralization. *Energy Procedia* **2014**, *63*, 4612–4619. [CrossRef]
26. Yang, Q.; Liu, Y.; Gu, A.; Ding, J.; Shen, Z. Investigation of calcium carbonate scaling inhibition and scale morphology by AFM. *J. Colloid. Interf. Sci.* **2001**, *240*, 608–621. [CrossRef] [PubMed]
27. Han, Y.; Zhang, C.; Wu, L.; Zhang, Q.; Zhu, L.; Zhao, R. Influence of alternating electromagnetic field and ultrasonic on calcium carbonate crystallization in the presence of magnesium ions. *J. Cryst. Growth* **2018**, *499*, 67–76. [CrossRef]
28. Floris, F.; Persico, M.; Tani, A.; Tomasi, J. Ab Initio effective pair potentials for simulations of the liquid state, based on the polarizable continuum model of the solvent. *Chem. Phys. Lett.* **1992**, *199*, 518–524. [CrossRef]
29. Busch, K.W.; Busch, M.A.; Parker, D.H.; Darling, R.E.; McAtee, J.L. Studies of a water treatment device that uses magnetic fields. *Corrosion* **1986**, *42*, 211–221. [CrossRef]
30. Tsuzuki, S.; Shinoda, W.; Saito, H.; Mikami, M.; Tokuda, H.; Watanabe, M. Molecular dynamics simulations of ionic liquids: Cation and anion dependence of self-diffusion coefficients of ions. *J. Phys. Chem. B* **2009**, *113*, 10641–10649. [CrossRef]
31. Henritzi, P.; Bormuth, A.; Klameth, F.; Vogel, M. A molecular dynamics simulations study on the relations between dynamical heterogeneity, structural relaxation, and self-diffusion in viscous liquids. *J. Chem. Phys.* **2015**, *143*, 164502. [CrossRef]
32. Fischler, M.A.; Bolles, R.C. Random sample consensus: A paradigm for model fitting with applications to image analysis and automated cartography. *Commun. Assoc. Comput.* **1981**, *24*, 381. [CrossRef]
33. Guo, B.; Han, H.; Chai, F. Influence of magnetic field on microstructural and dynamic properties of sodium, magnesium and calcium ions. *Trans. Nonferrous Met. Soc. China* **2011**, *21*, s494–s498. [CrossRef]
34. Wei, S.; Zhong, C.; Su-Yi, H. Molecular dynamics simulation of liquid water under the influence of an external electric field. *Mol. Simulat.* **2005**, *31*, 555–559. [CrossRef]
35. Bruneval, F.; Donadio, D.; Parrinello, M. Molecular dynamics study of the solvation of calcium carbonate in water. *J. Phys. Chem. B* **2007**, *111*, 12219–12227. [CrossRef]
36. Ying, T.Y.; Yang, K.L.; Yiacoumi, S.; Tsouris, C. Electrosorption of ions from aqueous solutions by nanostructured carbon aerogel. *J. Colloid. Interf. Sci.* **2002**, *250*, 18–27. [CrossRef]
37. Morimoto, T.; Hiratsuka, K.; Sanada, Y. Development and current status of electric double-layer capacitors. *MRS Proc.* **1995**, *393*, 397. [CrossRef]
38. Chen, L.; Zhang, H.; Qi, Z.-P.; Yao, W.; Wan, J.-J.; Shao, M.; Zhang, C.-C. Synthesis of calcium carbonate crystals by using bacteria. *Synth. React. Inorg. Met. Nano-Metal Chem.* **2012**, *42*, 972–975. [CrossRef]
39. Pach, L.; Hrabe, Z.; Komarneni, S.; Roy, R. Controlled crystallization of vaterite from viscous solutions of organic colloids. *J. Mater. Res.* **1990**, *5*, 2928–2932. [CrossRef]
40. Ricciardiello, F.; Roitti, S. The corrosion of Fe and Ag in S liquid at lowtemperature. Effect of S viscosity. *Corros. Sci.* **1972**, *12*, 651–659. [CrossRef]
41. Jada, A.; Verraes, A. Preparation and microelectrophoresis characterisation of calcium carbonate particles in the presence of anionic polyelectrolyte. *Colloids Surf. A Physicochem. Eng. Asp.* **2003**, *219*, 7–15. [CrossRef]

42. Domingo, C.; García-Carmona, J.; Loste, E.; Fanovich, A.; Fraile, J.; Gómez-Morales, J. Control of calcium carbonate morphology by precipitation in compressed and supercritical carbon dioxide media. *J. Cryst. Growth* **2004**, *271*, 268–273. [CrossRef]
43. Xie, A.J.; Zhang, C.Y.; Shen, Y.H.; Qiu, L.-G.; Xiao, P.P.; Hu, Z.Y. Morphologies of calcium carbonate crystallites grown from aqueous solutions containing polyethylene glycol. *Cryst. Res. Technol.* **2006**, *41*, 967–971. [CrossRef]
44. Hess, B. Determining the shear viscosity of model liquids from molecular dynamics simulations. *J. Chem. Phys.* **2002**, *116*, 209. [CrossRef]
45. Bakker, H.J. Structural dynamics of aqueous salt solutions. *Chem. Rev.* **2008**, *108*, 1456–1473. [CrossRef] [PubMed]
46. Jiao, D.; King, C.; Grossfield, A.; Darden, T.A.; Ren, P. Simulation of Ca^{2+} and Mg^{2+} solvation using polarizable atomic multipole potential. *J. Phys. Chem. B* **2006**, *110*, 18553–18559. [CrossRef]
47. Ohtaki, H.; Radnai, T. Structure and dynamics of hydrated ions. *Chem. Rev.* **1993**, *93*, 1157–1204. [CrossRef]
48. Han, G.; Fang, Z.; Chen, M. Modified Eyring viscosity equation and calculation of activation energy based on the liquid quasi-lattice model. *Sci. China Phys. Mech. Astron.* **2010**, *53*, 1853–1860. [CrossRef]
49. Ravi, K.R.; Pillai, R.M.; Amaranathan, K.R.; Pai, B.C.; Chakraborty, M. Fluidity of aluminum alloys and composites: A review. *J. Alloy. Compd.* **2008**, *456*, 201–210. [CrossRef]
50. Lee, W.J.; Chang, J.G.; Ju, S.P. Hydrogen-Bond structure at the interfaces between Water/Poly(Methyl Methacrylate), Water/Poly(Methacrylic Acid), and Water/Poly(2-Aminoethylmethacrylamide). *Langmuir* **2010**, *26*, 12640–12647. [CrossRef]
51. Pan, Z.; Chen, J.; Lü, G.; Geng, Y.Z.; Zhang, H.; Ji, Q. An Ab initio molecular dynamics study on hydrogen bonds between water molecules. *J. Chem. Phys.* **2012**, *136*, 164313. [CrossRef]
52. Li, Y.X.; Wei, Z.D.; Zhao, Q.L.; Ding, W.; Zhang, Q.; Chen, S.G. Preparation of Pt/Graphene catalyst and its catalytic performance for oxygen reduction. *Acta. Phys-Chim. Sin.* **2011**, *27*, 858–862. [CrossRef]
53. Lee, H.-S.; Tuckerman, M.E. Structure of liquid water at ambient temperature from ab initio molecular dynamics performed in the complete basis set limit. *J. Chem. Phys.* **2006**, *125*, 154507. [CrossRef] [PubMed]
54. Karmakar, A.; Choudhuri, J.R.; Yadav, V.K.; Mallik, B.S.; Chandra, A. A first principles simulation study of vibrational spectral diffusion in aqueous NaBr solutions: Dynamics of water in ion hydration shells. *Chem. Phys.* **2013**, *412*, 13–21. [CrossRef]

coatings

Review

Preparation and Application of Graphene and Derived Carbon Materials in Supercapacitors: A Review

Haiqiu Fang [1], Dongfang Yang [2], Zizhen Su [3], Xinwei Sun [4], Jiahui Ren [4], Liwei Li [5,*] and Kai Wang [4,*]

1 College of Chemistry and Chemical Engineering, China University of Petroleum (East China), Qingdao 266580, China
2 Xi'an Traffic Engineering Institute, Xi'an 710300, China
3 School of Materials Science and Engineering, Beihang University, Beijing 100190, China
4 College of Electrical Engineering, Weihai Innovation Research Institute, Qingdao University, Qingdao 266071, China
5 School of Control Science and Engineering, Shandong Unverisity, Jinan 250100, China
* Correspondence: liliwei@sdu.edu.cn (L.L.); wkwj888@163.com (K.W.)

Abstract: Graphene has recently attracted a wide range of research interests due to its rigorous two-dimensional structure and extraordinary electrical, thermal and mechanical properties. As a conductive agent, an activated carbon supercapacitor can obtain better performance. This paper summarizes the latest research progress, mainly from two aspects: (1) the preparation of an activated carbon base for a supercapacitor based on waste sugar solution and the relationship between pore structure and activation parameters, and (2) the application of the two-dimensional materials graphene and its composite materials in electric double-layer capacitors, graphene–polymer composite tantalum capacitors, graphene–transition metal oxide composite tantalum capacitors, and asymmetric super capacitors. The studies found that graphene and its composite materials have obvious advantages in improving the cycle efficiency, conversion rate, and energy density of supercapacitors, the overall energy efficiency of mechanical systems, and the chemical properties of nanoelectronics. Therefore, it is urgent to summarize these works in order to promote the next development. Graphene is expected to be effectively and environmentally quantified in the near future, and its application in supercapacitors will be further expanded and matured.

Keywords: supercapacitor; graphene; graphene composite; application; activated carbon

Citation: Fang, H.; Yang, D.; Su, Z.; Sun, X.; Ren, J.; Li, L.; Wang, K. Preparation and Application of Graphene and Derived Carbon Materials in Supercapacitors: A Review. *Coatings* **2022**, *12*, 1312. https://doi.org/10.3390/coatings12091312

Academic Editor: Marco Laurenti

Received: 11 July 2022
Accepted: 1 September 2022
Published: 8 September 2022

1. Introduction

In the 21st century, one of the biggest challenges in the world is energy consumption. However, the energy stored in nature is decreasing. Therefore, it is important to develop new energy storage and conversion systems. Under this premise, an electrode material with high conductivity, a large surface area, easy functionalization, and excellent electrochemical performance is required. Among these, carbon nanomaterials (for example, carbon nanotubes (CNTs), activated carbon, and graphene) have been widely studied for electrochemical energy storage systems, especially supercapacitors, due to their special physicochemical structures and excellent electrochemical properties [1–6].

Since 1985, when Kroto and his colleagues discovered fullerenes (C_{60}), several interesting carbon nanomaterials have been isolated. In 1991, Iijima discovered carbon nanotubes (CNTs), and Novoselov et al. isolated two-dimensional graphite graphenes in 2004 [7–9]. Due to its unique physical properties, graphene has become an interesting super material. This new type of two-dimensional carbon nanostructure has attracted great attention in research in physics and chemistry, as well as in materials science. Today, graphene is the most attractive nanomaterial, not only because it is the thinnest known material in the universe but also because it is the strongest material to date, due to its excellent electrical, thermal, mechatronic and optical properties. It has high specific surface area, high chemical stability, high light transmittance, high elasticity, high porosity, biocompatibility, an adjustable band gap, and easy chemical functionalization. These extraordinary properties of graphene can be used to fabricate many novel electronic devices such as field effect transistors (FETs), sensors, and supercapacitors [10–15]. Although the specific capacitance of graphene is much higher than that of activated carbon, activated carbon is much cheaper to make than graphene. Therefore, activated carbon has an advantage when the ratio capacitance is within an acceptable range. Therefore, the preparation and porosity of activated carbon and graphene and their composites have become the focus of research in recent years [16–21]. In addition, the method of constructing asymmetric supercapacitors to improve their performance has also been widely researched. The main research results are shown in Table 1.

Table 1. Main research results on the properties of graphene supercapacitors and porous activated carbon.

Researchers	Research Technique	Research Results
Zhu C et al. [2]	Three-Dimensional Hierarchical Graphene Aerogels with Periodic Macropores	Power densities (>4 kW kg^{-1})
Shao YL et al. [4]	Micro-Super capacitors and fiber-type Super capacitors	Proved that graphene material in wearable super capacitor application prospect
Yan J et al. [5]	Integrating Super capacitors with other applications	Developed multi-functional super capacitor
Liu L et al. [6]	Structure-designed fabrication of all-printed flexible	Higher energy density of asymmetric Super capacitors can be achieved (from 0.00177 mWh cm^{-2} to 0.00687 mWh cm^{-2})
Zhu S et al. [8]	Design and construction of three-dimensional CuO/polyaniline/rGO ternary hierarchical architectures	Energy density of 126.8 Wh kg^{-1} with a power density of 114.2 kW kg^{-1} at a current density of 1.0 A g^{-1}
Sun HT et al. [9]	Three-dimensional holey-graphene/niobia composite architectures for ultrahigh-rate energy storage	The highly interconnected graphene network in the 3D architecture provides excellent electron transport properties
Yu DS et al. [10]	Scalable synthesis of hierarchically structured carbon nanotube-graphene fibres	Energy density of similar to 6.3 mWh cm^{-3}
Zhang L et al. [12]	Freestanding three-dimensional reduced graphene oxide/MnO_2 on porous carbon/nickel foam as a designed hierarchical multihole Super capacitor electrode	Power density of 13.5 kW kg^{-1}
Wang X et al. [13]	Dielectric and optical properties of porous graphenes with uniform pore structures	Chemical synthesis for graphenes with uniform pore structures opens a new way for the precise modulation toward the performances of graphene-based materials.
Zhou YA et al. [14]	Transition-metal single atoms in nitrogen-doped graphenes as efficient active centers	The results shed light on the potential applications of TM and N co-doped graphenes as efficient single-atom bifunctional catalysts for water splitting

Specifically, the content is divided into five sections. In Section 2, we briefly introduce the double-layer capacitor, point out the relationship between its performance and the porous carbon structure, and analyze the relationship between the pore structure and the activation parameters. Section 3 introduces a graphene–polymer composite pseudocapacitor. Section 4 introduces the graphene–transition metal oxide composite pseudocapacitor. Asymmetrical supercapacitors are introduced in Section 5. Finally, a summary and prospects are proposed.

2. Electric Double-Layer Capacitor

The electric double-layer capacitor (EDLC) is a new type of energy storage device with the advantages of stable electrochemical performance, environmental protection, high power density, and a long cycle life [22,23]. The EDLC is a new capacitor based on the theory of the interfacial double-layer proposed by German physicist Helmholtz. As we all know, excess charges of opposite signs will appear on the surface of metal electrodes inserted into an electrolyte solution and on both sides of the liquid surface, resulting in a potential difference between phases. Then, if two electrodes are inserted into the electrolyte at the same time, and a voltage less than the decomposition voltage of the electrolyte solution is applied between them, then the positive and negative ions in the electrolyte will move rapidly to the poles under the action of the electric field, and form a tight charge layer on the surface of the two electrodes, namely the double electric layer. It is formed by the electric double layer and the traditional capacitor of the dielectric in the electric field under the action of the similar polarization charge, resulting in a capacitance effect which is closely similar to the electric double-layer plate condenser. However, due to the tighter-than-ordinary capacitor charge layer between the charge layer, the spacing distance is much smaller, and thus has greater capacity than the general capacitor [24–26]. The basic principle is as follows: when charging the electrode, the surface charge of the electrode in the ideal polarized electrode state will attract the opposite ions in the surrounding electrolyte solution, such that these ions will attach to the electrode surface to form a double charge layer, forming an electric double-layer capacitance. Because the distance between the two charge layers is very small (generally less than 0.5 nm), coupled with the special electrode structure, the surface area of the electrode is increased by tens of thousands of times, resulting in a great capacitance.

EDLC has been widely and successfully implemented in many areas, such as energy management/protection applications and day and night storage. The energy storage mechanism of EDLC is the electrostatic interaction at the electrode/electrolyte interface. Carbon-based electrode materials, especially graphene, are considered to be suitable electrode materials due to their high specific surface area (SSA), excellent electrical conductivity, good electrochemical stability, and inexpensive source of raw materials. SSA refers to the surface area per unit of mass of porous solid material. Because the outer surface area of a solid material is so small compared to the inner surface area that it can be ignored, this surface area usually refers to the inner surface area. The common unit is square meters per gram (m^3/g). It is essential to determine the type and properties of the material. In addition, the specific capacitance of the EDLC is primarily determined by the effective specific surface area (SSA), which is accessible by the electrolyte ions. The effective SSA depends mainly on the total SSA and pore size distribution (micro, meso and macropores [27,28]).

2.1. Research Progress of Activated Carbon Based on Waste Sugar Solution

Among the many active electrode materials, graphene shows the best performance. However, compared with activated carbon materials, the preparation cost of graphene is relatively high. Improving the production process of graphene and reducing the production cost has become one of the most important research topics. In many pieces of research, the production method using waste sugar solution (WSS) as a raw material has been widely investigated. In the current processing of many sugar products (xylose, glucose, galactose), the production of waste sugar solution is inevitable. It contains all kinds of substances that are harmful to the environment. Therefore, recycling waste sugar solution can not only effectively protect the environment but also greatly reduce the wastage of resources. Therefore, the experiment of using WSS as the material for the preparation of activated carbon is most often introduced.

The abundant O-containing functional groups (-COOH, -OH and C=O) of WSS are the basis for the preparation of high-performance supercapacitor materials. Hao et al. prepared porous carbon structures using dried WSS as precursors using conventional methods. However, its performance is not satisfactory when it is applied to supercapacitors [29]. In order to enhance the SSA and optimize the pore structure, Wu et al. successfully fabricated a porous carbon mechanism with a relatively high specific capacity (342 F/g) by the direct high-temperature decomposition of WSS using the nitrogen atom doping method [30]. However, under the condition of a high temperature, the utilization of nitrogen atoms is not high. In order to solve this problem, Lin et al. and Hao et al. improved the production process, changing the high-temperature production method to a hydrothermal method, greatly improving the availability of nitrogen atoms [31]. The relationship between the pore structure and the activation parameters was investigated by using KOH as an activator to increase the SSA of carbon spheres (CS). At the same time, the oxygen-containing groups on the surface of the porous carbon spheres (PCS) generated during the activation of the Heat Transfer Compound (HTC) and KOH can enhance the specific capacitance by improving the wettability of the electrodes, and in the form of a tantalum capacitor. In summary, it is meaningful and important to prepare carbon electrode materials by recycling WSS for EDLC. The current studies on the preparation of porous carbon structures by WSS are shown in Table 2.

Table 2. Research progress and properties of porous carbon structures prepared by WSS.

Researchers	Method	Specific Capacitance
Hao et al. [29].	Typical carbonation/activation method	240 F/g
Wu et al. [30].	Nitrogen doping method	342 F/g
Lin et al. [31].	Hydrothermal treatment	406 F/g
Hao et al. [29].	Hydrothermal treatment	296 F/g

2.2. Research Progress of Hollow Carbon Spheres

Among various carbon materials, hollow carbon spheres have attracted more and more attention due to their unique structure and potential applications in supercapacitors. Hollow carbon spheres differ from active carbon in that they can control not only shell thickness but also surface characteristics. Therefore, they are more flexible than active carbon [32–34]. At present, there are two ways to prepare hollow carbon spheres: the hard-templating and soft-templatingg routes. The hard-templating method mainly includes the coating of the template with the carbon precursor, pyrolysis, and the removal of the template. Soft-templating routes are very attractive because of the easier removal of the templates. This method is usually adopted by the template for the colloidal system (emulsion, droplets, micelles and vesicles, and gas bubbles) [35].

The hollow carbon-sphere structure can act through "ion-buffering desorption", such that the distance between the ions in the electrolyte and the pores in the shell is greatly reduced. Therefore, use the hollow carbon sphere as the electrode material of the supercapacitor shows high specific capacitance, according to Portet et al. and Murali et al. The hollow carbon structures proposed by Bhattacharjya et al. are 146 F/g and 122 F/g, respectively [36,37]. The paper presented a kind of core mesoporous shell carbon capsules, the specific capacity of which was 162 F/g. Under the same conditions, their performance is better than that of activated carbon. Similarly to the activated carbon material, the nitrogen doping method is also suitable for hollow carbon structures. It can not only make it have higher surface wetness but also increase SSA.

2.3. Research Progress of Graphene

Graphene is a well-known one-atom-thick sheet; it has 2-D monolayers composed of all-sp2 hybridized carbon atoms in a polyaromatic crystal lattice with a honeycomb structure. It has the advantages of active carbon and hollow carbon spheres; as such, it has become the first choice for energy storage devices such as supercapacitors [38,39]. In the traditional method of obtaining graphene, first, graphite is oxidized to produce graphite oxide; second, the graphite oxide is completely stripped by simple ultrasonic treatment; graphene is obtained by the reducing agent [40,41]. A graphene-based supercapacitor with ionic liquid electrolytes has an energy density value of 31.9 Wh/kg and a specific capacity of 135 F/g.

The effective capacity of graphene depends on the number of layers, and the problem of restacking will inevitably occur during use. In order to avoid this phenomenon, combining graphene with porous carbon, carbon nanotubes and carbon nanofibers has become a good solution [42]. In many current studies, different electrolytes have been used to improve the performance of graphene-based supercapacitors. For ionic liquid electrolytes, the specific capacity can reach 75 F/g [43]. In aqueous electrolytes, the specific capacitance can reach 135 F/g [44]. This can be seen for 2D and 3D macroscopic structures after peeling the graphene using different materials for the organic electrolyte to reach a specific capacity of 205 F/g [45]. A higher surface area can be obtained. Doping graphene with chemicals with electron donors and acceptors is also an effective measure to improve the electrochemical performance of graphene electrodes [46].

3. Graphene–Polymer Composite Pseudo-Capacitor

Recent developments in graphene and polymer composites have shown the very promising features of these composites for supercapacitor applications. Graphene can be divided into three types: graphene, graphene oxide, and reduced graphene oxide. Pseudo-capacitors are a type of supercapacitor that can mimic the behavior of EDLC for energy storage and release through multiple rapid and highly reversible processes of Faradaic redox reactions on the electrode surface [47,48]. The capacitance properties depend on the porosity of the material and the insertion and exit of ions at the electrode/electrolyte interface. Compared with EDLC, the specific capacitance of a pseudo-capacitor is higher. However, the cycle stability is low due to volume changes during charging and discharging. Conductive polymers (CPs), metal sulfides, metal oxides, metal carbides and metal nitrides are the most common pseudo-capacitor materials, and CPs are the representative electrode material of a pseudo-capacitor. Among them, CPs are probably the most representative, showing excellent pseudo-capacitive performance through rapid, reversible redox reactions. However, there are some limitations to this type of SC. For example, the most significant disadvantage is the rapid decay cycle stability during charging and discharging. This is mainly due to considerable mechanical degradation (such as expansion and contraction) and irreversible structural changes. At the same time, because of its compact structure, the power density of CPs that contact the electrolyte only with limited properties is very low [49].

In order to overcome the shortcomings of pseudo-capacitors, it has become a research hotspot to combine carbon materials with pseudo-capacitors to make new hybrid supercapacitors [50,51]. The combination of CP/graphene to form new composites has been widely studied for its advantages of containing both CP and graphene. Due to its excellent electrochemical performance, environmental stability, and rapid Faraday reaction, the method of combining polyaniline with graphene has been widely studied in recent decades. Zhao et al. proposed a synthesis using in situ high-gravity chemical polymerization means in a rolenating-bed (RPB) polyaniline/graphene method. In the process of preparation, it was pointed out that the ammonium persulfate/aniline mole ratio, graphene dosage, reactor type, and aniline concentration play a crucial role in the performance of the synthesized electrode materials [52]. Polyaniline is another material with high specific capacity, which is widely used in the manufacture of flexible electronic equipment due to its flexibility. However, polyaniline has poor cyclic stability. In order to alleviate this defect, Jia et al. synthesized polyaniline/graphene materials by electrochemical synthesis. The experiment showed that the specific capacity of the improved composite material can reach 202 F/g [53]. The current research on binary composites is shown in Table 3.

Table 3. Electrochemical performance of some of the previously reported ternary composites.

Researchers	Materials	Specific Capacitance
Y. Xu et al. [53].	Nano graphene platelet/polyaniline	269 F/g
M. Xue et al. [54].	Graphene oxide patterns	970 F/g
Y. Meng et al. [55].	Graphene/polyaniline composite	385 F/g
L. Liu et al. [56].	Nanostructured graphene composite	224 F/g
Z.-S. Wu et al. [57].	Ultrathin printable grphene	348 F/g
J.W. Park et al. [58].	Graphene/polyselephene	293 F/g

Binary composites synthesized from conductive polymers and graphene have been widely proven to have good electrochemical properties. Inspired by this, binary composites are combined with other materials (metal sulfide, metal oxide and non-metal oxide) to form ternary composites. Finally, the overall electrochemical performance can be maximized. Wang et al. proposed a material that combines SiO_2, polyaniline and graphene simultaneously. The introduction of SiO_2 can inhibit the accumulation of graphene's hierarchical structure to a certain extent, and can effectively improve the ion exchange and interaction at the electrolyte/electrode interface. Moreover, the specific capacity reached an astonishing 727 F/g [59]. The current research on ternary composites is shown in Table 4.

Table 4. Electrochemical performance of some of the previously reported ternary composites.

Researchers	Materials	Specific Capacitance
Dywili, N.R. et al. [60].	Graphene Oxide Decorated Nanometal-Poly	227.2 F/g
Xu, Z. et al. [61].	zinc sulfide/reduced graphene oxide/conductive polymer	722 F/g
Golkhatmi, S.Z. et al. [62].	nickel oxide/graphene/Polyaniline hybrid	970.85 F/g
Azizi, E. et al. [63].	reduced graphene oxide/polyindole/gamma—Al_2O_3	308 F/g
Ramesh, S. et al. [64].	Co_3O_4graphene oxide/polyindole composite	680 F/g
Li, S. et al. [65].	TiO_2@PPy/rGO	462.1 F/g
Wang, H. et al. [66].	Graphene Hybrids Embedded with Silica	727 F/g

It is well known that polymeric binders are a very important part of forming supercapacitor electrodes. However, one disadvantage of using polymeric binders is that they are generally not electrically conductive, and they may reduce the energy density of the supercapacitors. In combination with the binder polymer, graphene will compensate for the undesirable characteristics of the insulating polymer (e.g., its insulating properties, low surface area, and low specific capacitance). On the other hand, when graphene is made of a composite material with a conductive polymer, it will greatly improve the cycle performance and specific capacitance. In addition, the flexible nature of the graphene/polymer film enables flexible, wearable, conformal energy storage devices [67,68]. Whilst graphene/polymer supercapacitor devices exhibit innovative concepts and technologies that are unique to current state-of-the-art technologies, graphene/polymer composite-based supercapacitors still present many challenges to their full potential. One of the main challenges is to find a viable method for the low-cost mass production of graphene/polymer supercapacitor electrodes without compromising the micro/nanostructure of graphene due to stacking or aggregation. In addition, the integration of graphene–polymer supercapacitors with other electronic devices (e.g., solar cells, batteries) remains a practical challenge. In summary, graphene–polymer composites have great potential for supercapacitor applications.

4. Graphene–Transition Metal Oxide Composite Pseudo-Capacitor

Among the many materials used in supercapacitors, metal oxides have been widely used because of their superior physical and chemical properties. Compared to metal sulfides, metal oxides are easy to prepare (available for mass production), chemically stable, environmentally friendly, and compatible with a wide range of electrolytes. These advantages are essential for the production of durable and cost-effective supercapacitors. In addition, metal oxides have a higher theoretical surface area due to the combined effect of a high electrode surface area and various oxidation states. Metal oxide is based on Faraday's interesting electrochemical redox reaction performance, but because of the restriction, it may not be in the form of the original as a supercapacitor electrode: (1) most of the metal oxides' conductivities are very poor, which leads to the slice resistance of the electrode and the charge transfer resistance increases, which in turn leads to the attenuation rate capacity and power density of supercapacitor devices; (2) during the cycle of metal-oxide-based supercapacitors, strain will be generated on the surface of metal oxides, resulting in the structural degradation of electrode materials and a poor cycle life, and adjusting some of the properties of metal oxides is tricky. Therefore, it is important to develop composite and/or hybrid electrode materials by combining metal oxides with carbonaceous materials or conductive polymers.

Combining metal oxides with graphene has great potential. (1) Due to metal oxide/graphene's heterozygous structure due to the injection of electrons from the graphene layer in the metal oxide, the hole concentration in graphene is increased, which can lead to high conductivity. (2) The high surface area of the metal oxide/graphene hybrid enhances the interaction between the electroactive material and the electrolyte, thereby improving the electrochemical performance. (3) In the metal oxide/graphene hybrid, the metal oxide nanosheets are sandwiched between the graphene nanosheets, which can effectively avoid the re-stacking of graphene layers. Zhang et al.'s MnO_2/graphene hybrid was prepared by electrostatic precipitation with a specific capacity of 188 F/g. Due to electrostatic interactions, MnO_2 nanosheets are dispersed on the surface of the graphene substrate. Compared with the original MnO_2 and graphene, the 2D MnO_2/graphene hybrid exhibits better electrochemical performance due to the simultaneous contribution of the excellent conductivity in graphene and the pseudo-capacitance properties of the MnO_2 nanosheets [69]. The current studies on the heterozygous structure of metal oxides and graphene are shown in Table 5.

Table 5. Study on the heterozygous structure of metal oxides and graphene.

Researchers	Materials	Specific Capacitance
Wang Y et al. [70].	CeO_2 nanoparticles/graphene	18 F/g
Dong X et al. [71].	3D Graphene-Cobalt oxide	1100 F/g
He G et al. [72].	Co_3O_4@graphene nanocomposite	415 F/g
Qu Q et al. [73].	2D sandwich-like sheets of iron oxide grown on graphene	349 F/g
Wang Z et al. [74].	Fe_2O_3–graphene nano composite	226 F/g
Wang Q et al. [75].	Fe_3O_4 nanoparticles grown on graphene	220 F/g
Peng L et al. [76].	MnO_2/Graphene	267 F/g
Zhao K et al. [77].	Mn_3O_4@N-doped carbon/graphene	456 F/g
Wang C et al. [78].	Hierarchical composite electrodes of nickel oxide nanoflake 3D graphene	1829 F/g
Wang W et al. [79].	Hydrous ruthenium oxide nanoparticles anchored to graphene	502.78 F/g
Li F et al. [80].	Graphene/SnO_2	43.4 F/g
Zhang Z et al. [81].	TiO_2—Graphene	206.7 F/g
Wang H et al. [82].	Three-dimensional Graphene/VO_2	426 F/g
Perera SD et al. [83].	Vanadium oxide nanowire—Graphene	80 F/g

4.1. Preparation of the PGO-Ni Electrode and the Application of Graphene

This paper reviews a method for obtaining vibration energy by using a hybrid system containing a porous graphene nickel oxide (PGO-Ni) electrode and potassium chloride (KCl) solution. The results of this work are of great significance to improve the overall energy efficiency of the mechanical system

The electromechanical energy conversion system is composed of KCl solution flowing through the PGO-Ni electrode. Here, firstly, the foam nickel is ultrasonically cleaned in ethanol and acetone solution for 15 min, then 3 mol^{-1} hydrochloric acid is used to remove the surface oxide layer, and the sample is cleaned with deionized water to remove the surface oxide layer of foam nickel. After that, the foam nickel was immersed in 3 mg mL^{-1} graphene oxide suspension to adsorb some graphene oxide sheets on the surface. The graphene oxide suspension was prepared by a modified Hummers method. Then, the foam nickel with graphene oxide on the surface was dried at 60 °C in a blast dryer. After three repetitions of soaking and drying, the platinum group metal nickel electrode was prepared, and some graphene oxide sheets were adsorbed on the electrode surface. The platinum group metal nickel electrode was prepared by this method [84–86].

The SEM image of the PGO-Ni electrode is shown in Figure 1b. Compared with the typical SEM images of foam nickel, the surface of the foam nickel was covered by some graphene oxide sheets after impregnation and drying. In addition, the colors of foam nickel and PGO-Ni are also different in the optical image, as shown in the illustrations of Figure 1a,b. The XRD (performed using a Rigaku X-ray diffractometer) and EDS (performed using the energy spectrum analyzer of a JSM-6390A) of the PGO-Ni electrode are shown in Figure 1c,d. There is a peak at 12° in the XRD pattern. The chemical elements of carbon and oxygen are found in the EDS pattern, which further confirms the existence of graphene oxide.

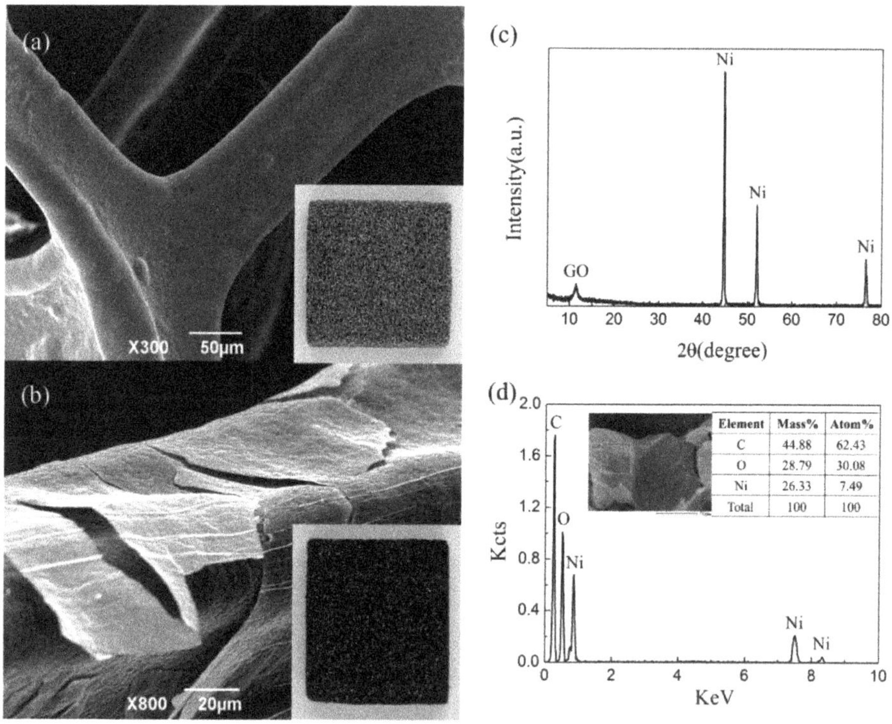

Figure 1. Characteristics of PGO-Ni electrodes. Reprinted with permission (**a**) SEM observations of the microstructure of nickel foam; (**b**) SEM observations of the micro-structure of PGO-Ni electrode; (**c**) XRD patterns of PGO-Ni electrode showing the peak of graphene oxide at 12° and the peaks of nickel at 44°, 52°, 76°. (**d**) EDS patterns of PGO-Ni electrode showing the elements C, O, Ni. from Ref. [84]. Copyright 2017, Elsevier B.V.

In order to further enhance its performance, transition metal oxide nanoparticles (TMO NPs) are hybridized with graphene. Some of the disadvantages of nanoparticles (NPs) can be offset by graphene. Common disadvantages of semiconductor nanoparticles

(NPs) include their relatively low conductivity and high recombination of photo generated electron–hole pairs. Additionally, during Li insertion–extraction processes in LIB applications, NPs cause large-volume expansions, which hinder their use in applications such as energy storage, sensing, advanced catalysis, solar cells, diodes, and biometrics. Therefore, the strongly coupled graphene–NP hybrid system appears to be promising to overcome these problems.

Among various graphene-based materials, graphene–transition metal oxide nanoparticles have recently attracted great research interest due to their unique structural advantages. In reference [84], the paper mainly introduces the influence of pH on the structural, optical and electrical properties of graphene–vanadium oxide nanoparticle (rGO/VO-NPs) nanocomposites. This study reveals the high crystallinity of rGO/VO-NPs nanocomposites and the graphene layer formed on the surface of vanadium oxide nanoparticles, ensuring good electrical contact and higher conductivity. In addition, UV-Vis (absorbance) confirmed an improvement in optical properties. Therefore, the growth conditions have a great influence on the synthesis of the final material. The use of these nanocomposites is a promising way to develop technology applications, particularly for energy storage devices that use lithium-ion batteries.

4.2. The Effect of Vanadium Oxide on Graphene

The experimental results demonstrate the pH effect of vanadium oxide growth on the graphene layer. FTIR spectroscopy revealed the formation of new bonds which, due to the functionalization of oxygen functional groups on the rGO surface, indicate the reoxidation of rGO nanosheets after coating with vanadium oxide nanoparticles. The control of the NaOH concentration has an important influence on the formation of VO-NPs in the composite. The optical properties indicate that the absorbance intensity of rGO/VO-NPs nanocomposites increases in the acidic range and gradually decreases in the alkaline range, which may be due to the stoichiometry and crystallinity of vanadium oxide nanoparticles. In addition, the mobility of charge carriers increases with the increasing pH, indicating that a large number of charge carriers can be used for conduction. Thus, the uniform coating of the graphene layer around the surface of the vanadium oxide nanoparticles ensures good electrical contact and exhibits higher conductivity under the effect of pH. In general, this is a new insight into controlling the preparation of such mixtures and tailoring such material properties to improve their efficiency and performance in many applications, particularly as storage devices.

4.3. The Effect of Polyoxometalate-Based Complexes on Graphene

Polyoxometalate has emerged as a promising candidate for the design of nanocomposites with unique properties and improved functionality. Polyoxometalate is an inorganic cluster composed of oxygen and early transition-metal atoms, usually in the highest oxidation state, with d^0 or d^1 electron configurations. Polyoxometalate is typically an extensive library of electrons that can easily perform rapid multielectron redox reactions and transfer multiple electrons per molecule while maintaining their intrinsic stability. Therefore, they are considered to be promising electroactive materials for energy storage applications. However, the inherent low conductivity and high solubility of these metal oxides in many solvents seriously hinder their direct use as electrode materials in supercapacitors. Polyoxometalate clusters are usually immobilized on chemically stable, high-surface-area substrates. Therefore, anchoring paraformaldehyde, which exhibits high redox activity, into capacitive carbon electrodes (e.g., GO, CNTS, AC) can increase the net specific energy and operating voltage range of storage devices. Various properties of polyoxometalate-based hybrid materials have been studied, including electrochemical adsorption and catalytic capacity. Sukanya Maity et al. doped different masses of $K_2H_5[NiV_{14}O_{40}](NiV_{14})$ into porous activated carbon, and found that the resulting complex had the electrochemical enhancement function. When applied to supercapacitors, the specific capacitance reaches 375 F/g [87]. In order to further improve its electrochemical performance, two metaloxates ($[PVMo_{11}O_{40}]^{4-}$ and $[PV_2Mo_{10}O_{40}]^{5-}$) were integrated with AC in a new way. The composite shows a high specific capacitance of 450 F/g. In addition, metaloxate not only plays an important role in improving the electrochemical performance of AC but also improves the electrochemical performance to a large extent when the metaloxate is applied to graphene materials [88]. Decavanadate–graphene oxide nanocomposite is an electrode material for electrochemical capacitors. It uses Fourier transform infrared spectroscopy, powder X-ray diffraction, scanning electron microscopy and transmission electron microscopy. The electrochemical behavior of the composite electrode was studied using a neutral 1M sodium sulfate (Na_2SO_4) solution in a three-electrode cyclic voltammetry (CV) system, which showed a specific capacitance of 306 F/g at a scan rate of 5 mV/s [89]. In addition, $MnV_{14}O_{40}$ is impregnated on a high-surface-area substrate of GO; the synergistic effect between GO and MnV_{14} provides a better route for ion transport to the interface, thus improving the conductivity and diffusion ability of the nanohybrid. Oxygen-containing functional groups in GO act as active sites anchored to the MnV_{14} cluster. Therefore, the surface modification of the MnV14 cluster improves the specific capacitance of the nanohybrid material, and has significant electrical and mechanical stability [90].

5. Asymmetric Super Capacitors Based on Graphene

The energy density of supercapacitors is much lower than the energy density of batteries, thus limiting their practical application. The method by which to construct high-energy-density supercapacitors is not particularly important without sacrificing power density and cycle life. Currently, two effective strategies are widely used to increase the energy density of supercapacitors. One is to enhance the specific capacitance of the electrode material; the other is to increase the operating voltage window by developing an asymmetric supercapacitor device with a battery-type Faraday electrode and a capacitor-type electrode. This paper summarizes the second method; the hollow $NiCo_2O_4$ material with a high surface area is introduced, and its influence on the energy density of asymmetric supercapacitor is analyzed [91–94]. An asymmetric supercapacitor was assembled by using the hollow $NiCo_2O_4$ nanosphere electrode as the positive and the AC electrode as the negative, which manifests high energy density with excellent cycling stability.

Figure 2a shows a scanning electron microscope (SEM) image of SiO_2 nanospheres. SiO_2 nanospheres with a smooth surface and a highly uniform diameter of about 140–170 nm were obtained by the improved St ö BER method. Subsequently, the prepared SiO_2 nanospheres were used as hard templates, and hollow $NiCo_2O_4$ nanospheres were prepared by simple hydrothermal process and post-annealing treatment. Figure 2b shows a hollow $NiCo_2O_4$ nanosphere with a diameter of 200–220 nm assembled from a number of $NiCo_2O_4$ nanosheets. The hollow nanostructures are clearly visible from the figure. This hollow nanostructure can provide more active sites for ion access and transport, and can thus shorten the diffusion distance of ions and electrons, which is very important for tantalum capacitors. For comparison, $NiCo_2O_4$ nanospheres with a diameter of 4.7 μm were also prepared under similar synthesis conditions, as shown in Figure 2c. The crystal phases and structures of the hollow $NiCo_2O_4$ nanospheres and $NiCo_2O_4$ microspheres were characterized by X-ray diffraction (XRD), as shown in Figure 2d. Through the experiment, no residue or contaminant was detected, indicating that the purity of the sample was high.

The Brunauer-Emmett-Teller (BET) surface area of the hollow $NiCo_2O_4$ nanospheres and $NiCo_2O_4$ microspheres was further investigated by nitrogen adsorption–desorption measurements. The increase in the BET surface area of hollow $NiCo_2O_4$ nanospheres can be explained by their fractional hollow nanostructures. The pore size distribution curve is shown in the inset of Figure 3, in which the average pore diameters of the hollow $NiCo_2O_4$ nanospheres and $NiCo_2O_4$ microspheres were 3.8 nm and 4.3 nm, respectively (calculated by the Barrett-Joyner-Halenda ((BJH) method). The larger BET surface area of the hollow $NiCo_2O_4$ nanospheres can provide more electrochemical reaction sites for efficient electrolyte ion transport, thereby improving the supercapacitor's performance.

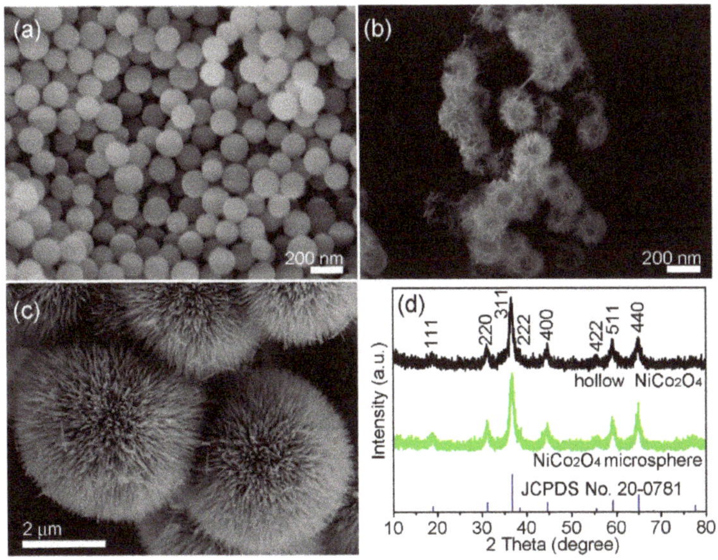

Figure 2. SiO_2 nanospheres and $NiCo_2O_4$ nanospheres. SEM images of (**a**) SiO_2 nanospheres, (**b**) hollow $NiCo_2O_4$ nanospheres, (**c**) $NiCo_2O_4$ microspheres, (**d**) XRD patterns of hollow $NiCo_2O_4$ nanospheres and $NiCo_2O_4$ microspheres. Reprinted with permission from Ref. [86]. Copyright 2017, Elsevier Inc.

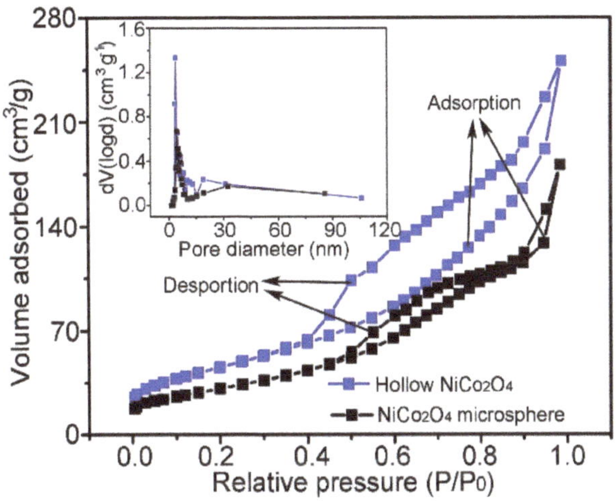

Figure 3. Nitrogen adsorption–desorption isotherms of hollow $NiCo_2O_4$ nanospheres and $NiCo_2O_4$ microspheres. Reprinted with permission from Ref. [86]. Copyright 2017, Elsevier Inc.

The electrochemical properties of the hollow $NiCo_2O_4$ nanospheres and $NiCo_2O_4$ microspheres were measured using a three-electrode configuration with the Pt plate counter electrode and the SCE reference electrode in a 1 M KOH aqueous electrolyte. Figure 4a shows cyclic voltammetry (CV) curves for hollow $NiCo_2O_4$ nanospheres at various scan rates in the range of 10–50 mV/s. The CV curve shows a well-defined redox peak indicating the tantalum capacitance of the electrode during charge storage.

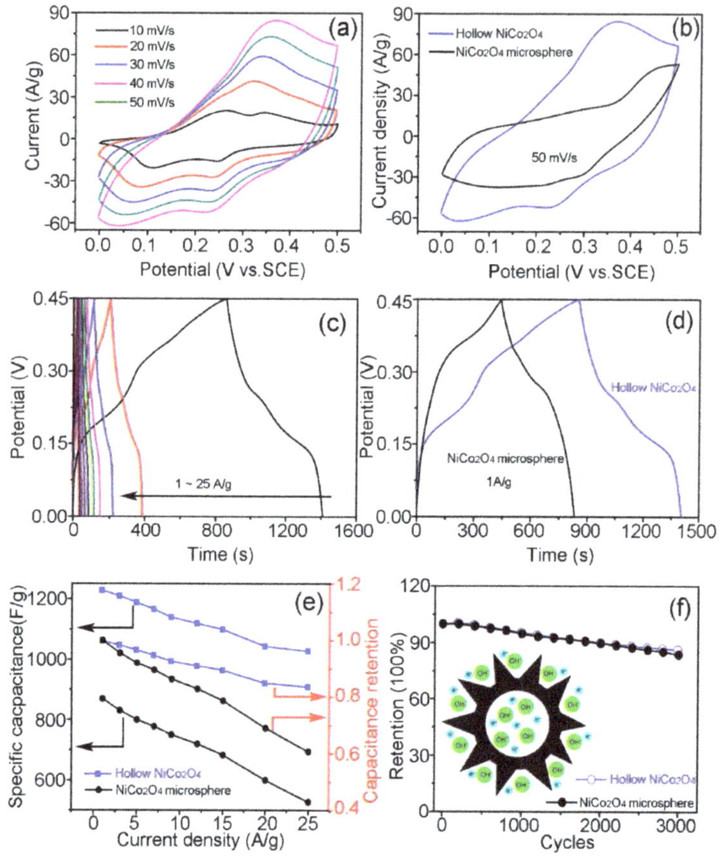

Figure 4. Hollow $NiCo_2O_4$ nanospheres' characteristic curve. (**a**) CV curves of the hollow $NiCo_2O_4$ nanospheres at varied scan rates, (**b**) CV curves of the hollow $NiCo_2O_4$ nanospheres and $NiCo_2O_4$ microspheres at a scan rate of 50 mV/s, (**c**) CD curves of the hollow $NiCo_2O_4$ nanospheres at different current densities, (**d**) CD curves of the hollow $NiCo_2O_4$ nanospheres and $NiCo_2O_4$ microspheres at a current density of 1 A/g, (**e**) The specific capacitances and capacitance retentions of the hollow $NiCo_2O_4$ nanospheres and $NiCo_2O_4$ microspheres, (**f**) Cycle performance of the hollow $NiCo_2O_4$ nanospheres and $NiCo_2O_4$ microspheres at 50 mV/s for 3000 cycles. Inset is schematic of the charge storage advantage of the hollow $NiCo_2O_4$ nano-spheres. Reprinted with permission from Ref. [86]. Copyright 2017, Elsevier Inc.

It is worth noting that as the scanning rate increases, no significant changes are observed in the shape and position of the oxidation and reduction peaks, indicating that the hollow $NiCo_2O_4$ nanosphere electrode has good capacitance characteristics and a high rate of performance. Figure 4b compares the CV curves of a hollow $NiCo_2O_4$ nanosphere electrode and a $NiCo_2O_4$ microsphere electrode at a scan rate of 50 mV/s. It can be seen that the CV integral area of the hollow $NiCo_2O_4$ nanosphere electrode is significantly larger than the CV integral area of the $NiCo_2O_4$ microsphere electrode, indicating that the electrochemical capacitance of the hollow $NiCo_2O_4$ nanosphere electrode is significantly improved. Figure 4c shows the constant current charge and discharge (CD) curves for hollow $NiCo_2O_4$ nanosphere electrodes between 0 and 0.45 V at different current densities. The charging and discharging platforms can be clearly seen in the CD curve, indicating the presence of the Faraday process. In addition, these CD curves are approximately symmetrical, indicating that the hollow $NiCo_2O_4$ nanosphere electrode has good electrochemical reversibility. Figure 4d shows a comparison of the CD curves for hollow $NiCo_2O_4$ nanosphere electrodes and $NiCo_2O_4$ microsphere electrodes at the same current density of 1 A/g. Obviously, the hollow $NiCo_2O_4$ nanosphere electrode has a longer discharge time and a higher capacity than the $NiCo_2O_4$ microsphere electrode. As shown in Figure 4e, the specific capacitance of the hollow $NiCo_2O_4$ nanosphere electrode and the $NiCo_2O_4$ microsphere electrode was calculated. The hollow $NiCo_2O_4$ nanosphere electrode produced a significant specific capacity of 1229 F/g at 1 A/g, which is much larger than that of the $NiCo_2O_4$ microsphere electrode (869 F/g). In addition, the hollow $NiCo_2O_4$ nanosphere electrode showed a significantly better rate of performance than the $NiCo_2O_4$ microsphere electrode. When the capacitance density of the hollow $NiCo_2O_4$ nanosphere electrode increased from 1 to 25 A/g, 83.6% of the capacitance was retained. The cyclic stability of the hollow $NiCo_2O_4$ nanosphere electrode and the $NiCo_2O_4$ microsphere electrode was 3000 cycles at 50 mV/s (Figure 4f). After 3000 cycles, the total permittivity retention of the hollow $NiCo_2O_4$ nanosphere electrode was 86.3%, while the permittivity retention of the $NiCo_2O_4$ microsphere electrode was 83.7%. The remarkable electrochemical properties of hollow $NiCo_2O_4$ nanospheres can be attributed to their uniform fractionated hollow nanostructures with high specific surface areas, which can provide more electroactive sites for rapid ion implantation in the entire electrode material, thereby improving the electrode material's utilization rate [17–19,95–97].

In order to further evaluate the practical application effect of the hollow $NiCo_2O_4$ nanosphere electrode, a hollow $NiCo_2O_4$ nanosphere electrode was used as the positive electrode, the AC electrode was used as the negative electrode, and the asymmetric supercapacitor was assembled by the separator. Figure 5a shows the CV curve of a $NiCo_2O_4$//AC-ASC device collected at a scan rate of 50 mV/s in different voltage windows. As expected, the stable potential window of the fabricated device can be extended to 1.5 V. A typical CV curve for a $NiCo_2O_4$//AC-ASC device is shown in Figure 5b. These CV curves exhibit similar shapes, indicating that the asymmetric supercapacitor device has excellent capacitive performance. Based on the total mass of the active material of the two electrodes, the specific capacitance of the $NiCo_2O_4$//AC-ASC device is shown in Figure 5c. Figure 5d shows the energy and power density of a $NiCo_2O_4$//AC-ASC device. It is worth noting that these results are superior to other previously reported asymmetric supercapacitors. Figure 5e shows the Nyquist impedance spectrum of the fabricated $NiCo_2O_4$//AC-ASC device. It is worth noting that the equivalent series resistance (ESR, the intercept on the real axis) is estimated to be 0.6 Ω, indicating that the internal resistance of the $NiCo_2O_4$//AC-ASC device is low. The cycle performance of the fabricated asymmetric supercapacitor device was evaluated by repeating the CD test for 2000 cycles at a large current density of 5 A/g. As shown in Figure 5f, the total specific capacitance can remain approximately 87.8% even after 2000 cycles. In addition, it maintains significant coulombic efficiency close to 100% in continuous cycles.

In summary, hollow $NiCo_2O_4$ nanospheres exhibit excellent electrochemical performance. The specific capacitance and rate of performance of the hollow $NiCo_2O_4$ nanosphere electrode are much better than those of the $NiCo_2O_4$ microsphere electrode. In addition, the asymmetric supercapacitor is assembled with hollow $NiCo_2O_4$ nanosphere electrodes and AC electrodes, and the stable potential window can be extended to 1.5V. Importantly, the device has a high energy density of 21.5 Wh/kg, an significant cycle performance and coulombic efficiency.

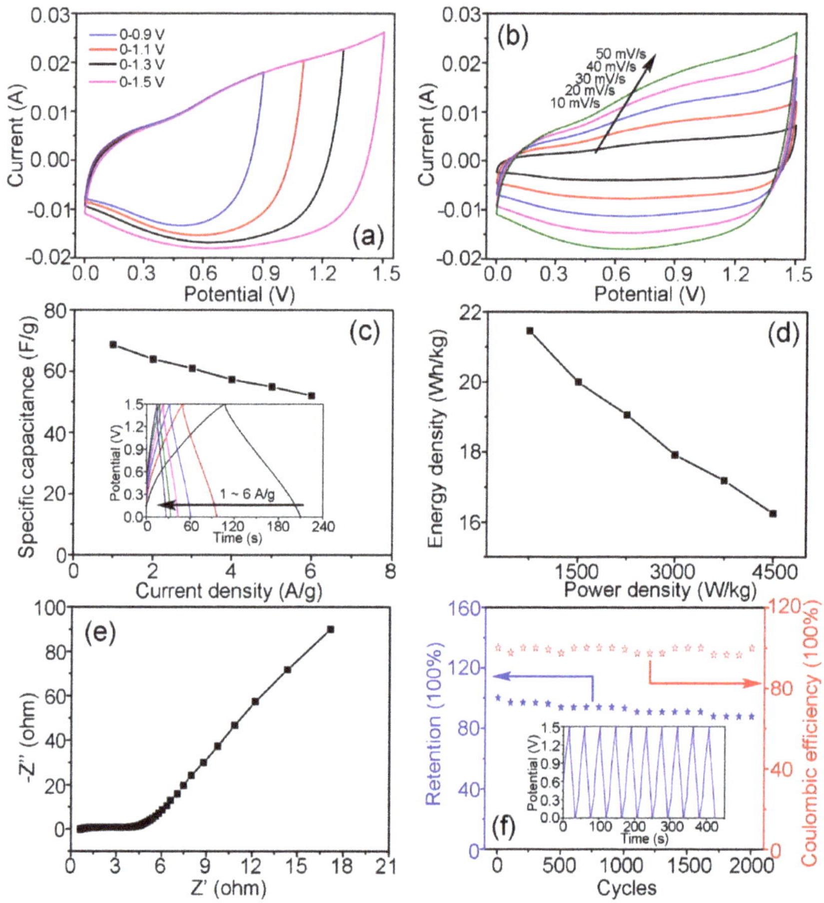

Figure 5. $NiCo_2O_4$//AC-ASC characteristic chart. (**a**) CV curves of $NiCo_2O_4$//AC-ASC at different scan voltage windows, (**b**) CV curves, (**c**) The specific capacitances (based on the total mass of active materials), (**d**) Ragone plots, (**e**) Nyquist plots, (**f**) Cycle performance and Coulombic efficiency of $NiCo_2O_4$//AC-ASC device. Inset in (**c**,**f**) is CD curves of $NiCo_2O_4$//AC-ASC at different current densities and the first 10th CD curves of $NiCo_2O_4$//AC-ASC device, respectively. Reprinted with permission from Ref. [86]. Copyright 2017, Elsevier Inc.

6. Summary and Outlook

Carbon material, metal oxide, and chlorinated paraffin are three commonly used electrode materials for supercapacitors. Metal oxides have higher specific capacity, but their low conductivity, high cost and environmental pollution limit their application in supercapacitors. Graphene is one of the ideal supercapacitor materials due to its superior electrochemical properties (such as high conductivity) and high specific surface area.

It is expected that the remarkable electronic properties of graphene will bring a new era to nanoelectronics. The excellent electronic properties of graphene are critical for many device applications. Graphene—with its good electron conductivity, large specific surface area, and suitable porosity—is an ideal supercapacitor electrode material. It can effectively improve the power and energy density of supercapacitors: on the one hand, it can effectively avoid the agglomeration problem of graphene; on the other hand, it can improve the shortcomings of the poor conductivity, low specific capacitance and poor power characteristics of traditional materials. The most direct application of graphene may be in the design of tough, lightweight materials by spreading a small amount of graphene in the polymer. The composite is electrically conductive and can withstand much higher temperatures than polymers.

At present, supercapacitors have high power density but still have the problem of low energy density. The low ionic conductivity (especially for the solid electrolytes used in flexible supercapacitors and organic electrolytes) and narrow potential windows of electrolytes limit the performance of supercapacitors in terms of power density and energy density. This problem can be solved by using novel electrolytes with high ionic conductivity and a large electrochemical potential window. Recently, there have been new efforts to manufacture hybrid energy storage devices, such as lithium/sodium ion hybrid capacitors, which consist of both battery-type and capacitor-type electrodes, and can have both high energy and power density. Graphene and its composite materials have shown great advantages as electrodes for supercapacitors. However, their electrochemical properties largely depend on how the two components interact and the resulting structure/morphology. In this regard, the structure and interface of electrode materials must be optimized in order to provide more electroactive sites and greatly improve ion transport kinetics. One of the main goals in the design of new electrode materials is to improve their rate performance at high current densities, which requires the electrode materials to have high conductivity.

However, regarding how to give full play to the excellent electrical properties of graphene and accelerate its industrialization process, this paper believes that we should also pay attention to the following aspects:

(1) Improving the uniformity of the distribution of nanoparticles on the surface of graphene, controlling the morphology and structure of the nanoparticles, and increasing the specific capacitance need further study.
(2) We should improve the synergistic effect of graphene composites, prevent graphene agglomeration, and improve the wettability of composite electrode materials and electrolytes without affecting the conductivity of graphene.
(3) We should simplify the electrode material synthesis process, improve its production efficiency, and reduce its production costs.
(4) The general graphene-based composite electrode materials only discuss mass-to-capacitance capacitance, and there is very little research on volume-to-capacitance.

In summary, graphene supercapacitors also need to improve relevant theoretical research and develop new synthetic processes. In the near future, it is foreseeable that with the efficient and environmentally-friendly production of graphene, its application in supercapacitors will mature.

Funding: This paper has been supported by the project of Industry University Cooperation and Collaborative Education of the Ministry of Education (No. 202102152006).

Institutional Review Board Statement: Not applicable.

Informed Consent Statement: Not applicable.

Data Availability Statement: Not applicable.

Conflicts of Interest: The authors declare no conflict of interest.

References

1. Yu, X.; Kang, Y.; Park, H.S. Sulfur and phosphorus co-doping of hierarchically porous graphene aerogels for enhancing Super capacitor performance. *Carbon* **2016**, *101*, 49–56. [CrossRef]
2. Zhu, C.; Liu, T.; Qian, F.; Han, T.Y.J.; Duoss, E.B.; Kuntz, J.D.; Spadaccini, C.M.; Worsley, M.A.; Li, Y. Super capacitors based on three-dimensional hierarchical graphene aerogels with periodic macropores. *Nano Lett.* **2016**, *16*, 3448–3456. [CrossRef] [PubMed]
3. Gao, Z.; Yang, W.; Wang, J.; Song, N.; Li, X. Flexible all-solid-state hierarchical $NiCo_2O_4$/porous graphene paper asymmetric Super capacitors with an exceptional combination of electrochemical properties. *Nano Energy* **2015**, *13*, 306–317. [CrossRef]
4. Shao, Y.; El-Kady, M.; Wang, L. Graphene-based materials for flexible Super capacitors. *Chem. Soc. Rev.* **2015**, *44*, 3639–3665. [CrossRef]

5. Yan, J.; Li, S.; Lan, B.; Wu, Y.; Lee, P.S. Rational Design of Nanostructured Electrode Materials toward Multifunctional Super capacitors. *Adv. Funct. Mater.* **2019**, *30*, 1902564. [CrossRef]
6. Liu, L.; Feng, Y.; Liang, J.; Li, S.; Tian, B.; Yao, W.; Wu, W. Structure-designed fabrication of all-printed flexible in-plane solid-state Super capacitors for wearable electronics. *J. Power Sources* **2019**, *425*, 195–203. [CrossRef]
7. Jang, S.; Kang, J.; Kwak, S.; Seol, M.L.; Meyyappan, M.; Nam, I. Methodologies for Fabricating Flexible Super capacitors. *Micromachines* **2021**, *12*, 163. [CrossRef]
8. Zhu, S.; Wu, M.; Ge, M.H.; Zhang, H.; Li, S.K.; Li, C.H. Design and construction of three-dimensional CuO/polyaniline/rGO ternary hierarchical architectures for high performance Super capacitors. *J. Power Sources* **2016**, *306*, 593–601. [CrossRef]
9. Sun, H.; Mei, L.; Liang, J.; Zhao, Z.; Lee, C.; Fei, H.; Ding, M.; Lau, J.; Li, M.; Wang, C.; et al. Three-dimensional holey-graphene/niobia composite architectures for ultrahigh-rate energy storage. *Science* **2017**, *356*, 599–604. [CrossRef]
10. Yu, D.; Goh, K.; Wang, H.; Wei, L.; Jiang, W.; Zhang, Q.; Dai, L.; Chen, Y. Scalable synthesis of hierarchically structured carbon nanotube–graphene fibres for capacitive energy storage. *Nat. Nanotechnol.* **2014**, *9*, 555–562. [CrossRef]
11. Hou, J.; Cao, C.; Idrees, F.; Ma, X. Hierarchical porous nitrogen-doped carbon nanosheets derived from silk for ultrahigh-capacity battery anodes and Super capacitors. *ACS Nano* **2015**, *9*, 2556–2564. [CrossRef] [PubMed]
12. Zhang, L.; Li, T.; Ji, X.; Zhang, Z.; Yang, W.; Gao, J.; Li, H.; Xiong, C.; Dang, A. Freestanding three-dimensional reduced graphene oxide/MnO_2 on porous carbon/nickel foam as a designed hierarchical multihole Super capacitor electrode. *Electrochim. Acta* **2017**, *252*, 306–314. [CrossRef]
13. Wang, X.; Ma, X.; Zhang, L.; Jiang, G.; Yang, M. Dielectric and optical properties of porous graphenes with uniform pore structures. *J. Mol. Model.* **2019**, *25*, 266. [CrossRef]
14. Zhou, Y.; Gao, G.; Li, Y.; Chu, W.; Wang, L.W. Transition-metal single atoms in nitrogen-doped graphenes as efficient active centers for water splitting: A theoretical study. *Phys. Chem. Chem. Phys.* **2019**, *21*, 3024–3032. [CrossRef] [PubMed]
15. Wang, Y.; Wang, W.; Zhu, S.; Yang, G.; Zhang, Z.; Li, P. Theoretical studies on the structures and properties of doped graphenes with and without an external electrical field. *RSC Adv.* **2019**, *9*, 11939–11950. [CrossRef]
16. Cao, Y.; Liu, C.; Jiang, J.; Zhu, X.; Zhou, J.; Ni, J.; Zhang, J.; Pang, J.; Rummeli, M.H.; Zhou, W.; et al. Theoretical Insight into High-Efficiency Triple-Junction Tandem Solar Cells via the Band Engineering of Antimony Chalcogenides. *Sol. RRL.* **2021**, *5*, 2000800. [CrossRef]
17. Cao, Y.; Zhu, X.; Tong, X.; Zhou, J.; Ni, J.; Zhang, J.; Pang, J. Ultrathin microcrystalline hydrogenated Si/Ge alloyed tandem solar cells towards full solar spectrum conversion. *Front. Chem. Sci. Eng.* **2020**, *14*, 997–1005. [CrossRef]
18. Wang, Y.; Pang, J.; Cheng, Q.; Han, L.; Li, Y.; Meng, X.; Ibarlucea, B.; Zhao, H.; Yang, F.; Liu, H.; et al. Applications of 2D-Layered Palladium Diselenide and Its van der Waals Heterostructures in Electronics and Optoelectronics. *Nano-Micro Lett.* **2021**, *13*, 143. [CrossRef]
19. Sun, B.; Pang, J.; Cheng, Q.; Zhang, S.; Li, Y.; Zhang, C.; Sun, D.; Ibarlucea, B.; Li, Y.; Chen, D.; et al. Synthesis of Wafer-Scale Graphene with Chemical Vapor Deposition for Electronic Device Applications. *ADV. Mater. Technol.* **2021**, *6*, 2000744. [CrossRef]
20. Cui, Z.; Kang, L.; Li, L.; Wang, L.; Wang, K. A combined state-of-charge estimation method for lithium-ion battery using an improved BGRU network and UKF. *Energy* **2022**, *259*, 124933. [CrossRef]
21. Liu, C.; Li, D.; Wang, L.; Li, L.; Wang, K. Strong Robustness and High Accuracy Remaining Useful Life Prediction on Supercapacitors. *APL Mater.* **2022**, *10*, 061106. [CrossRef]
22. Cui, Z.; Dai, J.; Sun, J.; Li, D.; Wang, L.; Wang, K. Hybrid methods using neural network and Kalman filter for the state of charge estimation of lithium-ion battery. *Math. Probl. Eng.* **2022**, *2022*, 9616124. [CrossRef]
23. Yi, Z.; Zhao, K.; Sun, J.; Wang, L.; Wang, K.; Ma, Y. Prediction of the Remaining Useful Life of Supercapacitors. *Math. Probl. Eng.* **2022**, *2022*, 7620382. [CrossRef]
24. Sun, H.; Sun, J.; Zhao, K.; Wang, L.; Wang, K. Data-driven ICA-Bi-LSTM-combined lithium battery SOH estimation. *Math. Probl. Eng.* **2022**, *2022*, 9645892. [CrossRef]
25. Li, D.; Wang, L.; Duan, C.; Li, Q.; Wang, K. Temperature prediction of lithium-ion batteries based on electrochemical impedance spectrum: A review. *Int. J. Energy Res.* **2022**, *46*, 10372–10388. [CrossRef]
26. Li, D.; Li, S.; Zhang, S.; Sun, J.; Wang, L.; Wang, K. Aging state prediction for supercapacitors based on heuristic kalman filter optimization extreme learning machine. *Energy* **2022**, *250*, 123773. [CrossRef]
27. Hao, Z.; Cao, J.; Zhao, X.; Wu, Y.; Zhu, J.; Dang, Y.; Zhuang, Q.; Wei, X. Preparation of porous carbon spheres from 2-keto-l-gulonic acid mother liquor by oxidation and activation for electric double-layer capacitor application. *J. Colloid Interface Sci.* **2017**, *513*, 20–27. [CrossRef]
28. Ni, Z.; Wang, Y.; Yu, T.; Shen, Z. Raman spectroscopy and imaging of graphene. *Nano Res.* **2008**, *1*, 273–291. [CrossRef]
29. Hao, Z.; Cao, J.; Wu, Y.; Zhao, X.; Zhou, L.; Fan, X.; Zhao, Y.; Wei, X. Preparation of porous carbons from waste sugar residue for high performance electric double-layer capacitor. *Fuel Process Technol.* **2017**, *162*, 45–54. [CrossRef]
30. Wu, Y.; Cao, J.; Zhou, Z.; Zhao, X.; Zhuang, Q.; Wei, Y.; Zhao, M.; Zhao, Y.; Bai, H. Transforming waste sugar solution into N-doped hierarchical porous carbon for high performance Super capacitors in aqueous electrolytes and ionic liquid. *Int. J. Hydrog. Energy* **2020**, *45*, 31367–31379. [CrossRef]
31. Lin, Y.; Chen, Z.; Yu, C.; Zhong, W. Heteroatom-doped sheet-like and hierarchical porous carbon based on natural biomass small molecule peach gum for high-performance Super capacitors. *ACS Sustain. Chem. Eng.* **2019**, *7*, 3389–3403. [CrossRef]

32. Li, D.; Yang, D.; Li, L.; Wang, L.; Wang, K. Electrochemical Impedance Spectroscopy Based on the State of Health Estimation for Lithium-Ion Batteries. *Energies* **2022**, *15*, 6665. [CrossRef]
33. Li, Q.; Li, D.; Zhao, K.; Wang, L.; Wang, K. State of health estimation of lithium-ion battery based on improved ant lion optimization and support vector regression. *J. Energy Storage* **2022**, *50*, 104215. [CrossRef]
34. Cui, Z.; Wang, L.; Li, Q.; Wang, K. A comprehensive review on the state of charge estimation for lithium-ion battery based on neural network. *Int. J. Energy Res.* **2022**, *46*, 5423–5440. [CrossRef]
35. Li, S.; Pasc, A.; Fierro, V.; Celzard, A. Hollow carbon spheres, synthesis and applications—A review. *J. Mater. Chem. A* **2016**, *4*, 12686. [CrossRef]
36. Portet, C.; Yang, Z.; Korenblit, Y.; Gogotsi, Y.; Mokaya, R.; Yushin, G. Electrical Double-Layer Capacitance of Zeolite-Templated Carbon in Organic Electrolyte. *J. Electrochem. Soc.* **2009**, *156*, A1–A6. [CrossRef]
37. Murali, S.; Dreyer, D.R.; Valle-Vigón, P.; Stoller, M.D.; Zhu, Y.; Morales, C.; Fuertes, A.B.; Bielawski, C.W.; Ruoff, R.S. Mesoporous carbon capsules as electrode materials in electrochemical double layer capacitors. *Phys. Chem. Chem. Phys.* **2011**, *13*, 2652–2655. [CrossRef]
38. Liu, C.; Zhang, Y.; Sun, J.; Cui, Z.; Wang, K. Stacked bidirectional LSTM RNN to evaluate the remaining useful life of supercapacitor. *Int. J. Energy Res.* **2022**, *46*, 3034–3043. [CrossRef]
39. Liu, C.; Li, Q.; Wang, K. State-of-charge estimation and remaining useful life prediction of supercapacitors. *Renew. Sustain. Energy Rev.* **2021**, *150*, 111408. [CrossRef]
40. Kang, L.; Du, H.; Deng, J.; Jing, X.; Zhang, S.; Zhang, Y. Synthesis and Catalytic Performance of a New V-doped CeO_2-supported Alkali-activated-steel-slag-based Photocatalyst. *J. Wuhan Univ. Technol. Mater. Sci. Ed.* **2021**, *36*, 209–214. [CrossRef]
41. Xu, H.; Du, H.; Kang, L.; Cheng, Q.; Feng, D.; Xia, S. Constructing Straight Pores and Improving Mechanical Properties of Gangue-Based Porous Ceramics. *J. Renew. Mater.* **2021**, *9*, 2129–2141. [CrossRef]
42. Matsumoto, H.; Imaizumi, S.; Konosu, Y.; Ashizawa, M.; Minagawa, M.; Tanioka, A.; Lu, W.; Tour, J.M. Electrospun composite nanofiber yarns containing oriented graphene nanoribbons. *ACS Appl. Mater. Interfaces* **2013**, *5*, 6225–6231. [CrossRef] [PubMed]
43. Vivekchand, S.; Rout, C.; Subrahmanyam, K.; Govindaraj, A. Graphene-based electrochemical Super capacitors. *J. Chem. Sci.* **2008**, *120*, 9–13. [CrossRef]
44. Stoller, M.; Park, S.; Zhu, Y.; An, J.; Ruoff, R. Graphene-based ultracapacitors. *Nano Lett.* **2008**, *8*, 3498–3502. [CrossRef] [PubMed]
45. Wang, Y.; Shi, Z.; Huang, Y.; Ma, Y. Super capacitor devicesbased on graphene materials. *J. Phys. Chem. C* **2009**, *113*, 13103–13107. [CrossRef]
46. Cai, Y.; Wang, Y.; Deng, S.; Chen, G.; Li, Q.; Han, B. Graphene-tungsten oxides composite for Super capacitor electrode. *Ceram. Int.* **2014**, *40*, 4109–4116. [CrossRef]
47. Ran, H.; Du, H.; Ma, C.; Zhao, Y.; Feng, D.; Xu, H. Effects of A/B-Site Co-Doping on Microstructure and Dielectric Thermal Stability of $AgNbO_3$ Ceramics. *Sci. Adv. Mater.* **2021**, *13*, 741–747. [CrossRef]
48. Ma, C.; Du, H.; Liu, J.; Kang, L.; Du, X.; Xi, X.; Ran, H. High-temperature stability of dielectric and energy-storage properties of weakly-coupled relaxor (1-x)$BaTiO_3$-xBi($Y_{1/3}Ti_{1/2}$)O_3 ceramics. *Ceram. Int.* **2021**, *47*, 25029–25036. [CrossRef]
49. Cai, X.; Sun, K.; Qiu, Y.; Jiao, X. Recent Advances in Graphene and Conductive Polymer Composites for Supercapacitor Electrodes: A Review. *Crystals* **2021**, *11*, 947. [CrossRef]
50. Zhou, W.; Du, H.; Kang, L. Microstructure Evolution and Improved Permeability of Ceramic Waste-Based Bricks. *Materials* **2022**, *15*, 1130. [CrossRef]
51. Feng, D.; Du, H.; Ran, H.; Lu, T.; Xia, S.; Xu, L.; Wang, Z.; Ma, C. Antiferroelectric stability and energy storage properties of Co-doped $AgNbO_3$ ceramics. *J. Solid State Chem.* **2022**, *310*, 123081. [CrossRef]
52. Zhao, Y.; Arowo, M.; Wu, W.; Zou, H.; Chen, J.; Chu, G. Polyaniline/graphene nanocomposites synthesized by in situ high gravity chemical oxidative polymerization for Super capacitor. *J. Ind. Eng. Chem.* **2015**, *25*, 280–287. [CrossRef]
53. Xu, Y.; Schwab, M.G.; Strudwick, A.J.; Hennig, I.; Feng, X.; Wu, Z.; Mellen, K. Screen-Printable Thin Film Super capacitor Device Utilizing Graphene/Polyaniline Inks. *Adv. Energy Mater.* **2013**, *3*, 1035–1040. [CrossRef]
54. Xue, M.; Li, F.; Zhu, J.; Song, H.; Zhang, M.; Cao, T. Structure-Based Enhanced Capacitance: In Situ Growth of Highly Ordered Polyaniline Nanorods on Reduced Graphene Oxide Patterns. *Adv. Funct. Mater.* **2012**, *22*, 1284–1290. [CrossRef]
55. Meng, Y.; Wang, K.; Zhang, Y.; Wei, Z. Hierarchical Porous Graphene/Polyaniline Composite Film with Superior Rate Performance for Flexible Super capacitors. *Adv. Mater.* **2013**, *25*, 6985–6990. [CrossRef]
56. Liu, L.; Niu, Z.; Zhang, L.; Zhou, W.; Chen, X.; Xie, S. Nanostructured Graphene Compsite Papers for Highly Flexible and Foldable Super capacitors. *Adv. Mater.* **2014**, *26*, 4855–4862. [CrossRef]
57. Wu, Z.; Liu, Z.; Parvez, K.; Feng, X.; Mellen, K. Ultrathin Printable Graphene Superca pacitors with AC Line-Filtering Performance. *Adv. Mater.* **2015**, *27*, 3669–3675. [CrossRef]
58. Park, J.; Park, S.; Kwon, O.; Lee, C.; Jang, J.; Jang, J.; Park, J.; Park, S.; Lee, C.; Kwon, O. In Situ Synthesis of Graphene/Polyselenophene Nanohybrid Materials as Highly Flexible Energy Storage Electrodes. *Chem. Mater.* **2014**, *26*, 2354–2360. [CrossRef]
59. Wang, H.; Liu, R.; Liu, X.; Wu, L.; Li, Y.; Zhang, X. Improved Electrochemical Performances of Graphene Hybrids Embedded with Silica as the Functional Connection Layer for Super capacitors. *J. Energy Storage* **2021**, *36*, 102315. [CrossRef]

60. Dywili, N.R.; Ntziouni, A.; Ikpo, C.; Ndipingwi, M.; Hlongwa, N.; Yonkeu, A.; Masikini, M.; Kordatos, K.; Iwuoha, E. Graphene Oxide Decorated Nanometal-Poly(Anilino-Dodecylbenzene Sulfonic Acid) for Application in High Performance Super capacitors. *Micromachines* **2019**, *10*, 115. [CrossRef]
61. Xu, Z.; Zhang, Z.; Yin, H.; Hou, S.; Lin, H.; Zhou, J.; Zhuo, S. Investigation on the role of different conductive polymers in Super capacitors based on a zinc sulfide/reduced graphene oxide/conductive polymer ternary composite electrode. *RSC Adv.* **2020**, *10*, 3122–3129. [CrossRef] [PubMed]
62. Golkhatmi, S.; Sedghi, A.; Miankushki, H.; Khalaj, M. Structural properties and supercapacitive performance evaluation of the nickel oxide/graphene/Polyaniline hybrid ternary nanocomposite in aqueous and organic electrolytes. *Energy* **2021**, *214*, 118950. [CrossRef]
63. Azizi, E.; Salimi, J.A.A.; Lee, J. Fabrication of an asymmetric Super capacitor based on reduced graphene oxide/polyindole/gamma—Al_2O_3 ternary nanocomposite with high-performance capacitive behavior. *Polymer* **2020**, *195*, 122429. [CrossRef]
64. Ramesh, S.; Yadav, H.; Bathula, C.; Shinde, S.; Sivasamy, A.; Kim, H.; Kim, J. Cubic nanostructure of Co_3O_4@nitrogen doped graphene oxide/polyindole composite efficient electrodes for high performance energy storage applications. *J. Mater. Res. Technol.* **2020**, *9*, 11464–11475. [CrossRef]
65. Li, S.; Zhang, L.; Zhou, J.; Zhou, H.; Chen, X.; Tang, T. The in situ construction of three-dimensional core-shell-structured TiO_2@PPy/rGO nanocomposites for improved Super capacitor electrode performance. *New J. Chem.* **2021**, *45*, 1092–1099. [CrossRef]
66. Yuan, D.; Zhang, C.; Tang, S.; Wang, Z.; Sun, Q.; Zhang, X.; Jiao, T.; Zhang, Q. Ferric ion-ascorbic acid complex catalyzed calcium peroxide for organic wastewater treatment: Optimized by response surface method. *Chin. Chem. Lett.* **2021**, *32*, 3387–3392. [CrossRef]
67. Wang, G.; Gao, E.; Dai, Z. Degradation and recovery of graphene/polymer interfaces under cyclic mechanical loading. *Compos. Sci. Technol.* **2017**, *149*, 220–227. [CrossRef]
68. Huang, J.; Wang, J.; Wang, C. Hierarchical porous graphene carbon-based Super capacitors. *Chem. Mater.* **2015**, *27*, 2107–2113. [CrossRef]
69. Kandasamy, M.; Sahoo, S.; Nayak, S.; Chakraborty, B.; Rout, C. Recent ADVances in engineered metal oxide nanostructures for supercapacitor applications: Experimental and theoretical aspects. *J. Mater. Chem. A* **2021**, *9*, 17643. [CrossRef]
70. Wang, Y.; Guo, C.; Liu, J.; Chen, T.; Yang, H.; Li, C. CeO_2 nanoparticles/graphene nanocomposite-based high performance Super capacitor. *Dalton Trans.* **2011**, *40*, 6388–6391. [CrossRef]
71. Dong, X.; Xu, H.; Wang, X.; Huang, Y.; Chan, M.; Zhang, H. 3D Graphene-Cobalt Oxide Electrode for High-Performance Super capacitor and Enzymeless Glucose Detection. *ACS Nano* **2012**, *6*, 3206–3213. [CrossRef]
72. He, G.; Li, J.; Chen, H. Hydrothermal Preparation of Co_3O_4@graphene Nanocom posite for Super capacitor with Enhanced Capacitive Performance. *Mater. Lett.* **2012**, *82*, 61–63. [CrossRef]
73. Qu, Q.; Yang, S.; Feng, X. 2D sandwich-like sheets of iron oxide grown on graphene as high energy anode material for Super capacitors. *Adv. Mater.* **2011**, *23*, 5574–5580. [CrossRef] [PubMed]
74. Wang, Z.; Ma, C.; Wang, H.; Liu, Z.; Hao, Z. Facilely synthesized Fe_2O_3–graphene nano composite as novel electrode materials for Super capacitors with high performance. *J. Alloys Compd.* **2013**, *552*, 486–491. [CrossRef]
75. Wang, Q.; Jiao, L.; Du, H.; Wang, Y.; Yuan, H. Fe_3O_4 nanoparticles grown on graphene as Advanced electrode materials for Super capacitors. *J. Power Sources* **2014**, *245*, 101–106. [CrossRef]
76. Peng, L.; Peng, X.; Liu, B.; Wu, C.; Xie, Y.; Yu, G. Ultrathin Two-Dimensional MnO_2/Graphene Hybrid Nanostructures for High-Performance, Flexible Planar Super capacitors. *Nano Lett.* **2013**, *13*, 2151–2157. [CrossRef]
77. Zhao, K.; Lyu, K.; Liu, S.; Gan, Q.; He, Z.; Zhou, Z. Ordered porous Mn_3O_4@N-doped carbon/graphene hybrids derived from metal-organic frameworks for Super capacitor electrodes. *J. Mater. Sci.* **2017**, *52*, 446–457. [CrossRef]
78. Wang, C.; Xu, J.; Yuen, M. Hierarchical composite electrodes of nickel oxide nanoflake 3D graphene for high-performance pseudo capacitors. *Adv. Funct. Mater.* **2014**, *24*, 6372–6380. [CrossRef]
79. Wang, W.; Guo, S.; Guo, I. Hydrous Ruthenium Oxide Nanoparticles An chored to Graphene and Carbon Nanotube Hybrid Foam for Super capacitors. *Sci. Rep.* **2015**, *4*, 4452. [CrossRef]
80. Li, F.; Song, J.; Yang, H. One-step synthesis of graphene/SnO_2 nanocompo sites and its application in electrochemical Super capacitors. *Nanotechnology* **2009**, *20*, 455602. [CrossRef]
81. Zhang, Z.Z.; Xiao, X.F.; Guo, G.Y.; Wang, W.S.; Liu, L.Y. One-Pot Self-Assembled Three-Di mensional TiO_2-Graphene Hydrogel with Improved Adsorption Capacities and Photocatalytic and Electrochemical Activities. *ACS Appl. Mater. Interfaces* **2013**, *5*, 2227–2233. [CrossRef] [PubMed]
82. Wang, H.; Yi, H.; Chen, X.; Wang, X. One-step strategy to three-dimensional graphene/VO_2 nanobelt composite hydrogels for high performance Super capacitors. *J. Mater. Chem. A* **2014**, *2*, 1165–1173. [CrossRef]
83. Perera, S.; Liyanage, A.; Nijem, N.; Ferraris, J.; Chabal, Y.; Balkus, K. Vanadium oxide nanowire-Graphene binder free nanocomposite paper electrodes for Super capacitors: A facile green approach. *J. Power Sources* **2013**, *230*, 130–137. [CrossRef]
84. Zhang, C.; Dang, F.; Chen, Y. Vibration-to-electric energy conversion with porous graphene oxide-nickel electrode. *J. Power Sources* **2017**, *368*, 73–77. [CrossRef]
85. Derkaoui, I.; Khenfouch, M.; Mothudi, B. PH effect on the optoelectronic properties of graphene vanadium oxides nanocomposites. *J. Mater. Sci. Mater. Electron.* **2017**, *28*, 17710–17718. [CrossRef]

86. Xu, K.; Yang, J.; Hu, J. Synthesis of hollow $NiCo_2O_4$ nanospheres with large specific surface area for asymmetric Super capacitors. *J. Colloid Interface Sci.* **2018**, *511*, 456–462. [CrossRef] [PubMed]
87. Maity, S.; Neethu, B.; Kella, T.; Shee, D.; Das, P.; Mal, S. Activated carbon- supported Vanado-nickelate (IV) based hybrid materials for energy application. *J. Energy Storage* **2021**, *40*, 102727. [CrossRef]
88. Maity, S.; Vannathan, A.; Kella, T.; Shee, D.; Das, P.; Mal, S. Electrochemical performance of activated carbon-supported vanadomolybdates electrodes for energy conversion. *Ceram. Int.* **2021**, *47*, 27132–27141. [CrossRef]
89. Maity, S.; Das, P.; Mal, S. Decavanadate-graphene oxide nanocomposite as an electrode material for electrochemical capacitor. *Mater. Technol.* **2022**, *37*, 1129–1139. [CrossRef]
90. Guo, Y.; Yu, P.; Zhu, C.; Zhao, K.; Wang, L.; Wang, K. A state-of-health estimation method considering capacity recovery of lithium batteries. *Int. J. Energy Res.* **2022**. [CrossRef]
91. Peng, Z.; Liu, X.; Meng, M.H. Design and tailoring of the 3D macroporous hydrous RuO_2 hierarchical architectures with a hard-template method for high-performance Super capacitors. *ACS Appl. Mater. Interfaces* **2016**, *9*, 4577–4586. [CrossRef] [PubMed]
92. Zhao, Y.; Ran, W.; He, J. High-performance asymmetric Super capacitors based on multilayer MnO_2/graphene oxide nanoflakes and hierarchical porous carbon with enhanced cycling stability. *Small* **2015**, *11*, 1310–1319. [CrossRef] [PubMed]
93. Li, M.; Cheng, J.; Wang, J. The growth of nickel-manganese and cobalt-manganese layered double hydroxides on reduced graphene oxide for super capacitor. *Electrochim. Acta.* **2016**, *206*, 108–115. [CrossRef]
94. Sangermano, M.; Chiolerio, A.; Veronese, G. Graphene-epoxy flexible transparent capacitor obtained by graphene-polymer transfer and UV-induced bonding. *Macromol Rapid Commun.* **2014**, *35*, 355–359. [CrossRef]
95. Rehmen, J.; Pathirana, T.; Garcia-Quintana, L. Structuring PEDOT Hollow Nanosphere Electrodes for High Specific Energy Li-Metal | Polymer Thin-Film Batteries. *ACS Appl. Nano Mater.* **2020**, *3*, 3820–3828. [CrossRef]
96. Cui, Z.; Kang, L.; Li, L.; Wang, L.; Wang, K. A hybrid neural network model with improved input for state of charge estimation of lithium-ion battery at low temperatures. *Renew. Energy* **2022**, *98*, 1328–1340. [CrossRef]
97. Sun, H.; Yang, D.; Wang, L.; Wang, K. A method for estimating the aging state of lithiumion batteries based on a multi-linear integrated model. *Int. J. Energy Res.* **2022**. [CrossRef]

coatings

Article

Metakaolin-Reinforced Sulfoaluminate-Cement-Solidified Wasteforms of Spent Radioactive Resins—Optimization by a Mixture Design

Jiaqian Xu, Mengzhou Wang, Cong Li, Mengxing Han, Qi Wang and Qina Sun *

Hebei Key Laboratory of Heavy Metal Deep-Remediation in Water and Resource Reuse, School of Environmental and Chemical Engineering, Yanshan University, Qinhuangdao 066004, China
* Correspondence: sunqn@ysu.edu.cn

Abstract: Cement solidification is a main technique for radioactive waste treatment to reduce its risk to the environment and human health. However, this method underperforms when dealing with spent radioactive ion-exchange resin, taking much space, and costing much money for final disposal. In this work, simulated spent radioactive resin was solidified using a metakaolin-reinforced sulfoaluminate cement system, which was optimized by a mixture design based on the effects of components and parameters, and the durability of solidified wasteforms was assessed in terms of strength and Cs(I) leaching. Solidified by an optimized formula of 40 wt.% spent resin, 55.8 wt.% sulfoaluminate cement, 2.2 wt.% metakaolin, and 2 wt.% water reducer, the resin loading in wasteforms reached 64% and the compressive strength 13.7 MPa. The dominant mineral phases of hydration products were ettringite crystalline of acicular and columnar morphology, with small amounts of scattered amorphous clusters of aluminum gels and C–S–H gels. Metakaolin, a source of aluminum, promoted the growth of ettringite, which facilitated (1) the encapsulation of resin beads with high strengths, even in acidic environments or during frequent freezing-thawing, and (2) the retention of Cs(I), with a 42 day leaching rate of 2.3×10^{-4} cm/day. This work offers a technical justification for spent resin solidification in the metakaolin-reinforced sulfoaluminate cement system, which is an applicational solution for the efficient treatment of radioactive waste.

Keywords: spent radioactive resin; cement solidification; sulfoaluminate cement; metakaolin; nuclides leaching; mixture design

Citation: Xu, J.; Wang, M.; Li, C.; Han, M.; Wang, Q.; Sun, Q. Metakaolin-Reinforced Sulfoaluminate-Cement-Solidified Wasteforms of Spent Radioactive Resins—Optimization by a Mixture Design. *Coatings* **2022**, *12*, 1466. https://doi.org/10.3390/coatings12101466

Academic Editor: Maria Bignozzi

Received: 31 August 2022
Accepted: 29 September 2022
Published: 4 October 2022

1. Introduction

Radioactive waste streams come from a wide range of sources, including the nuclear fuel cycle, nuclear technology utilization, and reactor accidents [1–3]. One of the commonly used purification processing for radioactive waste streams is ion exchange, in which nuclide ions are extracted from aqueous streams and enriched onto ion exchange resin beads [4–6]. When the ion exchange sites of resins are saturated with radionuclides, they lose purification capacity and are discharged from treatment columns, thus forming spent radioactive resins to be solidified and disposed of. Cement solidification, with the advantages of low cost, operability, and good radiation resistance of the solidified wasteforms, is currently the most widely used solidification technique for treating spent resin before final disposal [7–15]. New cement compositions besides ordinary Portland cement (OPC) offers fresh perspectives on radioactive waste solidification, especially for wastes that are challenging for OPC to solidify [16,17]. Sulfoaluminate cement (SAC) is one of the emerging cementitious materials produced by calcining limestone, bauxite, and gypsum in a certain composition ratio configuration at relatively low temperatures [7,18–21]. The clinker minerals of SAC are mainly anhydrous calcium sulfoaluminate ($3CaO{\cdot}3Al_2O_3{\cdot}CaSO_4$), dicalcium silicate ($2CaO{\cdot}SiO_2$), and iron phase [7,18,20]. After blending the binder with water, hydration products of hydrated calcium sulfoaluminate (ettringite, $3CaO{\cdot}Al_2O_3{\cdot}3CaSO_4{\cdot}32H_2O$), aluminum gel ($Al_2O_3 \cdot nH_2O$), and hydrated calcium silicate hydrate gels (C–S–H gels) generate early strength, frozen resistance, and impermeability, etc. Radioactive wastes, such as filter cartridges from gas purification, membranes, and evaporator concentrates from liquids treatment, are solidified by SAC and the wasteforms exhibit high compressive strengths and good resistance to immersion, indicating SAC as being a promising cementitious material for radioactive waste treatment and disposal [7,20,22–24].

However, similar to OPC, the biggest problem with spent resin wasteforms solidified by SAC is that they cannot withstand prolonged water immersion [21,25,26]. This is because, during the curing process, the water in the wet resin is thoroughly customed in the hydration reaction, thus reducing the diameter of the resin beads and encapsulating them in a comparable size cavity. When the wasteforms are subjected to prolonged water immersion, water enters through the pores of wasteforms, the resin absorbs water, and the beads swell and become larger in diameter, thus generating static pressure on the cavity walls [5]. With volume loadings of resin in the wasteforms of 30%~40%, it may cause the wasteforms to crack, thus making it easier for water to penetrate [7]. Resin swelling force and water chemistry result in the wasteforms breaking and disintegrating. In this case, free nuclide ions may migrate into the water; broken fragments and dislodged resin beads may also diffuse, causing radioactive contamination. To overcome this drawback, researchers have investigated many approaches, such as reducing the resin loadings and wrapping the wasteforms with hydrophobic film, which increases the waste to be disposed of and the process complexity [7].

Metakaolin ($Al_2O_3 \cdot 2SiO_2 \cdot 2H_2O$) is a typical clay material with the main components of SiO_2 and Al_2O_3; Al_2O_3 content is 40% and more, which is higher than other clay minerals [27–30]. Commercial metakaolin is produced by kaolinite calcination at a temperature of 600~800 °C [28,30,31]. The crystalline kaolinite is mainly in irregular squamous sheets bonded by Van der Waals bonds. By calcination, the lamellar structure of kaolinite remains unchanged, but the crystal size decreases [28,31]. Meanwhile, the particle gap reduces, forming metakaolin with a poorly crystallized transition phase with more agglomerated particles [27,29]. The formation of unsaturated ligands during calcination derives a thermodynamically metastable state of metakaolin, which possesses a high-pozzolanic reactivity as an artificial cementitious additive for cements [27,32–34]. Pozzolanic reaction of metakaolin occurs at the early stage of cement hydration, thus promoting the early hydration of clinker minerals, shortening the hydration induction period and early hydration acceleration period of medium- and low-heat Portland cement [31,32]. Afterwards, the ongoing pozzolanic reaction substantially increases hydration products as the secondary hydration reaction, in which the active aluminum component participates and generates C–S–H gels [33]. Moreover, the ultra-fine metakaolin powder fills the pores, decreases pore size, and reduces the total pore volume and porosity [31,32]. These reactions and effects lead to a denser microstructure and higher mechanical properties of the hardened cement [31–33,35]. Because of these advantages for cement mechanics, metakaolin has been widely used in the construction industry; therefore, it is reasonable to expect that metakaolin will play a beneficial role in solidifying radioactive wastes. However, metakaolin has been less studied as a solidification additive for radioactive waste treatment. The performance of metakaolin-reinforced solidification of spent resins in sulfoaluminate cementitious systems has never been reported.

In order to improve the performance of solidification of spent radioactive resins, in this work, we solidified simulated spent resin using a metakaolin-reinforced sulfoaluminate cementitious system and investigated the effects of solidifying components and parameters on the wasteforms. The solidification was optimized by a mixture design to improve the comprehensive performance of wasteforms, including compressive strength, durability, and nuclide leaching property. The microscopic morphology and hydration products of the cured bodies were characterized by SEM and XRD to analyze the curing mechanism of the metakaolin-reinforced sulfoaluminate cement system.

2. Materials and Methods

2.1. Materials

Sulfoaluminate cement was purchased from Polar Bear Special Cement Co., Tangshan, China, and the main mineral components are anhydrous calcium sulfoaluminate ($3CaO{\cdot}3Al_2O_3{\cdot}CaSO_4$), dicalcium silicate ($2CaO{\cdot}SiO_2$), and $CaSO_4$ and $CaCO_3$ blended during production (Figure S1, see Supplementary Materials). Metakaolin and naphthalenesulfonate water reducer were purchased from Beichen Admixture Co., Tianjin, China. Cesium nitrate (A.R.) was purchased from Kaitong Chemical Reagent Co., Tianjin, China. Polystyrene ion exchange resin was offered by Zhengguang Resin Co., Zhejiang, China, and the main composition and total exchange capacity are listed in Table S1. Simulated radioactive resins were prepared by immersing polystyrene ion-exchange spent resin in cesium nitrate solution and stirring continuously for 48 h to ensure adsorption equilibrium for Cs(I). Afterwards, the wet resin with a water content of 68.8% was solidified as the simulated resin without further treatment.

2.2. Solidification and Performance Tests

A typical out-of-vessel mixing method is used for the solidification operation [23,36]. The sulfoaluminate cement, metakaolin, and naphthalenesulfonate water reducer were put to the vessel and mixed at low speed. Then the simulated spent resin is added, and all materials are stirred into a well-mixed paste. The paste was slowly injected into Ø50 mm × 50 mm cylindrical molds and tapped to discharge large air bubbles. The upper surface was scraped and then cured at 25 ± 5 °C with relative humidity ≥ 90%.

The effects of metakaolin, spent resin loadings in wasteforms, water-binder (w/b) ratios, and water reducer dosages on the solidification performance were investigated by the controlled variable method. The compressive strength of demolded wasteforms was tested after 7 days curing on a fully automatic pressure tester (SY-200, Jinan Testing Machine Co., Jinan, China), and the arithmetic mean from four parallel samples was used for further assessment. The initial and final setting times of the wasteforms were determined using the Vicat needle test by a Vicat apparatus (ISO standard, Wuxi Construction Instrument and Material Machinery Co., Wuxi, China), according to the Chinese Standard GB/T 1346-2011. The fluidities were measured in accordance with the Chinese Standard GB/T 2419-2005 based on the spread diameter in a cone test, using a cement mortar fluidity tester (NLD-3, Zhongluda Instrument Technology Co., Tianjin, China).

2.3. Solidification Optimization by Mixture Design

A mixture design was employed to optimize the solidification of spent resin in metakaolin-reinforced cement. The factors in the mixture design, denoted as X_i, are the components of a cementitious system with the constraints shown in Equation (1) [37–39],

$$\begin{cases} \sum_{i=1}^{n} X_i = 1 \\ a_i \leq X_i \leq b_i \end{cases} \quad (1)$$

where, a_i and b_i are the lower and upper limits of factor X_i, respectively.

Fluidity, resin loading by volume, final setting time, and 7 day compressive strength are the responses, noted as Y_1, Y_2, Y_3, and Y_4, respectively. A quadratic polynomial model including only primary and interaction terms is used to describe the relationship between the response values and the factors, and to predict and optimize the response values, as shown in Equation (2) [37–40]. The experiments were arranged by the software Design Expert, the results were analyzed, and the model regression coefficients were computed. The multiobjective optimization was performed using the satisfaction function method [37,41].

$$Y = \sum_{i=1}^{n} \beta_i X_i + \sum_{i=1}^{n-1} \beta_{ij} X_i X_j \ (1 \leqq \mathrm{i} \leqq \mathrm{n}, \quad j = i+1) \quad (2)$$

2.4. Solidified Wasteforms Durability Tests

Solidified wasteforms durability tests include evaluation of compressive strength loss after immersion or freeze–thaw cycles and Cs(I) leaching behavior [23]. Compressive strength as control was measured after 28 days of curing. In the immersion tests, solidified samples were immersed in deionized water without pH adjustment at 25 ± 5 °C for 30 days, 90 days and 180 days, respectively, and different pHs of 3, 5, 7, 9, and 11 were used in the 90 day immersion. The change in compressive strength before and after immersion was used to evaluate the strength loss. In a freeze–thaw cycle, the cured bodies were frozen at −20 °C for three hours and thawed at 20 °C for four hours [42]. The samples were evaluated for strength loss after undergoing an infinitesimal number of cycles.

The Cs(I) leaching test was conducted using semidynamic experimental methods with reference to the Chinese National Standard GB7023-1986 [23,42]. The mass of cesium contained in each standard curing body was 66.14 mg. Deionized water 1.6 L at 25 ± 5 °C at different pH was used as the leachant, and the ratio of leachant volume to the surface area of the waste form V_L/F was 13.35. After the 1st, 3rd, 7th, 10th, 14th, 21st, 28th, 35th and 42nd days of leaching, the deionized water was replaced. The mass of Cs(I) in the leachant was calculated from Cs(I) concentration that was determined by an atomic absorption spectrophotometer (TAS-990 AFG, PuXi General Instrument Co., Beijing, China).

The leaching rate R and cumulative leaching fraction P are determined in a semidynamic leaching test and calculated from the mass of Cs(I) leached into the leachant from the wasteform, as shown in Equations (3) and (4), respectively [8,23,42],

$$R_n = \frac{a_n/A_0}{F/V} \cdot \frac{1}{t_n} \quad (3)$$

$$P_t = \frac{\sum a_n/A_0}{F/V} \quad (4)$$

where t is the durative days of immersion, d, $t = \Sigma t_n$, and in the nth interval of t_n days, the leaching rate and leached mass of Cs(I) are R_n, cm/day, and a_n, g, respectively; A_0 is the average initial Cs(I) mass in a sample, 66.14 mg in this case; F is the surface area of a sample, cm^2; V is the volume of a sample, cm^3; P_t is the cumulative fraction leached of Cs(I) on the tth day, cm.

2.5. Characterization of Solidified Wasteforms

The morphology and mineral phases of the hydration products of the solidified wasteforms were characterized and analyzed by scanning electron microscopy (FESEM, S-4800, Hitachi, Tokyo, Japan) and X-ray diffraction (XRD, D/max 2550 PC, Rigaku, Tokyo, Japan). The samples were crushed and ground, immersed in anhydrous ethanol to terminate the hydration, and dried. The SEM samples were observed after plating with gold films. XRD analysis was performed using samples ground to fine powders with a scan speed of 7° 2θ/min, step size of 0.02°, and a scan range of 5° to 70° 2θ.

3. Results and Discussion

3.1. Factors Affecting the Solidification Performance

The effects of metakaolin, spent resin loadings in wasteforms, water-binder (w/b) ratios, and water reducer dosages on the solidification performance were investigated and the results are shown in Figure 1. Metakaolin dosage is the major concerning factor (Figure 1a) for its pozzolanic reactivity. The 7 day compressive strength increased with the increase in metakaolin dosage, indicating that the secondary hydration of the metakaolin improved the microstructure and then enhanced the mechanical properties and durability of the solidified wasteforms at the macroscopic level [30]. At the same time, metakaolin reduced the fluidity significantly and shortened both the initial and final setting times of the paste. The trends of strength, fluidity, and setting times indicate that metakaolin boosted the hydration reaction rate of the sulfoaluminate cementitious system to facilitate the solidification. In terms of operating parameters, with no metakaolin addition, the fluidity reached 300 mm, which makes the resin beads subject to a decreasing fluid resistance in the paste and keep floating up and gathering in the upper part of the paste. The floating resin beads were delaminated with the cement during a period around 100 min before the initial setting, which is one of the reasons for the low compressive strength in this case. Owing to appropriate metakaolin dosages, it was also facile to blend the paste uniformly and fill the mold without large bubbles with fluidity between 150 mm and 200 mm, which meets the needs of solidifying procedures. The setting times for paste containing metakaolin were also within a reasonably acceptable range, with the initial setting time for sufficient stirring being no less than 40 min and the final setting time for minimizing delay on subsequent processing steps and personnel exposure risk being no more than 90 min.

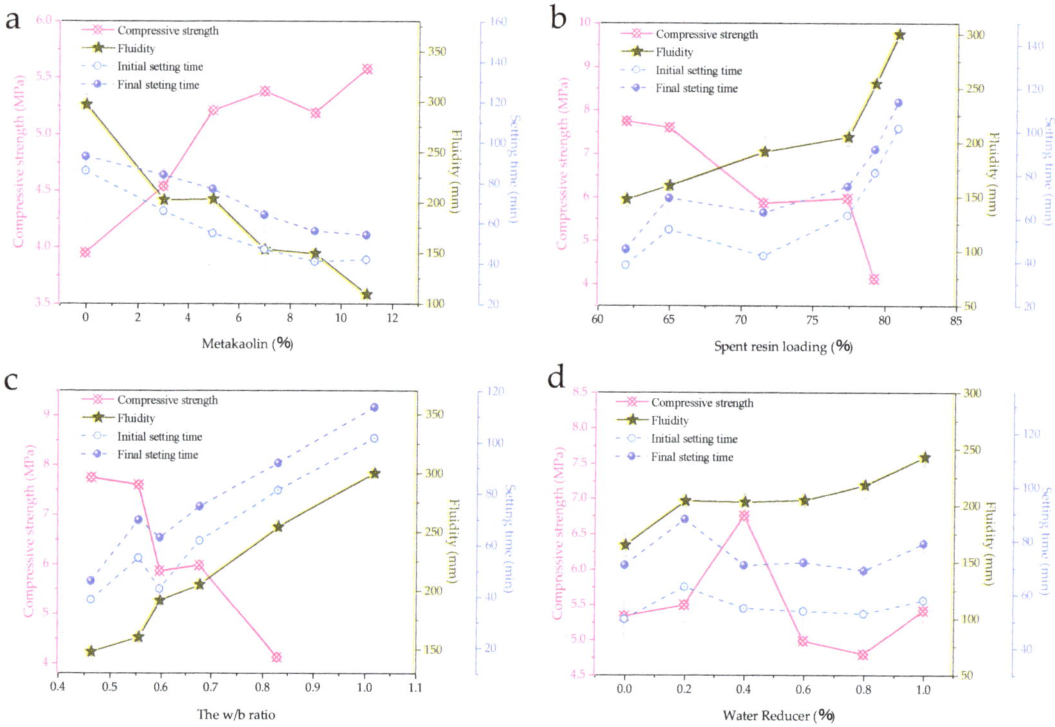

Figure 1. Solidification performance affected by (**a**) metakaolin (spent resin loading 77.4%, w/b ratio 0.68, water reducer 0.4%), (**b**) resin loading (metakaolin 5%, water reducer 0.6%~0.7%), (**c**) w/b ratio (metakaolin 5%, water reducer 0.6%~0.7%), and (**d**) water reducer (metakaolin 5%, spent resin loading 77.4%, w/b ratio 0.68).

Generally speaking, more resin added in the solidification system results in a larger resin loading in the final wasteforms, despite cementitious components affecting the composition of hydration products and the pore structure of the solidified wasteforms, resulting in differences in their volumes [25]. More practically, resin loading indicates whether the smallest waste package is formed for the same resin volume [21]. Therefore, resin loading is used as an indicator to examine the effects of spent resins on solidifying performance, as shown in Figure 1b. With increased resin loading, the wasteforms' compressive strength fell noticeably and rapidly, reaching ~5.8 MPa with a loading higher than 70%. High resin loadings also resulted in more free water and, therefore, caused an increase in fluidity and prolonged setting times. There was a modest rise in fluidity with a 64% loading, but when the loading was increased to 80%, the fluidity increased to 300 mm, and the resin floated up, which was one of the causes of the low compressive strength. To further illustrate the role of free water in the system, extra water was added in addition to the 68.8% water contained in the resin. The w/b ratio effects on the solidifying performance are shown in Figure 1c. The w/b ratio curve resembles the spent resin loading curve. The fluidity increased almost linearly with increasing w/c ratios. Although it benefited the operation, the excess water in the system over that required for the hydration reaction greatly retarded the solidification. Thus, there is a noticeable decline in the compressive strength curve. Therefore, water reducers providing additional fluidity without introducing excess water are essential for solidification. In the experimental range (Figure 1d), the water reducer increased the fluidity well, without drastically weakening both setting times and compressive strength. This is due to the effect of naphthalenesulfonate on the dispersion and surface lubrication of binder particles [43,44].

Through controlled variable experiments, we determined the effectiveness of metakaolin on the reinforcement of sulfoaluminate cement systems. We set a reasonable range of solidifying components based on the results in Figure 1, as shown in Table 1, for optimizing the overall performance of the solidified wasteforms through a mixture design.

Table 1. Upper and lower limits of factors.

Factors	Components	a_i (wt.%)	b_i (wt.%)
X_1	Spent resin	35	48
X_2	Sulfoaluminate cement	47	57.8
X_3	Metakaolin	2.2	5.9
X_4	Naphthalenesulfonate water reducer	2	2.5

3.2. Mixture Design Optimization

A $p = 4$ mixture design was performed with factors in Table 1 using the software Design Expert. The experimental runs were arranged, and the responses were 7 day compressive strength, final setting time, fluidity, and resin loading, as shown in Table S2. The data in Table 2 were fitted by the quadratic polynomial model of Equation (2), and the fitting coefficients were computed (Table S3). The model fitness assessment and analysis of variances results are presented in Table 2 and Table S4, respectively. The fluidities exceeded the 300 mm measuring range in runs 2 and 6 and were not measured; hence, they were removed from the fitting of Y_3. Runs 2, 3, 5, and 6 failed to record specific compressive strengths due to resin uplift and delamination, and these results were disregarded for fitting Y_1.

Table 2. Model fitness for responses.

Responses	R^2	$R^2{}_{adj}$	$R^2{}_{pred}$	S/N
Y_1	0.8588	0.8323	0.788	16.237
Y_2	0.9099	0.8289	0.809	10.521
Y_3	0.9652	0.9338	0.6735	15.745
Y_4	0.9478	0.938	0.9203	28.725

For each response in Table 2, both the coefficients of determination R^2 and the adjusted coefficients of determination $R^2{}_{adj}$ show that the model's predictions and the experimental data match well [37,38]. This is notably evident for the responses Y_3 and Y_4. The experimental data were well-predicted by the model since the coefficients of determination for the prediction, $R^2{}_{pred}$, were quite close to $R^2{}_{adj}$ [37,38]. The lowest signal-to-noise ratio S/N of the regressions was 10.521 (Y_2). All the $Pr > F$ values of the four fittings in the ANOVA (Table S4) were less than 0.0004, the lack-of-fit terms were all insignificant, and the fitting results were free of outliers, demonstrating that the model significantly fit the experimental data [45]. The model fitness assessment and analysis of variances results suggest that the quadratic polynomial model in Equation (2) is valid for regression analysis and prediction of the experimental data of simulated spent-radioactive-resin solidification by metakaolin-reinforced sulfoaluminate cement.

Based on the quadratic polynomial model, multi-objective optimization using the satisfaction function for solidification was conducted to enhance the overall performance of the solidified wasteforms [37]. The optimized mixture composition, model predictions, and validation experiment data are listed in Table 3. The experimental data of this optimized solidification formula matched the predicted values, and a w/b ratio of 0.445 provided smooth solidifying operation, high resin loading, and compressive strength of the wasteforms.

Table 3. Optimized mixture composition and results of prediction and validation experiments.

Mixture Composition (wt.%)				Predicted Responses				Experimental Data			
X_1	X_2	X_3	X_4	$Y_{1'}$ (MPa)	$Y_{2'}$ (min)	$Y_{3'}$ (mm)	$Y_{4'}$ (%)	Y_1 (MPa)	Y_2 (min)	Y_3 (mm)	Y_4 (%)
40	55.8	2.2	2	9.93	66.14	232.8	73.23	11.6	70	227	64

3.3. *Durability of Solidified Wasteforms*

After 28 days of curing, the optimized wasteforms were tested by immersion and freeze–thaw cycles, and the compressive strengths are depicted in Figure 2. The compressive strength did not reduce after being immersed in deionized water for 30, 90, or 180 days; rather, it increased by 6.3% to 15.6% compared to the control (Figure 2a). The sulfoaluminate cement continued to hydrate for 30 to 180 days in the water, ensuring the mechanical stability and resin beads sequestration in the wasteforms with a 64% resin loading. Whereas the pH of the immersion system impacted the durability of the wasteforms. Figure 2b shows that 90 days of immersion in neutral and alkaline systems increased compressive strength, while acidic systems with pHs of 3 and 5 significantly decreased compressive strength by 21.3% and 15.7%, respectively. This is because the sulfoaluminate hydration products disintegrated in acid media, weakening the strengths [21,46]. However, it is noteworthy that the compressive strength was 10.79 MPa, which is still higher than the 7 MPa required by the Chinese national standard GB14569.1-201 [42] for wasteforms to enter the final disposal sites, even after 90 days of immersion under unfavorable conditions at pH = 3. On the other hand, the durability was more negatively impacted by the repeated freeze–thaw cycles than by immersion. After 5 freeze–thaw cycles, the compressive strength was dramatically reduced with a loss of 16.7% before leveling off, while 20 cycles resulted in a 24.3% strength reduction. The optimized wasteforms performed better than the GB14569.1-2011 demands during freeze–thaw cycles, losing no more than 25% of their strength over five cycles. Additionally, their property satisfied practical applications because the permafrost layer thickness is typically taken into account when choosing disposal sites, preventing frequent freeze–thaw alternation of wasteforms [47–49].

The leaching behavior of typical nuclide Cs at different pHs was further investigated because of the significant effect of immersing pH on the strength of wasteforms. Nonradioactive isotopes were used instead of the radionuclide ^{137}Cs since they are chemically equivalent in solution [50]. Figure 3 shows the trends of leaching rate R and cumulative leached fraction P changes of Cs(I) in 42 days of leaching experiments, which exhibit similar trends at all pHs. R decreased with prolonged leaching time, while P increased; however, the extents of R and P altered at different stages. The initial 10 days corresponded to the rapid leaching phase of Cs(I). On the 10th day, P reached 42.5–46.8% of the total P for 42 days, while R reached a high order of magnitude of 10^{-2} cm/day. After that, the leaching of Cs(I) slowed down significantly, R stabilized at 10^{-3} and 10^{-4} cm/day orders of magnitude, and P accumulated the remaining ~50% during the following 32 days. R and P were not significantly affected by the varied pH values, notably pH = 3 and pH = 5, implying that the solidified package sequestered Cs(I) effectively in the acidic leaching solution. The highest P of 0.467 cm was observed under the alkaline condition of pH = 11, not considerably different from 0.416 cm at pH = 3, which was also related to the fact that Cs(I) is soluble as CsOH at alkalinity [50,51]. These results suggest that the solidified wasteforms can wrap the resin particles to maintain the whole with high resin loadings and sequester the soluble cations in a strongly acidic environment, effectively preventing the radionuclides from diffusing into and further migrating in the environment.

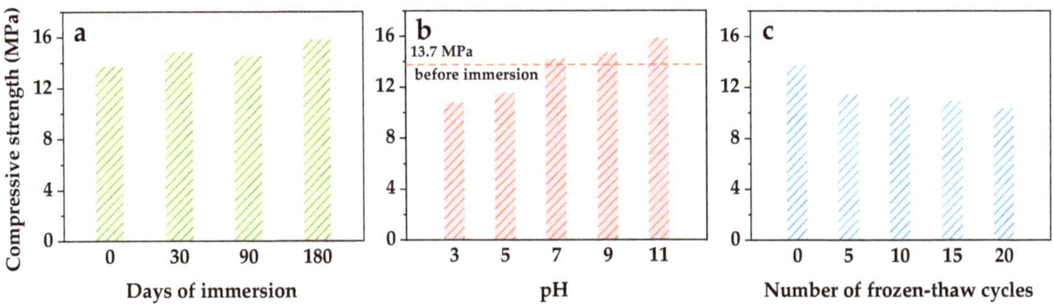

Figure 2. Compressive strength of optimized solidified wasteforms after (**a**) immersion of different days (pH = 7, 25 ± 5 °C), (**b**) 90 day immersion under different pH (25 ± 5 °C), and (**c**) freeze–thaw cycles, with the mixture composition of spent resin 40 wt.%, sulfoaluminate cement 55.8 wt.%, metakaolin 2.2 wt.%, and naphthalenesulfonate water reducer 2 wt.%.

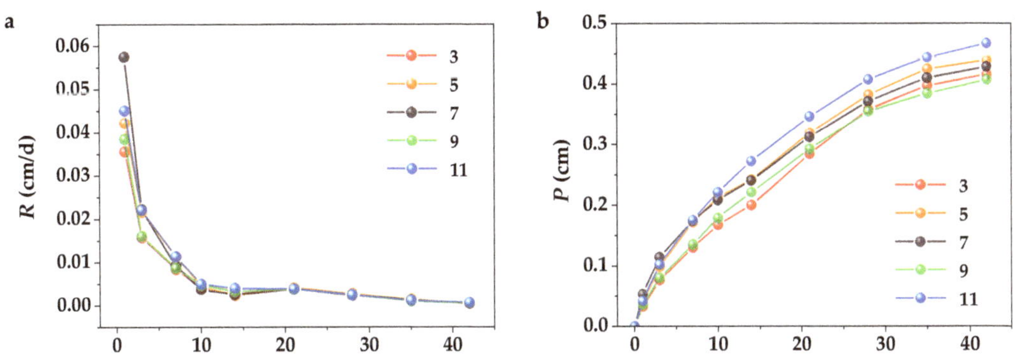

Figure 3. Leaching behavior of the simulated nuclide Cs^+: (**a**) Leaching rates and (**b**) cumulative leaching fractions; V_L = 1.6 L, V_L/F = 13.35, 25 ± 5 °C.

3.4. Hydration Products of Metakaolin-Reinforced Wasteforms

The encapsulation of resins in the metakaolin-reinforced wasteforms depends on its hydration products. The SEM images and XRD spectra in Figure 4 reveal the microscopic morphology and mineral content of the hydration products after 28 days of curing. As seen in Figure 4a, a stable crystalline skeletal structure scattered with amorphous products is formed by many acicular and columnar crystals interspersed throughout. The active ingredients in the sulfoaluminate clinker, such as anhydrous calcium thiosulfate, calcium aluminate, and dicalcium silicate, react when the cement comes into contact with water to create ettringite and C–S–H gels [19,20,52]. Following the abovementioned reactions, the generated $Ca(OH)_2$ interacts with calcium sulfate and aluminum gels to produce calcium alumina (Equation (5)) [51]. The basic structural unit of ettringite is $\{Ca_3[Al(OH)_6]\cdot 12H_2O\}^{3+}$, which belongs to the tripartite crystal system and presents the typical columnar structure in Figure 4a [18,21,53–56]. This microstructure also supports the effect of metakaolin dosage on the wasteform performance in Figure 1a. On the one hand, Cs(I) retention in the wasteforms derived from Cs(I) immobilization in the mineral lattice of fully grown ettringite crystalline. On the other hand, the integrity and durability of the optimized wasteforms through the response surface methodology contributed to the physical encapsulating of radioactive wastes.

$$3Ca(OH)_2 + 3CaSO_4 + Al_2O_3{\cdot}3H_2O + 26H_2O \rightarrow 3CaO{\cdot}Al_2O_3{\cdot}3CaSO_4{\cdot}32H_2O \quad (5)$$

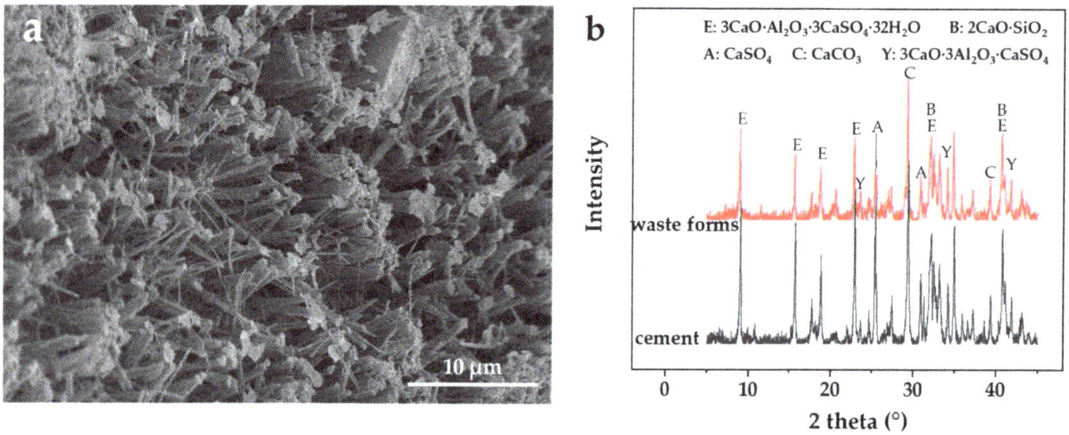

Figure 4. Characterization of metakaolin-reinforced wasteforms: (**a**) SEM image and (**b**) XRD spectra (red), with the mixture composition of spent resin 40 wt.%, sulfoaluminate cement 55.8 wt.%, metakaolin 2.2 wt.%, and naphthalenesulfonate water reducer 2 wt.%; the hardened cement XRD spectra (black) as comparison, with the mixture composition of water 40 wt.%, sulfoaluminate cement 55.8 wt.%, metakaolin 2.2 wt.%, and naphthalenesulfonate water reducer 2 wt.%.

Moreover, in the sulfoaluminate cementitious system reinforced by metakaolin, the amorphous Al_2O_3 in thermodynamically metastable metakaolin is very easily dissolved in an alkaline environment, which further offers the aluminum supply for hydration in Equation (5), in addition to the aluminum gels from sulfoaluminate clinker [27,29–32]. As a result, the reaction of Equation (5) and the creation of ettringite were both promoted, and the microstructure of the solidified wasteforms was strengthened. Furthermore, the amorphous products in Figure 4a are mainly C–S–H gels, aluminum gels, and some residual clinkers that were not entirely hydrated [23,25,26].

These hydration products exhibit characteristic peaks in the XRD spectra of wasteforms (red) and harden cement samples (black) without resins in Figure 4b, further qualitatively demonstrating that the mineral composition of hydration products was mainly ettringite, aluminum gels, and a small amount of C–S–H gels.

4. Conclusions

Metakaolin is a commonly used additive to improve the performance of cement in the construction industry and is a developing assistant in cementitious systems to solidify radioactive wastes. In this work, metakaolin was used to reinforce the sulfoaluminate-cement solidification of simulated spent radioactive resins, and the overall performance of the solidified wasteforms was optimized by the mixture design. With a high content of pozzolanic-reactive Al_2O_3, metakaolin facilitated the hydration of sulfoaluminate cement. The reinforced wasteforms, with a resin loading of 64%, met the disposal standards of GB14569.1-2011 for a compressive strength of 13.71 Mpa and the immersion and freeze–

thaw resistances. The wasteforms exhibited a good retention ability of Cs^+. The microscopic morphology was acicular and columnar crystals interspersed with amorphous gels and particles. Ettringite, aluminum gels, and C–S–H gels comprised the bulk of the mineral phases of the hydration products. The simulated spent radioactive resin was successfully solidified by the metakaolin-reinforced sulfoaluminate cementitious system, providing new insights into the safe treatment and disposal of radioactive wastes.

Supplementary Materials: The following supporting information can be downloaded at: https://www.mdpi.com/article/10.3390/coatings12101466/s1. Figure S1: Main mineral components of the sulfoaluminate cement used in the solidification experiment; Table S1: Main composition and property of the ion exchange resin used in the solidification experiment; Table S2: Mixture design and experimental data; Table S3: Regression model coefficients; Table S4: ANOVA results of the regression model.

Author Contributions: Writing—original draft preparation, J.X.; visualization, M.W.; methodology and validation, C.L.; investigation, M.H.; software and data curation, Q.W.; supervision and writing—reviewing and editing, Q.S. All authors have read and agreed to the published version of the manuscript.

Funding: This work was financially supported by the National Natural Science Foundation of China (No. 21301151), the Natural Science Foundation of Hebei Province (No. B2021203036 and No. B2018203331), and the University Science and technology Research Projects of Hebei Province (No. ZD2021103).

Institutional Review Board Statement: Not applicable.

Informed Consent Statement: Not applicable.

Data Availability Statement: Data sharing is not applicable to this article.

Conflicts of Interest: The authors declare no conflict of interest.

References

1. Burns, P.C.; Ewing, R.C.; Navrotsky, A. Nuclear fuel in a reactor accident. *Science* **2012**, *335*, 1184–1188. [CrossRef] [PubMed]
2. Wang, J.L.; Liu, X.J. Forward osmosis technology for water treatment: Recent advances and future perspectives. *J. Clean. Prod.* **2021**, *280*, 124354. [CrossRef]
3. Zhang, X.Y.; Gu, P.; Liu, Y. Decontamination of radioactive wastewater: State of the art and challenges forward. *Chemosphere* **2019**, *215*, 543–553. [CrossRef]
4. Wang, Z.; Guo, S.; Wu, Z.; Fan, H.; Guan, G.; Hao, X. A smart potential-responsive ion exchange nanomaterial with superparamagnetism for cesium ion separation and recovery. *Sep. Purif. Technol.* **2017**, *187*, 199–206. [CrossRef]
5. Wang, J.; Wan, Z. Treatment and disposal of spent radioactive ion-exchange resins produced in the nuclear industry. *Prog. Nucl. Energy* **2015**, *78*, 47–55. [CrossRef]
6. Rao, S.V.; Mani, A.G.; Karua, S.; Cheralathan, M.; Reddy, A.; Khandelwal, S.K.; Paul, B. Treatment of liquid wastes using composite resins. *J. Radioanal. Nucl. Chem.* **2016**, *307*, 463–469. [CrossRef]
7. Li, J.; Chen, L.; Wang, J. Solidification of radioactive wastes by cement-based materials. *Prog. Nucl. Energy* **2021**, *141*, 103957. [CrossRef]
8. Eskander, S.B.; Bayoumi, T.A.; Saleh, H.M. Leaching behavior of cement-natural clay composite incorporating real spent radioactive liquid scintillator. *Prog. Nucl. Energy* **2013**, *67*, 1–6. [CrossRef]
9. García-Gutiérrez, M.; Missana, T.; Mingarro, M.; Morejón, J.; Cormenzana, J.L. Cesium diffusion in mortars from different cements used in radioactive waste repositories. *Appl. Geochem.* **2018**, *98*, 10–16. [CrossRef]
10. Rakhimova, N.R.; Rakhimov, R.Z.; Morozov, V.P.; Potapova, L.I.; Osin, Y.N. Mechanism of solidification of simulated borate liquid wastes with sodium silicate activated slag cements. *J. Clean. Prod.* **2017**, *149*, 60–69. [CrossRef]
11. Deneanu, N.; Dulama, M.; Baboescu, E.; Teoreanu, I. The conditioning in Portland cement of oil radioactive wastes. *Rev. Chim.* **2004**, *55*, 966–970.
12. Lanier, S.; Davy, C.A.; Albert-Mercier, C.; Farcy, O.; Cau-Dit-Coumes, C.; Lambertin, D. Novel Portland cement matrices incorporating a gamma -MnO_2/Ag_2O hydrogen/tritium getter -structure changes and trapping performance. *J. Nucl. Mater.* **2022**, *567*, 153819. [CrossRef]
13. Phillip, E.; Khoo, K.S.; Yusof, M.A.; Rahman, R.O. Mechanistic insights into the dynamics of radionuclides retention in evolved POFA-OPC and OPC barriers in radioactive waste disposal. *Chem. Eng. J.* **2022**, *437*, 135423. [CrossRef]
14. Shaaban, I.; Assi, N. Measurement of the leaching rate of radionuclide Cs-134 from the solidified radioactive sources in Portland cement mixed with microsilica and barite matrixes. *J. Nucl. Mater.* **2011**, *415*, 132–137. [CrossRef]

15. Nicu, M.; Ionascu, L.; Dragolici, F.; Neacsu, E. The influence of chemical composition of the secondary radioactive waste on cement matrix conditioning. In Proceedings of the Energy and Clean Technologies Conference Proceedings, SGEM 2016, Vienna, Austria, 2–5 November 2016; Volume I, pp. 49–55, ISBN 978-619-7105-63-6.
16. Kononenko, O.A.; Milyutin, V.V.; Nekrasova, N.A. Composite binders for solidification of spent ion-exchange resins. *At. Energy* **2019**, *125*, 257–261. [CrossRef]
17. Kononenko, O.A.; Milyutin, V.V.; Makarenkov, V.I.; Kozlitin, E.A. Immobilization of NPP evaporator bottom high salt-bearing liquid radioactive waste into struvite-based phosphate matrices. *J. Hazard. Mater.* **2021**, *416*, 125902. [CrossRef]
18. Shi, C.J.; Jimenez, A.F.; Palomo, A. New cements for the 21st century: The pursuit of an alternative to Portland cement. *Cem. Concr. Res.* **2011**, *41*, 750–763. [CrossRef]
19. Luz, C.A.; Rocha, J.C.; Cheriaf, M.; Pera, J. Valorization of galvanic sludge in sulfoaluminate cement. *Constr. Build. Mater.* **2009**, *23*, 595–601. [CrossRef]
20. Zhu, J.P.; Chen, Y.; Zhang, L.; Guo, B.K.; Fan, G.X.; Guan, X.M.; Zhao, R.Q. Revealing the doping mechanism of barium in sulfoaluminate cement clinker phases. *J. Clean. Prod.* **2021**, *295*, 126405. [CrossRef]
21. Shi, C.; Spence, R. Designing of cement-based formula for solidification/stabilization of hazardous, radioactive, and mixed wastes. *Crit. Rev. Environ. Sci. Technol.* **2004**, *34*, 391–417. [CrossRef]
22. Zhang, W.; Wang, J. Leaching performance of uranium from the cement solidified matrices containing spent radioactive organic solvent. *Ann. Nucl. Energy* **2017**, *101*, 31–35. [CrossRef]
23. Sun, Q.N.; Hu, J.; Wang, J.L. Optimization of composite admixtures used in cementation formula for radioactive evaporator concentrates. *Prog. Nucl. Energy* **2014**, *70*, 1–5. [CrossRef]
24. Li, J.F.; Wang, J.L. Solidification of 30% TBP-OK Waste by calcium sulfoaluminate cement. *Adv. Mater. Res.* **2013**, *726–731*, 2782–2785. [CrossRef]
25. Li, J.F.; Wang, J.L. Advances in cement solidification technology for waste radioactive ion exchancre resins: A review. *J. Hazard. Mater.* **2006**, *135*, 443–448. [CrossRef]
26. Sun, Q.N.; Li, J.F.; Wang, J.L. Solidification of borate radioactive resins using sulfoaluminate cement blending with zeolite. *Nucl. Eng. Des.* **2011**, *241*, 5308–5315. [CrossRef]
27. He, C.; Osbaeck, B.; Makovicky, E. Pozzolanic reactions of six principal clay minerals: Activation, reactivity assessments and technological effects. *Cem. Concr. Res.* **1995**, *25*, 1691–1702. [CrossRef]
28. Murat, M.; Comel, C. Hydration reaction and hardening of calcined clays and related minerals III. Influence of calcination process of kaolinite on mechanical strengths of hardened metakaolinite. *Cem. Concr. Res.* **1983**, *13*, 631–637. [CrossRef]
29. Fernandez, R.; Martirena, F.; Scrivener, K.L. The origin of the pozzolanic activity of calcined clay minerals: A comparison between kaolinite, illite and montmorillonite. *Cem. Concr. Res.* **2011**, *41*, 113–122. [CrossRef]
30. Asteris, P.G.; Lourenco, P.B.; Roussis, P.C.; Adami, C.E.; Armaghani, D.J.; Cavaleri, L.; Chalioris, C.E.; Hajihassani, M.; Lemonis, M.E.; Mohammed, A.S.; et al. Revealing the nature of metakaolin-based concrete materials using artificial intelligence techniques. *Constr. Build. Mater.* **2022**, *322*, 126500. [CrossRef]
31. Antoni, M.; Rossen, J.; Martirena, F.; Scrivener, K. Cement substitution by a combination of metakaolin and limestone. *Cem. Concr. Res.* **2012**, *42*, 1579–1589. [CrossRef]
32. Bilal, H.; Chen, T.F.; Ren, M.; Gao, X.J.; Su, A.S. Influence of silica fume, metakaolin & SBR latex on strength and durability performance of pervious concrete. *Constr. Build. Mater.* **2021**, *275*, 122124. [CrossRef]
33. Nergis, D.D.; Vizureanu, P.; Sandu, A.V.; Nergis, D.P.; Bejinariu, C. XRD and TG-DTA study of new phosphate-based geopolymers with coal ash or metakaolin as aluminosilicate source and mine tailings addition. *Materials* **2022**, *15*, 202. [CrossRef] [PubMed]
34. Sabir, B.; Wild, S.; Bai, J. Metakaolin and calcined clays as pozzolans for concrete: A review. *Cem. Concr. Compos.* **2001**, *23*, 441–454. [CrossRef]
35. Zhang, P.; Wang, K.X.; Wang, J.; Guo, J.J.; Ling, Y.F. Macroscopic and microscopic analyses on mechanical performance of metakaolin/fly ash based geopolymer mortar. *J. Clean. Prod.* **2021**, *294*, 126193. [CrossRef]
36. Abdel Rahman, R.O.; Zin El Abidin, D.; Abou-Shady, H. Cesium binding and leaching from single and binary contaminant cement–bentonite matrices. *Chem. Eng. J.* **2014**, *245*, 276–287. [CrossRef]
37. Li, Z.P.; Lu, D.G.; Gao, X.J. Optimization of mixture proportions by statistical experimental design using response surface method—A review. *J. Build. Eng.* **2021**, *36*, 102101. [CrossRef]
38. Menchaca-Mendez, A.; Zapotecas-Martinez, S.; Garcia-Velazquez, L.M.; Coello, C.A. Uniform mixture design via evolutionary multi-objective optimization. *Swarm Evol. Comput.* **2022**, *68*, 100979. [CrossRef]
39. Wang, H.Y.; Zhang, C.Q. The mixture design threshold accepting algorithm for generating D-optimal designs of the mixture models. *Metrika* **2022**, *85*, 345–371. [CrossRef]
40. Scheffé, H. Experiments with mixtures. *J. R. Stat. Soc. Ser. B Methodol.* **1958**, *20*, 344–360. [CrossRef]
41. DeRousseau, M.A.; Kasprzyk, J.; Srubar, W.V. Computational design optimization of concrete mixtures: A review. *Cem. Concr. Res.* **2018**, *109*, 42–53. [CrossRef]
42. Zhang, W.; Li, J.; Wang, J. Solidification of spent radioactive organic solvent by sulfoaluminate and Portland cements. *J. Nucl. Sci. Technol.* **2015**, *52*, 1362–1368. [CrossRef]
43. Matsuzawa, K.; Atarashi, D.; Miyauchi, M.; Sakai, E. Interactions between fluoride ions and cement paste containing superplasticizer. *Cem. Concr. Res.* **2017**, *91*, 33–38. [CrossRef]

44. Liu, Y.; Li, H.; Wang, K.; Wu, H.; Cui, B. Effects of accelerator–water reducer admixture on performance of cemented paste backfill. *Constr. Build. Mater.* **2020**, *242*, 118187. [CrossRef]
45. Khaskhoussi, A.; Calabrese, L.; Bouhamed, H.; Kamoun, A.; Proverbio, E.; Bouaziz, J. Mixture design approach to optimize the performance of TiO_2 modified zirconia/alumina sintered ceramics. *Mater. Des.* **2018**, *137*, 1–8. [CrossRef]
46. Rahman, R.A.; Ojovan, M.I. Toward sustainable cementitious radioactive waste forms: Immobilization of problematic operational wastes. *Sustainability* **2021**, *13*, 11992. [CrossRef]
47. Helton, J.C.; Hansen, C.W.; Swift, P.N. Performance assessment for the proposed high-level radioactive waste repository at Yucca Mountain, Nevada. *Reliab. Eng. Syst. Saf.* **2014**, *122*, 1–6. [CrossRef]
48. Wang, J. Progress of geological disposal of high-level radioactive waste in China in the 21st century. *At. Energy Sci. Technol.* **2019**, *53*, 2072–2082.
49. Yu, X.; Ni, S.; Wang, Y.; Cai, G.; Xu, D. Experimental research on top cover of shallow-buried radioactive waste disposal repository. *Chin. J. Rock Mech. Eng.* **2009**, *28*, 1169–1176.
50. Li, J.W.; Xu, D.; Wang, W.L.; Wang, X.J.; Mao, Y.P.; Zhang, C.; Jiang, W.; Wu, C.L. Review on selection and experiment method of commonly studied simulated radionuclides in researches of nuclear waste solidification. *Sci. Technol. Nucl. Install.* **2020**, *2020*, 3287320. [CrossRef]
51. Alby, D.; Charnay, C.; Heran, M.; Prelot, B.; Zajac, J. Recent developments in nanostructured inorganic materials for sorption of cesium and strontium: Synthesis and shaping, sorption capacity, mechanisms, and selectivity—A review. *J. Hazard. Mater.* **2018**, *344*, 511–530. [CrossRef]
52. Xu, X.; Bi, H.; Yu, Y.; Fu, X.; Wang, S.; Liu, Y.; Hou, P.; Cheng, X. Low leaching characteristics and encapsulation mechanism of Cs+ and Sr2+ from SAC matrix with radioactive IER. *J. Nucl. Mater.* **2021**, *544*, 152701. [CrossRef]
53. Qian, G.R.; Shi, J.; Cao, Y.L.; Xu, Y.F.; Chui, P.C. Properties of MSW fly ash-calcium sulfoaluminate cement matrix and stabilization/solidification on heavy metals. *J. Hazard. Mater.* **2008**, *152*, 196–203. [CrossRef]
54. Shi, H.; Wu, Q.Y.; Yu, Z.Q.; Ma, J.; Shen, X.D. Properties of eco-friendly coral sand powder—Calcium sulfoaluminate cement binary system. *Construct. Build. Mater.* **2020**, *263*, 120181. [CrossRef]
55. Shi, J.Y.; Liu, B.J.; Tan, J.X.; Dai, J.D.; Chen, J.Z.; Ji, R.J. Experimental studies and microstructure analysis for rapid-hardening cement emulsified asphalt mortar. *J. Constr. Eng. Manag.* **2020**, *146*, 04020130. [CrossRef]
56. Wang, W.L.; Luo, Z.Y.; Shi, Z.L.; Cen, K.F. Experimental study on cement clinker co-generation in pulverized coal combustion boilers of power plants. *Waste Manag. Res.* **2006**, *24*, 207–214. [CrossRef] [PubMed]

coatings

Article

Domesticating a Halotolerant Bacterium of *Vibrio* sp. LY1024 with Heterotrophic Nitrification–Aerobic Denitrification Property for Efficient Nitrogen Removal in Mariculture Wastewater Treatment

Lu Wang, Yutong Fu, Shuaijie Wang *, Fei Ye, Enming Cui and Qina Sun

Hebei Key Laboratory of Heavy Metal Deep-Remediation in Water and Resource Reuse, School of Environmental and Chemical Engineering, Yanshan University, Qinhuangdao 066004, China
* Correspondence: ysushuaijie@163.com; Tel.: +86-0335-8387741; Fax: +86-0335-8061569

Abstract: Dealing with mariculture wastewater that contains high nitrogenous compounds with efficient biological nitrogen removal technology is challenging but meaningful. The key lies in developing an active microorganism that can spontaneously complete the nitrification–denitrification processes in the marine environment. Herein, a halotolerant heterotrophic nitrification–aerobic denitrification (HN-AD) bacterium of *Vibrio* sp. LY1024 with good nitrogen removal capacity is domesticated to achieve the aforementioned goal. As a result, ammonium (NH_4^+-N) and nitrate (NO_3^--N) removal rates of almost 100% and 98.5% are detected over *Vibrio* sp. LY1024 at the salinity of 3.5%, even further increasing the salinity of wastewater to 5.5%. Its removal capacity towards both NH_4^+-N and NO_3^--N can still maintain at almost 100% and 94.7%, respectively. Further combining these results with those of intermediate product determination, it can be speculated that the ammonium removal is according to the pathway of NH_4^+-N $\rightarrow$ NH_2OH $\rightarrow$ NO_3^--N $\rightarrow$ N_2O $\rightarrow$ N_2. Moreover, the influence of wastewater temperature on the nitrogen removal efficiency of *Vibrio* sp. LY1024 is also considered. The NH_4^+-N and NO_3^--N removal efficiency over *Vibrio* sp. LY1024 at a relatively low temperature of 15 °C is still up to 97.3% and 76.4%, respectively. Our work provides a promising halotolerant and low-temperature resistance microorganism for the treatment of mariculture wastewater.

Keywords: heterotrophic nitrification; aerobic denitrification; low-temperature resistance; salt tolerance; *Vibrio* sp. LY1024; mariculture wastewater treatment

Citation: Wang, L.; Fu, Y.; Wang, S.; Ye, F.; Cui, E.; Sun, Q. Domesticating a Halotolerant Bacterium of *Vibrio* sp. LY1024 with Heterotrophic Nitrification–Aerobic Denitrification Property for Efficient Nitrogen Removal in Mariculture Wastewater Treatment. *Coatings* **2022**, *12*, 1786. https://doi.org/10.3390/coatings12111786

Academic Editor: Michele Ferrari

Received: 16 October 2022
Accepted: 18 November 2022
Published: 21 November 2022

1. Introduction

The densely concentrated mariculture model leads to much wastewater that contains highly toxic nitrogenous compounds, such as ammonium (NH_4^+-N) and nitrite (NO_2^--N), etc., being discharged into the marine environment, which will inevitably pose a great threat to the marine ecological system [1–4]. Therefore, it is urgent to develop an effective nitrogen removal technique for tackling mariculture wastewater.

Among various nitrogen removal techniques, the biological method has been regarded as a technique that is feasible and economically reasonable [5–8]. The mechanism of the traditional biological nitrogen removal process relies mainly on the two-stage theory, that is, nitrification and denitrification processes [9]. During this process, NH_4^+-N is first oxidized to NO_2^--N and NO_3^--N under the action of nitrite bacteria and nitrate bacteria under aerobic conditions, and then NO_2^--N or NO_3^--N is in turn transformed into nitrogenous gas by denitrification bacteria in anoxic conditions. Although this has been supposed to be a relatively mature technology, the different requirements between nitrification and denitrification bacteria in the working atmosphere still perplex the nitrogen removal processes.

To conquer this deficiency, a new model of the heterotrophic nitrification–aerobic denitrification (HN-AD) process is proposed. In this process, HN-AD bacteria can simultaneously carry out nitrification and denitrification under aerobic conditions in the same reactor, which greatly simplifies the process flow, and concurrently compensates for the alkalinity consumed during the nitrification process by the denitrification process [10]. Moreover, HN-AD bacteria also preserve the advantages of high growth rate and nitrogen removal capacity. However, within the limits of our knowledge, seldom have reports focused on the application of HN-AD bacteria in the treatment of mariculture wastewater, which might be ascribed to the fact that the growth and reproduction of common HN-AD bacteria would be inhibited in an environment of high salinity and the relatively low temperature [11]. Therefore, the present study aims at domesticating a halotolerant and low-temperature tolerant HN-AD bacteria with high nitrogen removal capacity to treat mariculture wastewater that contains high nitrogenous compounds.

Taking these into consideration, a halotolerant and low-temperature resistant HN-AD bacterium (named *Vibrio* sp. LY1024) is domesticated and identified by 16S rRNA gene sequence analysis. The effects of carbon-to-nitrogen (C/N) ratio, initial pH, temperature, dissolved oxygen, and salinity on the nitrogen removal performance of *Vibrio* sp. LY1024 and its nitrogen removal mechanism under high-salinity and low-temperature conditions are subsequently investigated. In addition, the treatment performance of *Vibrio* sp. LY1024 on stimulated mariculture wastewater is also evaluated. The results obtained from our work will provide a scientific basis for mariculture wastewater treatment.

2. Materials and Methods

2.1. Mediums

All the chemicals were of analytical grade and used as obtained without further purification. The seawater sample (salinity of 3.5%) utilized in this study was sampled from the Jinmeng Bay beach in the city of Qinhuangdao (39°55′ N, 119°38′ E). All the mediums were autoclaved at 121 °C for 20 min before utilization.

Trace element solution included the following: 50 g of $C_{10}H_{14}N_2O_8Na_2{\cdot}2H_2O$, 3.92 g of $MgSO_4{\cdot}7H_2O$, 7.28 g of $CaCl_2{\cdot}H_2O$, 5.06 g of $MnCl_2{\cdot}4H_2O$, 5 g of $FeSO_4{\cdot}7H_2O$, 1.57 g of $CuSO_4{\cdot}5H_2O$, 1.61 g of $CoCl_2{\cdot}6H_2O$, and 1 L of distilled water.

The components of activation medium (pH 7.2) employed for activating HN-AD bacteria included the following: 40 g of glucose, 10 g of NaCl, 2 mL of trace element solution, and 1 L of distilled water.

The constituents of domestication medium (pH 7.2) used for halotolerant HN-AD bacteria domestication included the following: 0.382 g of NH_4Cl, 0.5 g of $NaNO_2$, 7.167 g of $C_6H_5NaO_7{\cdot}2H_2O$, 0.2 g of KH_2PO_4, 2 mL trace element solution, (1) 1 L seawater/distilled water solution (v/v, 1/1, labeled as domesticated Medium 1), and (2) 1 L of seawater (labeled as domesticated Medium 2).

The Luria–Bertani (LB) medium (pH 7.2) employed for halotolerant HN-AD bacteria isolation and purification comprised the following ingredients: 5 g of peptone, 1 g of yeast powder, 16 g of agar, 2 mL of trace element solution, and 1 L of seawater.

The heterotrophic nitrification and aerobic denitrification (HN-AD) medium used to determine the nitrogen removal performance of isolated halotolerant HN-AD bacteria included the following: appropriate amount of NH_4Cl, KNO_3, $NaNO_2$, and $C_6H_5NaO_7{\cdot}2H_2O$; 0.2 g of KH_2PO_4; 2 mL of trace element solution; and 1 L of seawater.

2.2. Domestication, Isolation, and Identification of Halotolerant HN-AD Bacteria

The HN-AD bacterial seed was obtained from Jiangnan Biotechnology Co., Ltd., Yichun, China. Before domestication, the bacterial seed was first added to the activation medium and aerobic cultured at 20 °C for 18 h. After two rounds of 48 h of domestication, the resultant bacterial suspension was spread on the solid LB plates. After 36 h of incubation at 30 °C, separate colonies were picked and purified by repeated streaking on fresh LB plates.

2.3. Identification and Phylogenetic Analysis of Isolates

The morphologies of the isolated halotolerant HN-AD strain were observed by scanning electron microscopy (SEM, EM-30AXN, OPTON, Beijing, China). Bacterial 16S rRNA genes were amplified with universal primers 27F (5′-AGAGTTTGATCMTGGCTCAG-3′) and lower primer 1492R (5′-TACGGYTACCTTGTTACGACT-3′) [12]. The polymerase chain reaction (PCR) protocol of Tamura [13] was followed. The amplified products were sequenced, and the sequence was compared with available 16S rRNA gene sequences in GenBank by BLAST. A phylogenetic tree was constructed using MEGA 5.0 by the neighbor-joining method.

2.4. Investigating the Influence of Different Operation Parameters on Nitrogen Removal over Strain LY1024

To investigate the nitrogen removal performance and related mechanism of the isolated strain, 5 mL of the suspension that contain strain LY1024 (10.7 log CFU/mL) was inoculated into 250 mL of triplicate flasks and contacted with 100 mL of HN-AD medium (170 mg/L of NH_4^+-N or 120 mg/L of NO_3^--N or 130 mg/L of NO_2^--N, COD ≈ 1900 mg/L). The flasks were shaken continuously at 150 rpm to maintain aerobic conditions, and the reaction temperature was kept at 30 °C. A volume of 2.5 mL of samples was taken from the flasks at each 3 h interval.

The concentrations of NH_4^+-N, NO_2^--N, and NO_3^--N chemical oxygen demand (COD) were determined according to the National Standards of the People's Republic of China (HJ 535-2009, HJ/T 346-2007, HJ/T 399-2007, GB/T 7493-87), and the growth of the strain was assayed by measuring the wavelength of 600 nm (OD_{600}) using a spectrophotometer (WFZ-24A, Spectrum Shanghai, Shanghai, China).

The influence of different culture conditions, including C/N ratios of 5, 10, 15, 20, and 25, initial pH values of 5, 6, 7, 8, and 9, different temperatures of 15, 20, 25, and 30 °C, dissolved oxygen concentrations (DO amount changed by setting shaking speed at 100, 130, 150, and 180 rpm), and salinity (0%, 3.5%, 5.5%, 8.5%, and 10%) on the nitrogen (NH_4^+-N ≈ 40 mg/L, NO_3^--N ≈ 25 mg/L) removal performances of strain LY1024 were further investigated.

2.5. Treatment of Simulated Mariculture Wastewater by Strain LY1024

The simulated mariculture wastewater was prepared by dissolving 5 g of grouper bait into 4 L of seawater, and then the three flasks that contained 1 L of simulated mariculture wastewater (DO ≈ 4.67 mg/L) were kept at 25 °C. The effect of strain LY1024 on the nitrogen and COD removal of the simulated mariculture wastewater was determined. The physical and chemical properties of the simulated mariculture wastewater were as follows: pH = 7.8, salinity = 35 g/L, NH_4^+-N = 31 mg/L, NO_3^--N = 4.5 mg/L, NO_2^--N = 0.2 mg/L, total nitrogen (TN) = 38.2 mg/L, and chemical oxygen demand (COD) = 526 mg/L.

3. Results and Discussion

3.1. Identification of Isolates

After domestication and isolation, a strain with high nitrogen removal capacity and high salinity tolerance ability was obtained and named LY1024. As shown in Figure 1a, the colonies of strain LY1024 on the LB plate display a round shape with a regular edge and a glossy surface, and the color of these colonies is faint yellow. The micromorphology of strain LY1024 was further characterized by SEM analysis. From Figure 1b, many arc-shaped bacteria with an average size of 0.8–1.6 μm in length and 0.2–0.3 μm in width can be observed, which are typical characteristics of *Vibrio* sp. Bacteria. The subsequent result of sequence comparative analysis reveals that strain LY1024 belongs to the genus *Vibrio*, because it exhibits the closest phylogenetic relative to *Vibrio* sp. A5-15, with a similarity of up to 99.66%, as illustrated in Table 1. In addition, the neighbor-joining phylogenetic tree of strain LY1024 also indicates that strain LY1024 and *Vibrio* sp. belong to the same group (Figure 2). Based on these results, the isolated strain is identified as *Vibrio* sp. LY1024.

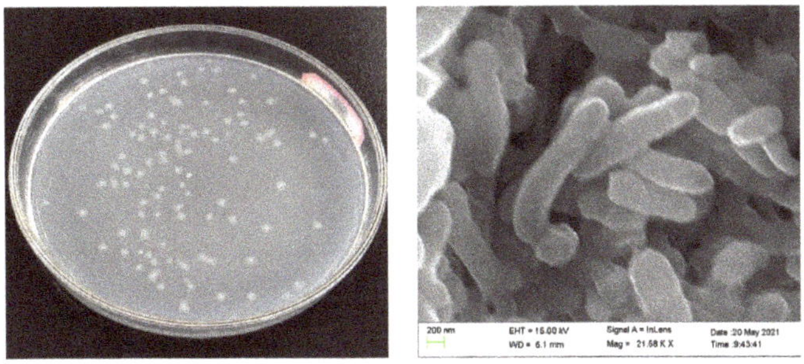

(**a**) (**b**)

Figure 1. (**a**) Colony morphology; (**b**) SEM image of strain LY1024.

Table 1. Sequence comparative analysis between strain LY1024 and other bacteria.

Description (16S Ribosomal RNA Gene Partial Sequence)	Per. Ident (%)
Vibrio sp. A5-15	99.66
Vibrio sp. B2-5-1	99.59
Vibrio sp. A5-5	99.52
Vibrio sp. B2-23	99.52
Vibrio sp. C4-5	99.52

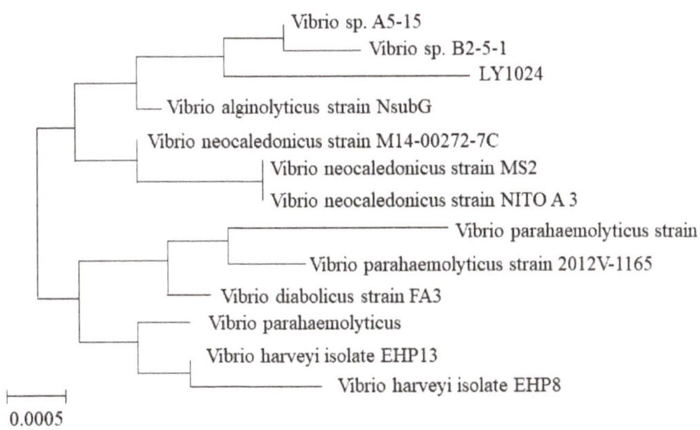

Figure 2. Phylogenetic tree of strain LY1024.

3.2. Heterotrophic Nitrification and Aerobic Denitrification Ability of Strain LY1024 at High Salinity

Ammonium (NH_4^+-N) was first selected as the sole nitrogen source to explore the heterotrophic nitrification and aerobic denitrification ability of strain LY1024 under high-salinity conditions, and the corresponding results are shown in Figure 3. It can be seen from the growth curve of strain LY1024 (Figure 3a) that, during the first 9 h of cultivation, the OD_{600} value of strain LY1024 displays a relatively low value of 0.28, indicating its growth rate was rather slow; correspondingly, relatively low removal rates of 11.5% and 21.5% toward NH_4^+-N and COD are observed on strain LY1024. Further prolonging the cultivation time to 27 h, the OD_{600} value rapidly increases from 0.28 to 1.6, demonstrating that the growth of strain LY1024 enters the logarithmic growth phase. As a result, both the concentrations of NH_4^+-N and COD are also degraded rapidly. Finally, approximately 98.8% of NH_4^+-N and 90.2% of COD are removed after 36 h of cultivation. The average ammonium removal rate was 4.74 mg/(L·h), which is higher than that of *Alcaligenes faecalis* C 16 (1.06 mg/(L·h)) [14] and *Rhodococcus* sp. CPZ 24 (3.4 mg/(L·h)) [15] and similar to that of *Alcaligenes faecalis* strain NR (4.35 mg/(L·h)) [16]. Moreover, by detecting the variation of nitrate (NO_3^--N) concentration during the NH_4^+-N removal process (Figure 3b), it can be found that the concentration of NO_3^--N is improved continuously, with a maximum accumulation concentration of 16.3 mg/L when the reaction time reaches 21 h, indicating that strain LY1024 has the ability of heterotrophic nitrification. Through this process, NH_4^+-N is transformed into NO_3^--N. Further prolonging the reaction time to 36 h, the concentration of NO_3^--N displays a gradual degradation trend, implying that strain LY1024 also presents the ability of aerobic denitrification. Almost no accumulation of NO_2^--N was observed during the whole culture process.

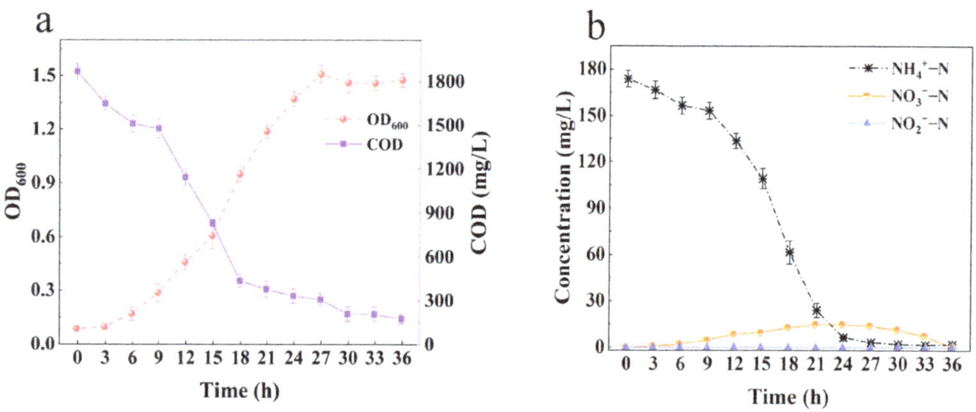

Figure 3. (**a**) The growth curve and COD removal capacity of strain LY1024; (**b**) heterotrophic nitrification performance of strain LY1024.

To verify the aforementioned inference, similar contrast experiments were carried out with NO_3^--N or NO_2^--N as the sole nitrogen source. As shown in Figure 4a, the growth trend of strain LY1024 with NO_3^--N as the nitrogen source is similar to the heterotrophic nitrification process, and 0–9 h is the adaptive stage of strain growth, during which the growth rate of LY1024 is slow. After 9 h, the strain enters the logarithmic growth stage, and the bacteria begin to grow and reproduce rapidly. In addition, the concentration of COD decreases rapidly with the growth of strain LY1024, and the removal rate reaches 90.1% after 36 h. In Figure 4b, NO_3^--N rapidly decreases from 121.58 mg/L to 0, and its removal rate is 100% after 24 h. There is almost no accumulation of NO_2^--N in the whole aerobic denitrification process. Therefore, it is preliminarily inferred that the intermediate binding of heterotrophic nitrification and aerobic denitrification of strain LY1024 is NO_3^--N.

The growth and COD removal capacity of strain LY1024 with NO_2^--N as the only nitrogen source are shown in Figure 4c. It follows that, after 6 h, strain LY1024 enters the logarithmic growth phase, and after 18 h, it enters the flat growth phase, and the OD_{600} value of the strain is stable at about 1.51, and the COD removal rate is 91.1% after 36 h. During the logarithmic growth phase, NO_2^--N is rapidly removed (Figure 4d). After 15 h, NO_2^--N is rapidly reduced from 132.5 mg/L to 0.06 mg/L, and the removal rate is nearly 100%. In the process of reaction, part of NO_3^--N is accumulated, the maximum amount of which is 8.58 mg/L, and then it is rapidly degraded.

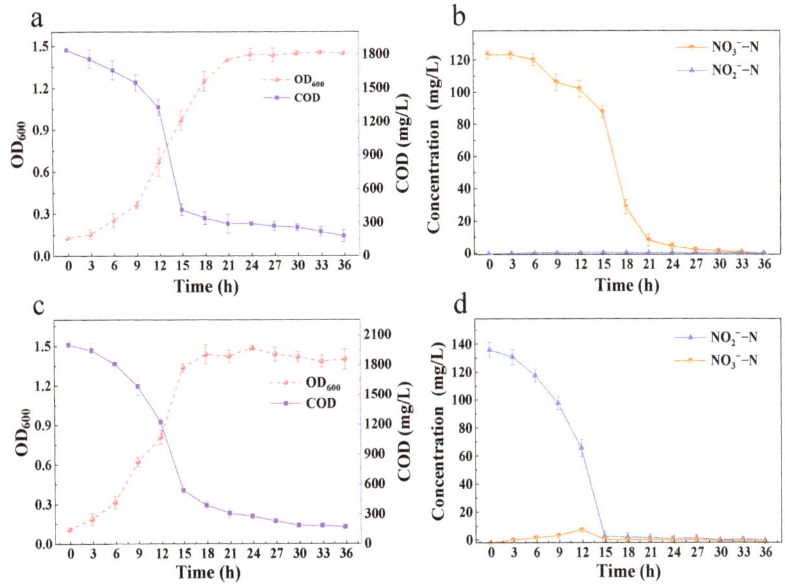

Figure 4. Aerobic denitrification characteristics of strain LY1024 with NO_3^--N (**a**,**b**) and NO_2^--N (**c**,**d**) as the solo nitrogen source.

3.3. *Influence of C/N Ratio, Initial pH Value, Temperature, DO, and Salinity on Nitrogen Removal over Strain LY1024*

Owing to the fact that carbon and nitrogen sources are the necessary energy sources for microbial growth and reproduction, to investigate the influence of the C/N ratio on the nitrogen removal capacity of HN-AD bacteria, a series of contrast experiments were conducted. As shown in Figure 5, it can be seen that when the C/N ratio is higher than 15:1 (including 15:1, 20:1, and 25:1), almost 100% of NH_4^+-N and more than 95% of NO_3^--N could be removed after 24 h of cultivation. However, when the C/N ratio reduces from 15:1 to 10:1 to 5:1, 92.7% and 51.5% of NH_4^+-N and 58.3% and 49.2% of NO_3^--N can be removed. It follows that the reduction of the C/N ratio has a greater effect on the removal rate of nitrate nitrogen than ammonia nitrogen. This can be ascribed to the lack of a carbon source for microbial growth and an electron donor for denitrification [9,17–19]. Moreover, it should be noted that with the C/N ratio increasing from 15 to 25, there is no significant difference in the NH_4^+-N removal rate. Therefore, to ensure sufficient NH_4^+-N removal, the C/N of 15 is selected in the following experiments.

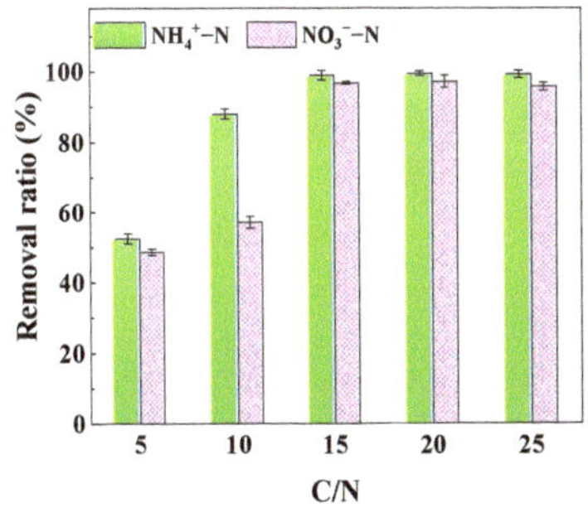

Figure 5. Influence of C/N on nitrogen compounds removal on *Vibrio* sp. LY1024.

To further explore the influence of the initial pH value on the NH_4^+-N and NO_3^--N degradation performances of strain LY1024, a group of contrast experiments were set at an initial pH value that varied from 5 to 6, 7, 8, and 9. From Figure 6, it can be observed that strain LY1024 presents excellent NH_4^+-N and NO_3^--N removal abilities at a wide pH value range from 6 to 9, where after 24 h of cultivation, 93.1%–99.2% of NH_4^+-N and 95.9%–99.1% of NO_3^--N can be degraded. In contrast, only 68.5% of NH_4^+-N and 46.4% of NO_3^--N are removed when the initial pH value is set at 5, suggesting that strain LY1024 is expected to display optimal nitrogen removal performance under a neutral or weak alkalinity solution. The pH values of all reaction systems increase after 24 h of cultivation (Table 2), further proving that the denitrification process is an alkali production process.

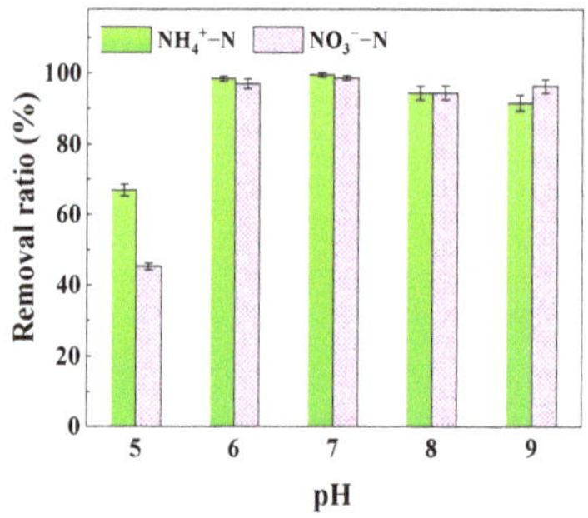

Figure 6. Influence of initial pH value on nitrogen compounds removal on *Vibrio* sp. LY1024.

Table 2. Variation of solution pH value before and after culture.

Initial pH Value	5	6	7	8	9
pH value after 24 h	5.12	6.80	7.75	8.67	9.05

The nitrogen compound removal properties of LY1024 at different temperatures are also investigated, and the related results are shown in Figure 7. After 24 h, the removal efficiencies of strain LY1024 towards NH_4^+-N at 15 °C, 20 °C, 25 °C, and 30 °C are 97.3%, 97.8%, 97.9%, and 98.5%, while its performance for NO_3^--N removal can also reach 76.4%, 95.3%, 96.3%, and 97.4%, respectively. These results demonstrate the strain LY1024 possesses excellent nitrogen removal ability in a wide temperature range, which is superior to most heterotrophic nitrification and aerobic denitrification bacteria already reported, as they were mainly carried out under mesophilic conditions (30–37 °C) [20–22]. The low-temperature-tolerance property endows strain LY1024 great potential to be employed to deal with practical mariculture wastewater, because the water temperature of mariculture wastewater treatment reactors is usually below 20 °C.

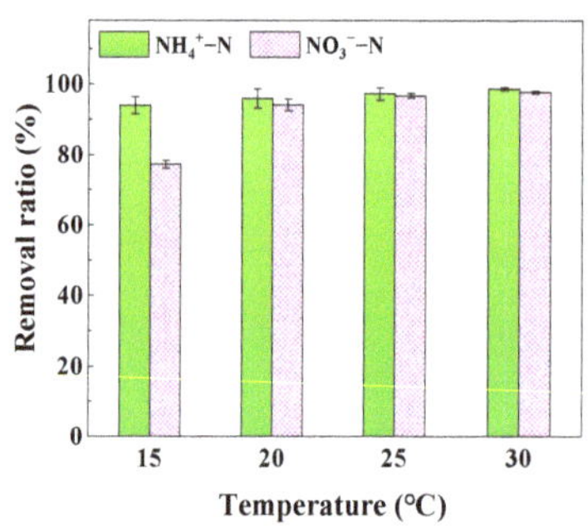

Figure 7. Influence of temperature on nitrogen compounds removal on *Vibrio* sp. LY1024.

In addition, the influence of contacted dissolved oxygen (DO) on the nitrogen removal activities of strain LY1024 is considered, and the degree of contacted DO is controlled by adjusting the shaking speed. From Figure 8, almost 100% of NH_4^+-N could be removed after 24 h of cultivation, regardless of whether the shaking speed is set at 100 rpm (DO $\approx$ 4.05 mg/L), 130 rpm (DO $\approx$ 4.67 mg/L), 150 rpm (DO $\approx$ 5.12 mg/L), or 180 rpm (DO $\approx$ 5.71 mg/L). Moreover, after 24 h of cultivation, the efficiencies of strain LY1024 towards NO_3^--N removal are 91.3%, 98.0%, 98.1%, and 98.0%, respectively. Thus, it can be concluded that an appropriate shaking speed is helpful to improve the nitrogen degradation rate owing to the increase in the mass transfer coefficient of oxygen. However, when the amount of DO reaches a certain level, an excessive increase in DO has no significant effect on the nitrogen removal rate.

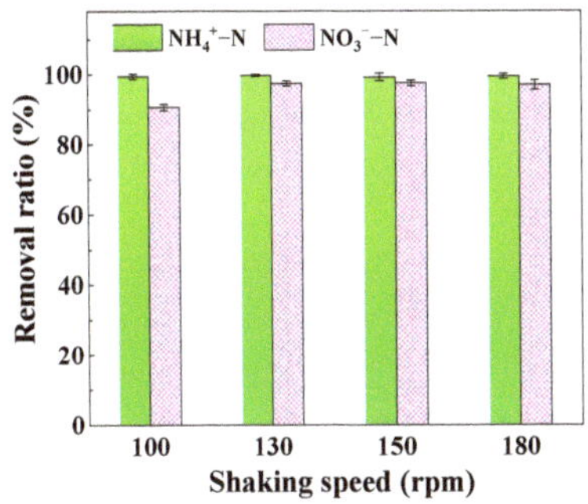

Figure 8. Influence of shaking speed on nitrogen compounds removal on *Vibrio* sp. LY1024.

Last but not the least, it is very necessary to investigate the salinity tolerance of LY1024 for achieving the goal of being employed to deal with mariculture wastewater, due to the latter usually containing high salinity. As shown in Figure 9, strain LY1024 exhibits relatively stable nitrogen removal efficiency with almost 100% of NH_4^+-N, and 81.1%–98.5% of NO_3^--N can be efficiently degraded in the salinity range of 0%–8.5% after 24 h of cultivation. However, further increasing the salinity to 10%, a sharp decrease in the NO_3^--N removal efficiency to 69.3% and a slight reduction of the NH_4^+-N removal rate to 86.4% are detected, respectively. These can be ascribed to the dehydration of cells caused by high salinity, which affects the biological activity of the microorganism [23]. The excellent nitrogen removal ability of strain LY1024 at the wide salinity range of 0%–8.5% greatly expands its application scope, which is also superior to most reported halotolerant strains, such as JR1, CL1502, etc. [24,25].

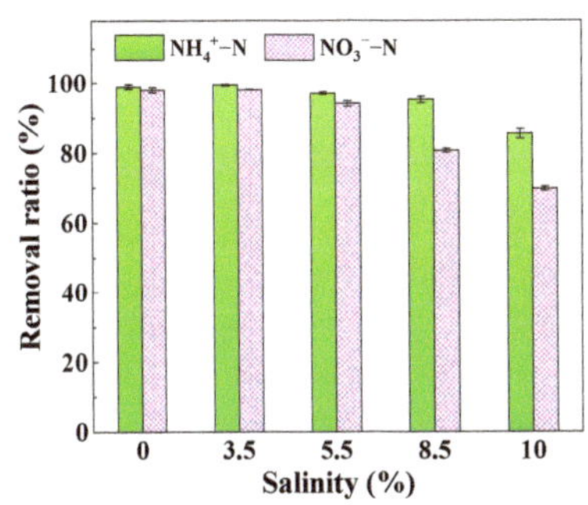

Figure 9. Influence of salinity on nitrogen compounds removal on *Vibrio* sp. LY1024.

3.4. Aerobic Nitrogen Removal Mechanism and Pathway by Strain LY1024

As mentioned above, when NH_4^+-N is used as the sole nitrogen source, the obvious accumulation of NO_3^--N and the significant decrease in COD appear during the reaction process, which indicates that obvious heterotrophic nitrification occurs in the process of nitrogen removal with organics as carbon sources. NO_3^--N accumulates to a certain extent and then gradually decreases under aerobic conditions, indicating that aerobic denitrification occurs in the process of nitrogen removal. The experimental results with NO_3^--N and NO_2^--N as the sole nitrogen source also well prove that *Vibrio* sp. LY1024 has good aerobic denitrification capacity.

At present, it is universally accepted that the aerobic denitrification process can be realized over bacteria through two main pathways [26], as depicted in Figure 10. One is NH_4^+-N is first converted into NH_2OH by the catalysis of ammonia monooxygenase (AMO), oxidized into NO_3^--N and NO_2^--N by hydroxylamine oxidase (HAO), reduced to NO and N_2O, and finally converted to N_2 released into the atmosphere [27–29]. The other is that NH_4^+-N is first converted into NH_2OH and then directly transformed into N_2O and N_2 without the conversion to intermediate, such as NO_3^--N and NO_2^--N [30–33].

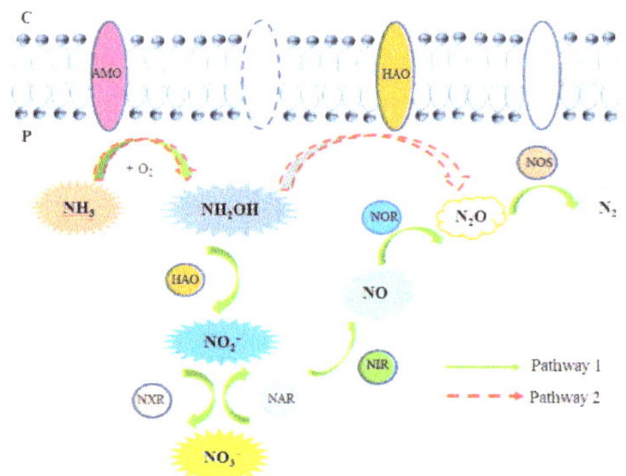

Figure 10. Possible pathways of aerobic nitrogen removal. Note: C: cytoplasm, P: periplasm, AMO: ammonia monooxygenase, HAO: hydroxylamine oxidase, NXR: nitrite oxidoreductase, NAR: nitrate reductase, NIR: nitrite reductase, NOR: nitric oxide reductase, NOS: nitrous oxide reductase.

Combining these theories with our determination results, *Vibrio* sp. LY1024 could oxidize NH_4^+-N under aerobic conditions, with obvious NO_3^--N and trace NO_2^--N accumulation appearing in the process of NH_4^+-N removal, meaning that the roles of AMO, HAO, and nitrite oxidoreductase (NXR) in the strain are expressed. In addition, *Vibrio* sp. LY1024 also displays the ability to degrade NO_3^--N and NO_2^--N under aerobic conditions, meaning that the roles of nitrate reductase (NAR) and nitrite reductase (NIR) in the strain are expressed. Therefore, the nitrogen removal mechanism over strain LY1024 is speculated via the pathway of NH_4^+-N $\rightarrow$ NH_2OH $\rightarrow$ NO_3^--N $\rightarrow$ N_2O $\rightarrow$ N_2.

3.5. Treatment of Stimulated Mariculture Wastewater by Vibrio sp. LY1024

Vibrio sp. LY1024 was finally employed to treat stimulated mariculture wastewater, and the corresponding result is shown in Figure 11. It can be seen from Figure 11 that strain LY1024 displays excellent activities on NH_4^+-N, NO_3^--N, and COD degradation. After 24 h of reaction, the removal rates of strain LY1024 towards NH_4^+-N, NO_3^--N, and COD reach 99.0%, 97.1%, and 88.5%, respectively. Moreover, almost no NO_2^--N accumulated in the process, further demonstrating that the nitrogen removal is mainly through the heterotrophic nitrification–aerobic denitrification (HN-AD) pathway. These experiment results imply that strain LY1024 has great potential in tackling the nitrogenous contaminants in practical mariculture wastewater.

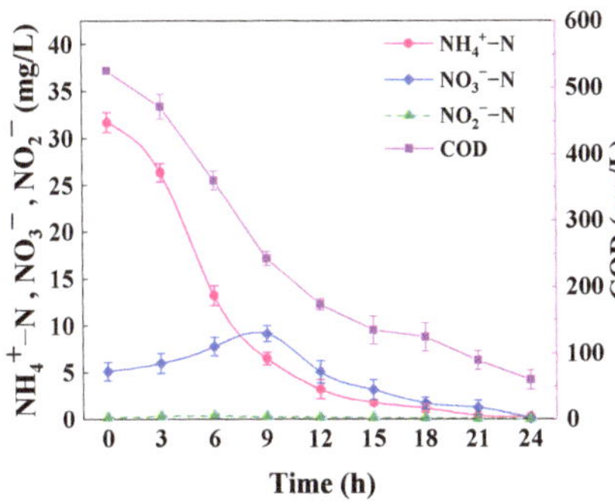

Figure 11. Treatment performance of simulated mariculture wastewater by *Vibrio* sp. LY1024.

The Monod model is usually used to describe the kinetics of microbial growth and substrate utilization. When it is used to describe the kinetics of substrate utilization, its formula can be expressed as:

$$v = (v_{\max} \times S)/(K_s + S) \quad (1)$$

where v is the specific substrate removal rate, $v_{\max}$ is the maximal specific substrate removal rate, S is the substrate concentration, and K_s is the Monod constant.

In this study, the Monod model is used to describe the nitrogen removal reaction kinetics of the simulated mariculture wastewater and the fitting results are shown in Figure 12. The fitting results show the nitrogen removal reaction kinetics can be well described by the Monod model with the correlation coefficient $R^2 = 0.973$. The maximal nitrogen (total inorganic nitrogen) removal rate v_{max} is calculated to be 2.17 mg/(L·h), which is comparable to most reported results [34,35], indicating the good denitrogenation performance of *Vibrio* sp. LY1024, and the Monod constant K_s is calculated to be 4.28 mg/L.

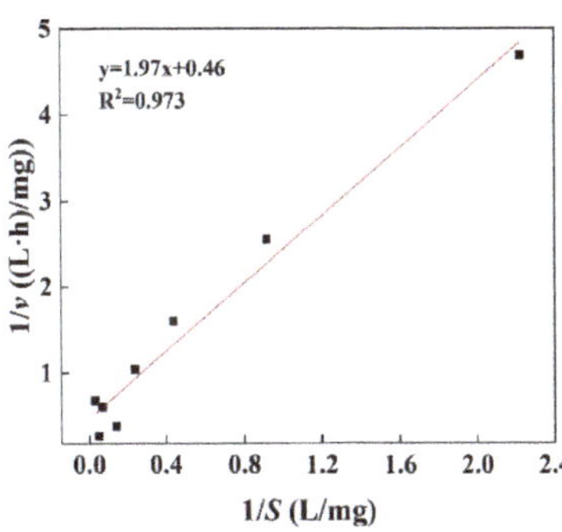

Figure 12. Fitting results of denitrification experimental data with Monod model.

4. Conclusions

Overall, a halotolerant HN-AD bacterium of *Vibrio* sp. LY1024 with high nitrogen removal capacity is domesticated and isolated. *Vibrio* sp. LY1024 exhibits excellent nitrogen removal performance under the conditions of a C/N ratio greater than 15, an initial pH value of 6–9, a temperature of 15–30 °C, a DO of 4.05–5.71 mg/L, and a salinity of 0%–10%. During the treatment of the simulated mariculture wastewater, 99.0%, 97.1%, and 88.5% of NH_4^+-N, NO_3^--N, and COD can be removed by *Vibrio* sp. LY1024 in 24 h of reaction time, and the nitrogen removal pathway is speculated as NH_4^+-N → NH_2OH → NO_3^--N → N_2O → N_2. Such a wide range of application conditions endows *Vibrio* sp. LY1024 with a great application potential in mariculture wastewater treatment.

Author Contributions: Methodology, formal analysis, data curation, L.W.; investigation, formal analysis, data curation, Y.F.; writing—review and editing, supervision, visualization, S.W.; conceptualization, writing—review and editing, F.Y.; data curation, investigation, E.C.; conceptualization, writing—review and editing, Q.S. All authors have read and agreed to the published version of the manuscript.

Funding: This research was funded by the Science and Technology Department of Hebei Province through the Natural Science Foundation of Hebei Province (No. E2021203038 and No. B2021203036), the Key Research and Development Program of Hebei Province (No. 22327309D) and by the Education Department of Hebei Province through the Demonstration Course Project for Postgraduates of Hebei Province (No. KCJSX2021026).

Institutional Review Board Statement: Not applicable.

Informed Consent Statement: Not applicable.

Data Availability Statement: Not applicable.

Conflicts of Interest: The authors declare no conflict of interest.

References

1. Piedrahita, R.H. Reducing the potential environmental impact of tank aquaculture effluents through intensification and recirculation. *Aquaculture* **2003**, *226*, 35–44. [CrossRef]
2. Kang, P.; Xu, S. The impact of mariculture on nutrient dynamics and identification of the nitrate sources in coastal waters. *Environ. Sci. Pollut. Res. Int.* **2016**, *23*, 1300–1311. [CrossRef] [PubMed]
3. Mousavi, S.R.; Asghari, M.; Mahmoodi, N.M. Chitosan-wrapped multiwalled carbon nanotube as filler within PEBA thin film nanocomposite (TFN) membrane to improve dye removal. *Carbohydr. Polym.* **2020**, *237*, 116128. [CrossRef]
4. Hosseini, F.; Sadighian, S.; Monfared, H.H.; Mahmoodi, N.M. Dye removal and kinetics of adsorption by magnetic chitosan nanoparticles. *Desalin. Water Treat.* **2016**, *57*, 24378–24386. [CrossRef]
5. Healy, M.G.; Ibrahim, T.G.; Lanigan, G.J.; Serrenho, A.J.; Fenton, O. Nitrate removal rate, efficiency and pollution swapping potential of different organic media in laboratory denitrification bioreactors. *Ecol. Eng.* **2011**, *40*, 198–209. [CrossRef]
6. Camargo, J.; Alonso, A. Ecological and toxicological effects of inorganic nitrogen pollution in aquatic ecosystems: A global assessment. *Environ. Int.* **2006**, *32*, 831–849. [CrossRef] [PubMed]
7. Mahmoodi, N.M.; Oveisi, M.; Bakhtiari, M.; Hayati, B.; Shekarchi, A.A.; Bagheri, A.; Rahimi, S. Environmentally friendly ultrasound-assisted synthesis of magnetic zeolitic imidazolate framework—Graphene oxide nanocomposites and pollutant removal from water. *J. Mol. Liq.* **2019**, *282*, 115–130. [CrossRef]
8. Mahmoodi, N.M.; Mokhtari-Shourijeh, Z. Preparation of PVA-chitosan blend nanofiber and its dye removal ability from colored wastewater. *Fibers Polym.* **2015**, *16*, 1861–1869. [CrossRef]
9. Zheng, H.Y.; Liu, Y.; Gao, X.Y.; Ai, G.M.; Miao, L.L.; Liu, Z.P. Characterization of a marine origin aerobic nitrifying-denitrifying bacterium. *J. Biosci. Bioeng.* **2012**, *114*, 33–37. [CrossRef]
10. Uygur, A. Specific nutrient removal rates in saline wastewater treatment using sequencing batch reactor. *Process Biochem.* **2006**, *41*, 61–66. [CrossRef]
11. Kong, Q.X.; Wang, X.W.; Jin, M.; Shen, Z.Q.; Li, J.W. Development and application of a novel and effective screening method for aerobic denitrifying bacteria. *FEMS Microbiol. Lett.* **2010**, *260*, 150–155. [CrossRef] [PubMed]
12. Jin, C.; Luo, P.; Zuo, H.; Chen, J.; Chem, M.; Wang, W. *Vibrio zhanjiangensis* sp. nov., isolated from sea water of shrimp farming pond. *Antonie Van Leeuwenhoek* **2010**, *101*, 743–751. [CrossRef] [PubMed]
13. Tamura, K.; Peterson, D.; Peterson, N.; Stecher, G.; Nei, M.; Kumar, S. MEGA5: Molecular evolutionary genetics analysis using maximum likelihood, evolutionary distance, and maximum parsimony methods. *Mol. Biol. Evol.* **2011**, *28*, 2731–2739. [CrossRef] [PubMed]
14. Liu, Y.; Wang, Y.; Li, Y. Nitrogen removal characteristics of heterotrophic nitrification-aerobic denitrification by *Alcaligenes faecalis* C16. *Chin. J. Chem. Eng.* **2015**, *23*, 827–834. [CrossRef]
15. Chen, P.; Ji, L.; Li, Q.X. Simultaneous heterotrophic nitrification and aerobic denitrification by bacterium *Rhodococcus* sp. CPZ24. *Bioresour. Technol.* **2012**, *116*, 266–270. [CrossRef]
16. Zhao, B.; Tian, M.; An, Q. Characteristics of a heterotrophic nitrogen removal bacterium and its potential application on treatment of ammonium-rich wastewater. *Bioresour. Technol.* **2016**, *226*, 46–54. [CrossRef]
17. Richardson, D.J.; Wehrfritz, J.M.; Keech, A.; Crossman, L.C.; Spiro, S. The diversity of redox proteins involved in bacterial heterotrophic nitrification and aerobic denitrification. *Biochem. Soc. Trans.* **1998**, *26*, 401–408. [CrossRef]
18. Lin, Y.; Kong, H.; Wu, D.; Li, C.; Wang, R.; Tanaka, S. Physiological and molecular biological characteristics of heterotrophic ammonia oxidation by *Bacillus* sp. LY. *World J. Microbiol. Biotechnol.* **2010**, *26*, 1605–1612. [CrossRef]
19. Yang, X.P.; Wang, S.M.; Zhang, D.W.; Zhou, L.X. Isolation and nitrogen removal characteristics of an aerobic heterotrophic nitrifying-denitrifying bacterium, *Bacillus subtilis* A1. *Bioresour. Technol.* **2011**, *102*, 854–862. [CrossRef]

20. Wang, T.; Wei, H.; Hu, Z.; Chai, H.; Zhao, H. Isolation and identification of a heterotrophic nitrifying and aerobic denitrifying strain and its denitrification characteristics. *Acta Scien. Circum.* **2017**, *37*, 945–953.
21. Chen, M.; Wang, W.; Feng, Y.; Zhu, X.; Zhou, H.; Tan, Z. Impact resistance of different factors on ammonia removal by heterotrophic nitrification-aerobic denitrification bacterium *Aeromonas* sp. HN-02. *Bioresour. Technol.* **2014**, *167*, 456–461. [CrossRef]
22. Fei, H.; Pan, L.; Na, L.; Tang, X. Characterization of novel *Bacillus* strain N31 from mariculture water capable of halophilic heterotrophic nitrification-aerobic denitrification. *J. Biosci. Bioeng.* **2017**, *124*, 564–571.
23. Lei, Y.; Wang, Y.; Liu, H.; Xi, C.; Song, L. A novel heterotrophic nitrifying and aerobic denitrifying bacterium, *Zobellella taiwanensis* DN-7, can remove high-strength ammonium. *Appl. Microbiol. Biotechnol.* **2016**, *100*, 4219–4229. [CrossRef] [PubMed]
24. Yang, J.R.; Wang, Y.; Chen, H.; Lyu, Y.K. Ammonium removal characteristics of an acid-resistant bacterium *Acinetobacter* sp. JR1 from pharmaceutical wastewater capable of heterotrophic nitrification-aerobic denitrification. *Bioresour. Technol.* **2019**, *274*, 56–64. [CrossRef] [PubMed]
25. Huang, G.; Ou, L.; Pan, F.; Wang, Y.; Fan, G.; Liu, G.; Wang, W. Isolation of a Novel Heterotrophic Nitrification–Aerobic Denitrification Bacterium *Serratia marcescens* CL1502 from Deep-Sea Sediment. *Environ. Eng. Sci.* **2017**, *34*, 453–459. [CrossRef]
26. Uygur, A.; Karg, F. Salt inhibition on biological nutrient removal from saline wastewater in a sequencing batch reactor. *Enzym. Microb. Technol.* **2004**, *34*, 313–318. [CrossRef]
27. Ren, Y.X.; Lei, Y.; Liang, X. The characteristics of a novel heterotrophic nitrifying and aerobic denitrifying bacterium, *Acinetobacter junii* YB. *Bioresour. Technol.* **2014**, *171*, 1–9. [CrossRef]
28. Teng, W.; Dang, Q.; Liu, C.; Yan, J.; Zhang, Y. Heterotrophic nitrogen removal by a newly-isolated alkalitolerant microorganism, *Serratia marcescens* W5. *Bioresour. Technol.* **2016**, *211*, 618–627.
29. Xiao, J.B.; Jiang, H.X.; Chu, S.Y. Denitrification characteristics of an aerobic denitrifying bacterium *Defluoibacter lusatiensis* str. DN7 using different sources of nitrogen. *Acta Ecol. Sin.* **2012**, *32*, 6463–6470. [CrossRef]
30. Qian, C.; Ni, J. Ammonium removal by *Agrobacterium* sp. LAD9 capable of heterotrophic nitrification-aerobic denitrification. *J. Biosci. Bioeng.* **2012**, *113*, 619–623.
31. Qian, C.; Ni, J. Heterotrophic nitrification-aerobic denitrification by novel isolated bacteria. *J. Ind. Microbiol. Biotechnol.* **2011**, *38*, 1305–1310.
32. Sun, Q.H.; Yu, D.S.; Zhang, P.Y. Identification and nitrogen removal characteristics of a heterotrophic nitrification-aerobic denitrification strain isolated from marine environment. *Environ. Sci.* **2016**, *37*, 1089–1097.
33. Zhao, B.; Yi, L.H.; Hughes, J.; Xiao, F.Z. Heterotrophic nitrogen removal by a newly isolated *Acinetobacter calcoaceticus* HNR. *Bioresour. Technol.* **2010**, *101*, 5194–5200. [CrossRef] [PubMed]
34. Zhang, Y.Q.; Liu, Y.; Shi, F.Y.; Cheng, B.; Song, B.B. Study on total ammonia nitrogen removal performance of marine biofilters and nitrification kinetics. *Chin. J. Environ. Eng.* **2010**, *4*, 1697–1703.
35. Hu, B.B.; Zhang, Y.C.; Wan, J.F. Effect of nitrate nitrogen concentration on denitrification efficiency of short cut denitrification. *Appl. Chem. Ind.* **2021**, *50*, 1468–1471+1477.

coatings

Article

Chemical Agglomeration to Enhance Blast Furnace Dust Capture Efficiency in Wet Electrostatic Precipitators

Yingying Han, Lichun Xiao * and Hongrui Chen

Hebei Key Laboratory of Heavy Metal Deep-Remediation in Water and Resource Reuse, School of Environmental and Chemical Engineering, Yanshan University, Qinhuangdao 066004, China
* Correspondence: xlc@ysu.edu.cn

Abstract: Wet electrostatic precipitators (WESPs) are increasingly used in iron and steel plants and coal-fired power plants due to their higher efficiency at capturing fine particles than conventional dry-type precipitators. In order to achieve ultra-pure purification of dust in steel plant gas, we propose an improved method that involves adding a chemical coagulant and a surfactant to a WESP. The effects of the type and concentration of chemical coagulant and surfactant on the agglomeration effect and dust removal efficiency of blast furnace dust were investigated. The results show that the addition of a chemical coagulant could promote the agglomeration of blast furnace dust particles, and the D_{50} of dust particles increased from 5.8 to 15.0 μm after the addition of xanthan gum (XTG). The best increase in the blast furnace's dust particle removal occurred at a concentration of 10 mg/L of XTG, and the dust removal efficiency reached 97.59%. The surfactant dodecyl trimethyl ammonium chloride (DTAC) improved the dust removal efficiency of blast furnace dust when added alone. The dust removal efficiency reached 97.82% when 10 mg/L of XTG and 9 mg/L of DTAC were added synergistically. The addition of a chemical coagulant and surfactant promoted the agglomeration of blast furnace dust and enhanced the dust capture effect of a WESP. We thus demonstrated that we can improve the efficiency of WESP in the future via chemical coagulation. The authors will further study the effect of multi-factor synergistic coupling on the chemical coagulation method in WESPs.

Keywords: blast furnace dust; wet electrostatic precipitator; chemical coagulant; synergy; dust removal

Citation: Han, Y.; Xiao, L.; Chen, H. Chemical Agglomeration to Enhance Blast Furnace Dust Capture Efficiency in Wet Electrostatic Precipitators. *Coatings* **2022**, *12*, 1937. https://doi.org/10.3390/coatings12121937

Academic Editor: Klemen Bohinc

Received: 31 October 2022
Accepted: 6 December 2022
Published: 9 December 2022

1. Introduction

The iron and steel industry is the second most polluting industry in China after the electric power industry, and it is an important source of fine particle pollution in the air in China [1–3]. With the continuous growth of steel production every year, the total emitted amount of pollutants has been increasing, which has had a great impact on the air quality [4–8]. According to statistical data, in the first half of 2021, the national crude steel production reached a total of 563 million tons, which represented an 11.80% year-on-year increase. Pig iron production reached 456 million tons, which represented a 4.0% year-on-year increase. Steel production reached 698 million tons, representing a cumulative year-on-year increase of 13.90% [9]. The total amount of waste gas emission in the first half of 2021 also increased compared with the same period of the previous year. Blast furnace gas generation reached 523,805 billion m^3, which represented a 6.80% year-on-year increase, and its recycling rate reached 97.95%. Converter gas generation reached 40.704 billion m^3, representing a 13.70% year-on-year increase, and its utilization rate reached 98.40%. Both blast furnaces and converters have improved their gas utilization rate compared with the same period of last year [10]. Although the gas recovery rate has increased in recent years with the improvement of emission standards, some of the gas that is not recovered and is discharged still has a negative impact on the environment [11]. Blast furnace gas is a by-product of blast furnace ironmaking. Blast furnace gas has the characteristics of high dust concentration, high pollution, low calorific value, and difficult recovery, and the poor hydrophilicity of the dust particles in the gas affects the recovery of the gas [12]. When blast furnace gas is used for power generation and other purposes, the ultra-pure purification of gas is required to reduce the dust content in the gas to less than 1 mg/m^3 (standard state). Therefore, this study proposes to enhance the hydrophilicity of blast furnace dust by adding a chemical coagulant and a surfactant to a wet electrostatic precipitator to improve the collection efficiency of blast furnace dust and to achieve the ultra-pure purification of gas [13].

The use of a chemical coagulant and a surfactant in wet electrostatic precipitators (WESPs) is based on adding one or more agents to the original dust removal equipment to enhance the agglomeration of fine dust particles to achieve a higher removal efficiency. In the 1990s, Durham et al. [14] sprayed a mixture containing atomized binder into an ESP gas stream and showed that the binder and fine particles effectively collected on the dust collection poles. In 2000, the U.S. Department of Energy and others developed new chemical conditioners for controlling fly ash resistivity and cohesion. In 2001, Johansen et al. [15] compared the morphology and particle size of agglomerates formed by PEG 20000 and PEG 3000 and concluded that PEG 20000 in powder form is the most effective agglomerate. Earlier research on the effect of chemical coagulants on the growth of ultrafine particles is described below. In 2003, Wei et al. [16] summarized the research on chemical agglomeration technology by foreign scientific researchers and proposed the use of a chemical coagulant to promote ultrafine particulate matter growth. In 2007, Zhao et al. [17] used coagulant spraying in a coal-fired flue to promote ultrafine particle growth. Later, researchers began to focus on the effects of temperature, pH, addition point, and coagulant aid when adding chemical coagulants. Rajniak et al. [18] investigated the effect of HPC in combination with other auxiliaries on the agglomeration of particles. In 2009, Li et al. [19] investigated the agglomeration effect of four chemical coagulants including PAM at different concentrations and under different pH conditions. In the same year, Carvalh et al. [20] investigated the physicochemical effects of using carrageenan as a coagulant. In 2011, Forbes et al. [21] investigated the effect of chemical coagulant droplets on the wetting and agglomeration of fine particulate matter at different temperatures. In 2013, Zhao et al. [22] analyzed the variation in dust concentration and particle size when using different coagulant addition points, temperatures, etc.

More and more researchers have begun to focus on the synergistic effect of chemical coagulation and electrocoagulation, the synergistic effect of chemical coagulation and turbulent coagulation, and the synergistic effect of chemical coagulants on the capture efficiency of fine particles. In 2014, Thonglek et al. [23] controlled fine particle agglomeration by pulsing the ESP to generate plasma inside the ESP. In the same year, Liu et al. [24] achieved a 30% increase in PM2.5 removal when adding chemical coagulants to a wet FGD system. In 2015, Balakin et al. [25] investigated the mechanism of agglomeration using capillary bridging in a multiphase flow agglomeration experiment. In 2016, Liu et al. [26] selected six chemical coagulants, including XTG, to study the growth of fine particles. In 2017, Guo et al. [27] evaluated the chemical agglomeration phenomenon by simulating the pressure, temperature, and velocity distributions in a chamber flow field. In 2018, Hu et al. [28] used water, pectin, and sodium alginate solution as coagulants to improve the fine particulate removal efficiency by more than 20%. In 2019, Sun et al. [29] proposed and investigated the effect of different chemical agglomeration and turbulent agglomeration coupling methods on the agglomeration effect and removal performance of ESP particles. Zhou et al. [30] investigated the agglomeration effect of sesbania gum (SBG) and styrene butadiene emulsion (SBE) on fine particulate matter using phase Doppler anemometry. In 2020, Li et al. [31] found that a carrageenan/Tween-80/NH_4Cl (KC/TW/NH_4Cl) three-component coagulant increased the average particle size of fine particulate matter from 2.8 μm to more than 10.0 μm. In 2021, Gao et al. [32] used chemical agglomeration and turbulent mixing to synergistically promote chemical agglomeration and greatly improve the effectiveness of electrostatic precipitation. Yang et al. [33] applied a new passivation coagulant to a 300 MW unit experimentally, and the exit dust concentration was reduced by about 32%. Therefore, the chemical agglomeration method has good prospects for application in electrostatic precipitators.

All of the above studies show that chemical coagulants have a good effect on the growth of fine particles of dust and can significantly improve dust removal efficiency. However, most of the particles selected in these studies were coal-fired dust, and the problems of high water consumption and high cost of coagulant dosage still exist, making it difficult to widely promote chemical agglomeration technology in dust removal applications. In this study, we focus on blast furnace dust with poor agglomeration performance. By adding a chemical coagulant to the original spraying system of a wet electrostatic precipitator, the dust removal efficiency can be improved without modifying the original equipment. In addition, this study applies a surfactant and a chemical coagulant simultaneously to investigate their synergistic effects and to identify a suitable concentration of chemical coagulant to promote dust agglomeration. This method could reduce economic costs and enhance the development of chemical agglomeration technology.

The purpose of this study is to investigate the effects of a chemical coagulant and a surfactant on the removal of blast furnace dust in wet scrubbers. The effects of chemical type and concentration on the agglomeration effect of blast furnace dust were systematically investigated to optimize the chemical agent type and concentration to improve the efficiency of WESP at capturing blast furnace dust. Our results may help improve the efficiency of WESPs in the future via chemical coagulation. The authors will further study the effect of multi-factor synergistic coupling on the chemical coagulation method in WESPs.

2. Materials and Methods

2.1. Experimental Setup

2.1.1. Wet Electrostatic Precipitators System

The wet type electric dedusting system used in this study is shown in Figure 1 and includes a low voltage control system, a high voltage power supply control system, a spraying system, and a fan system. The spraying system includes a circulating pool and a water pump. The electrode setup used a BS mansard cathode line and a 480C anode plate, and the dust collector size was 2.8 × 1.4 × 5.8 (L × W × H) (m). The dust inlet concentration was set to 80 mg/m^3, the power supply voltage was 72 kV, and the current was 0.3 mA. When the wet-type electric dedusting system was in operation, the water pump sent water from the circulating pool to the inside of the dedusting system through nozzle atomization.

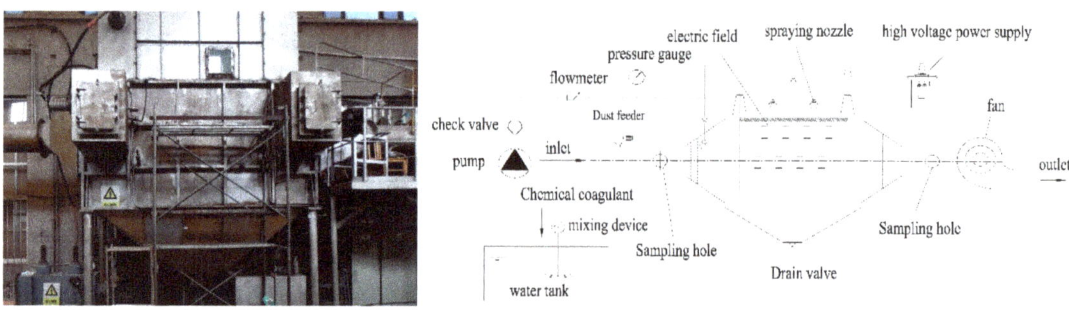

(**a**) (**b**)

Figure 1. Wet electrostatic precipitator system. (**a**) is the field physical diagram of WESP. (**b**) is the internal structure of WESP.

2.1.2. Sampling System

In this experiment, the isokinetic sampling method was used for the dust at the outlet of the WESP. The dust concentration obtained via this method was more accurate and closer to the dust concentration under real conditions. The following equipment was used in the sampling process: a filter cartridge (filter membrane), a sampling head, a drying bottle, a buffer bottle, a vacuum pump, and a flow meter. A diagram of the sampling system is shown in Figure 2. A high silica glass fiber filter cartridge was selected for dust sampling, and it was placed in the oven for drying before and after sampling to eliminate the influence of flue gas temperature and humidity. The total measurement error of the sampling system is less than 0.1 mg/m^3.

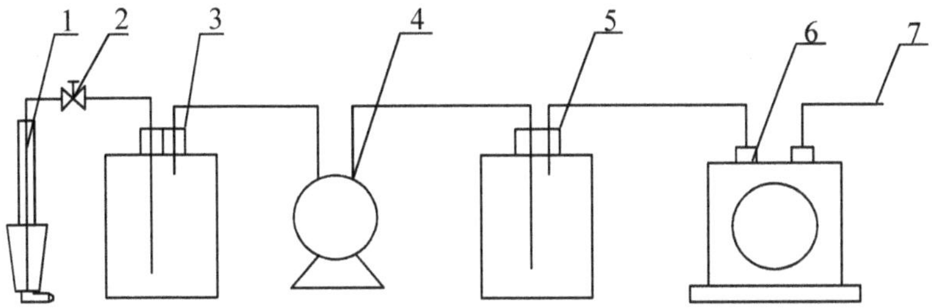

Figure 2. Sampling system. 1. Sampling tube. 2. Needle valves. 3. Dry bottles. 4. Pumping Pump. 5. Buffer bottles. 6. Wet gas flow meter. 7. Discharge tube.

2.1.3. Particle Size Distribution Measurement Device

A BT-9300H laser particle size analyzer was used to measure the dust particle size distribution. The device includes a sample cell, an ultrasonic disperser, a centrifugal circulation pump, and an electric stirrer.

2.1.4. Representation

In this study, a field emission scanning electron microscopc (SEM, S4800-II, Kyoto, Japan) equipped with an energy spectrometer system was used to observe the morphology and elemental composition of blast furnace dust.

2.2. *Experimental Materials*

The experimental conditions used in this study were as follows: the gas temperature was 18–25 °C and the relative humidity was 20%–40%. The dust used was blast furnace dust, the operating voltage of the WESP system was 40 kV, the field gas velocity was 1 m/s, and the nozzle pressure was 0.5 MPa. The blast furnace dust was dried before entering the WESP, and then a quantity of dust was uniformly transported from the feed inlet to the interior of the precipitator by a fan. Inside the dust collector, dust particles went through the stages of gas ionization, dust charging, dust collection, and spray cleaning. Most of the dust particles were trapped, and the uncaught dust escaped through the pipeline.

The drugs used in this experiment included a chemical coagulant and a surfactant, as is shown in Table 1 below. By changing the concentration and type of chemical coagulant and surfactant, the effects of different variables on the corona discharge effect, dust agglomeration effect, and removal efficiency of the wet-type dust collector were investigated. At the end of the experiment, the dust was collected to measure the particle size distribution, observe the dust particle growth effect, and calculate the dust removal efficiency.

Table 1. Chemical coagulants and surfactants used in the experiments.

Species	Name	Manufacturer
Chemical coagulant	PAM (polyacrylamide)	Tianjin Kemiou Chemical Reagent Co., Ltd., Tianjin, China
	KC (k-carrageenan)	Henan Wanbang Chemical Technology Co., Ltd., Zhengzhou, China
	KGM (konjac glucomannan)	Henan Wanbang Chemical Technology Co., Ltd., Zhengzhou, China
	XTG (xanthan gum)	Tianjin Guangfu Fine Chemical Research Institute, Tianjin, China
	Kieselguhr	Tianjin Dengfeng Chemical Reagent Factory, Tianjin, China
Surfactant	DTAC (Dodecyl trimethyl ammonium chloride)	Tianjin Kemiou Chemical Reagent Co., Ltd., Tianjin, China
	SDS (sodium dodecyl sulfate)	Tianjin Kemiou Chemical Reagent Co., Ltd., Tianjin, China
	TX-100 (Octylphenyl polyoxyethylene ether)	Tianjin Guangfu Fine Chemical Research Institute, Tianjin, China

3. Analysis and Discussion

3.1. Blast Furnace Dust Particle Size Distribution and Morphology

3.1.1. Particle Size Distribution Analysis

The particle size distribution of the blast furnace dust is shown in Figure 3. D_{10}, D_{50}, and D_{90} indicate the particle size when the cumulative particle size distribution number reached 10%, 50%, and 90%, respectively. D_{50} is also called xdmedian size or median particle size. It can be seen that the particle size of xdblast furnace dust is small, with a particle size distribution in the range of 0.2–40 μm, and D_{50} was 6.754 μm. We found that, 10% of the dust was less than 2.010 μm, 50% was less than 6.754 μm, and 90% was less than 17.33 μm. In addition, particles with a size below 2.5 μm accounted for about 20.14% of the total, and particles with a size below 10 μm accounted for about 79.77% of the total. These dust particles below the particle size can cause serious air pollution, which is why they are the key target of blast furnace dust capture.

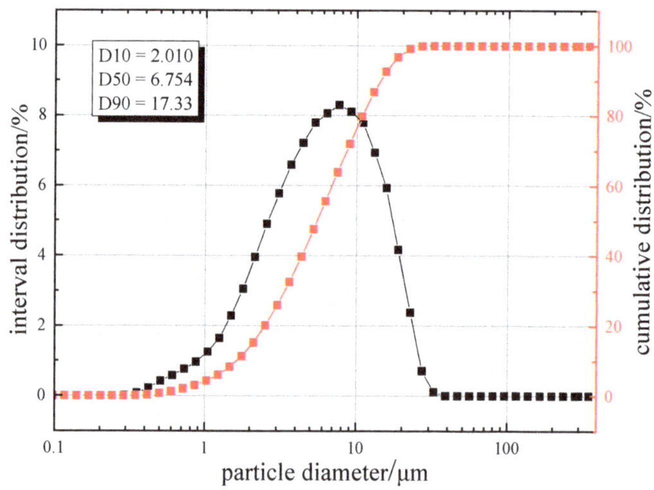

Figure 3. Particle size distribution of blast furnace dust.

3.1.2. Morphological Analysis

Figure 4a shows that the blast furnace dust particles are varied in size and showed an irregular shape and a porous structure. The particles were either connected or dispersed with each other, which may be due to the various chemical changes accompanying the combustion process, resulting in the particles bonding together to form larger particle clusters during cooling. Figure 4b shows that the metallic elements in the blast furnace dust were mainly Ca, Na, Fe, and Al, and the inorganic elements Si and C were the most abundant.

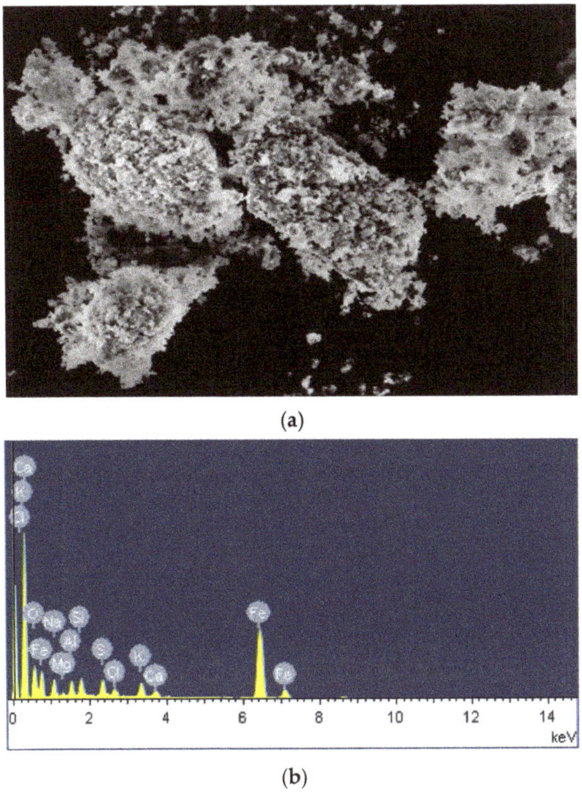

Figure 4. SEM morphology of blast furnace dust.

3.2. Influence of Chemical Coagulant on the Effect of Blast Furnace Dust Capture

3.2.1. Effect of Coagulant Type

Coagulation Effect

Different types of chemical coagulants have different effects on particle agglomeration. In this study, five coagulants, PAM, KGM, KC, XTG, and kieselguhr, were selected at a concentration of 15 mg/L, and the experimental results are shown in Figure 5. The D_{10} particle sizes of the five coagulants showed no significant differences, but the differences were more obvious for D_{50} and D_{90}. The sequence of the corresponding particle sizes of D_{90} were XTG, kieselguhr, PAM, KGM, and KC in descending order. XTG is a polymeric compound with strong hydrophilic properties, which dissolves in water and can form a highly viscous solution. XTG was added to the recirculation basin and sprayed into the dust collector through the nozzle as droplets, which collided with fine particles and bound together in a "bridging" manner to form larger particle clusters. The addition of a chemical coagulant makes the fine particles grow, increasing the probability of particle capture and improving the dust removal efficiency. Regarding the consumption of reagents, a simple calculation showed that less than 1.5 kg of agent was required to treat 100,000 m^3 of gas. Therefore, the cost of the reagents was low.

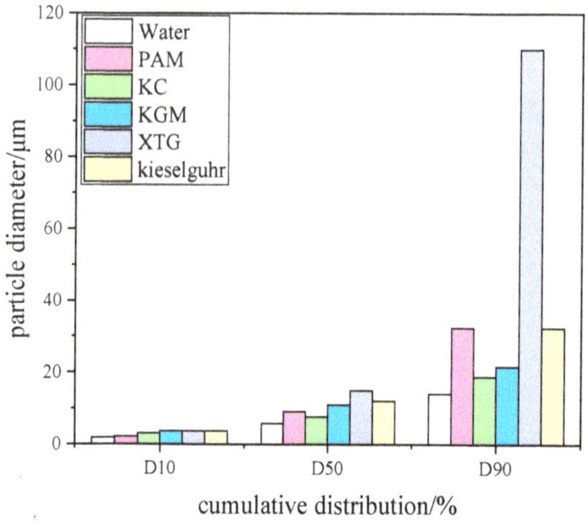

Figure 5. Histogram of particle size cumulative distribution of coagulant types.

Dust Removal Efficiency

To further investigate the effect of chemical coagulants on the removal of blast furnace dust in WESPs, this study applied sampling tubes to sample the inlet and outlet locations of the dust collector to calculate the dust removal efficiency of the WESP with different coagulants. The electric field wind speed was 1.0 m/s, the nozzle pressure was 0.5 MPa, the operating voltage was 40 kV, and the concentration of all five coagulants was 15 mg/L. Figure 6 shows that the dust removal efficiency increased after the addition of a coagulant. When the coagulant was XTG, the highest dust removal efficiency reached 97.59%. When the coagulant was KC, the dust removal efficiency was the lowest (95.68%). Among the five coagulants, XTG had the best agglomeration effect but the worst corona discharge performance and the highest dust removal efficiency. Similarly, the agglomeration effect of PAM was slightly worse than that of kieselguhr, but its corona discharge performance was better than that of kieselguhr. Therefore, it is assumed that the dust removal efficiency was more influenced by the agglomeration effect of particulate matter when the difference in corona discharge performance was not significant.

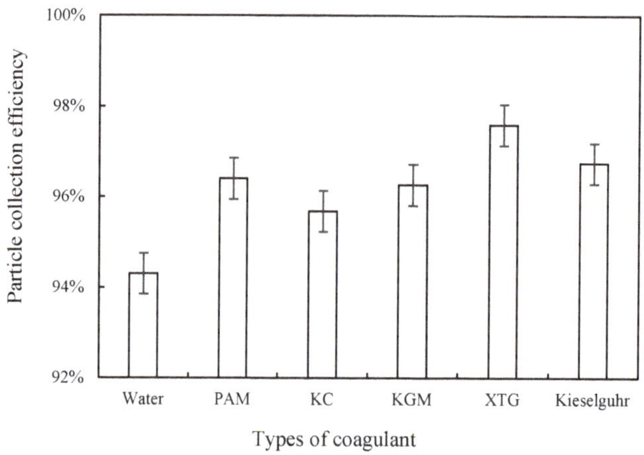

Figure 6. Effect of coagulant type on dust removal efficiency.

3.2.2. Effect of Coagulant Concentration

Coagulation Effect

The agglomeration effect of adding different concentrations of chemical coagulants was different. XTG, which had a better agglomeration effect, was selected to investigate the effect of concentration. Four different concentrations of XTG, 0 mg/L, 5 mg/L, 10 mg/L, and 15 mg/L, were added to the circulating pool for the experiment. As can be seen in Figure 7, the content of large-size particles gradually increased with an increase in coagulant concentration. Without XTG, 20.14% of $PM_{2.5}$, 79.77% of PM_{10}, and 20.23% of particles within 10–100 μm were found. When XTG was added, the $PM_{2.5}$ content was significantly reduced. A concentration of 5 mg/L of XTG can reduce the $PM_{2.5}$ content to 7.9% and the PM_{10} content to 62.33%, and the particulate matter content within 10–100 μm increased significantly to 37.67%. When the XTG concentration was 10 mg/L, the $PM_{2.5}$ content remained basically unchanged, the PM_{10} content continued to decrease, and the content of particulate matter within 10–100 μm was dominant. There was particulate matter with a particle size larger than 100 μm, and the content of particulate matter with a particle size larger than 100 μm was 0.2%. When the XTG concentration was 15 mg/L, the $PM_{2.5}$ content was reduced to 6.32% and the PM_{10} content was reduced to 41.9%. The dust particle size was concentrated between 10 μm and 100 μm, and the dust content in this particle size range was 45.85%. The proportion of particles larger than 100 μm increased to 12.25%. In general, chemical coagulant molecules contain chemical groups that can interact with the particle surface. When a chemical coagulant molecule collides with a particle, some groups in the chemical coagulant molecule will be adsorbed on the surface of the particle, while the rest of the particle can be in contact with other collision particles, resulting in secondary adsorption. In this way, the particles formed aggregates with each other by the action of chemical coagulant molecules [18,34]. Therefore, the particle growth was better with an increased concentration of the coagulant.

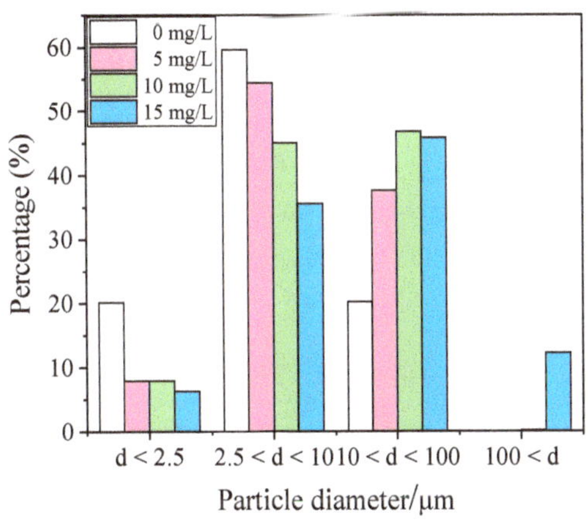

Figure 7. Histogram of particle size interval content of coagulant concentration.

Dust Removal Efficiency

In order to investigate the effect of chemical coagulant concentration on the removal of blast furnace dust in WESPs, dust at the inlet and outlet of the dust collector was collected through the sampling tube, and the dust removal efficiency of the five coagulants at different concentrations was calculated. As can be seen in Figure 8, the dust removal efficiency of the five chemical coagulants at different concentrations were XTG, kieselguhr, PAM, KGM, and KC in descending order, and the dust removal efficiency increases gradually with the coagulant concentration. The maximum efficiency of XTG was achieved at a concentration of 10 mg/L, and the efficiency did not change significantly when the concentration of XTG was increased. XTG has excellent physical and chemical properties. Compared with the several other agglomerating agents used, it has good dust coagulation and dust removal effects. XTG helical molecules can agglomerate fine particles into larger particles under the action of van der Waals forces, hydrogen bonds, hydrophobic interactions, etc. However, it has the disadvantage of poor solubility, and it can easily form a non-conductive coating layer on the electrode surface. When the concentration is high, the electrostatic precipitator cannot work normally. When the chemical coagulant is added in excess, the particle surface is saturated by the chemical coagulant molecules and there are no adsorption vacancies on the particle surface, so the chemical coagulant loses its bridging effect. At the same time, due to the effect of polymer adsorption membrane vacancy resistance or mutual repulsion between the particles, the particles were again in a stable state of dispersion. In addition, continuing to increase the concentration of the coagulant leads to corona discharge performance degradation, resulting in a decrease in the capture efficiency of blast furnace dust. Therefore, the amount of chemical coagulant should be moderate, as more does not necessarily mean better.

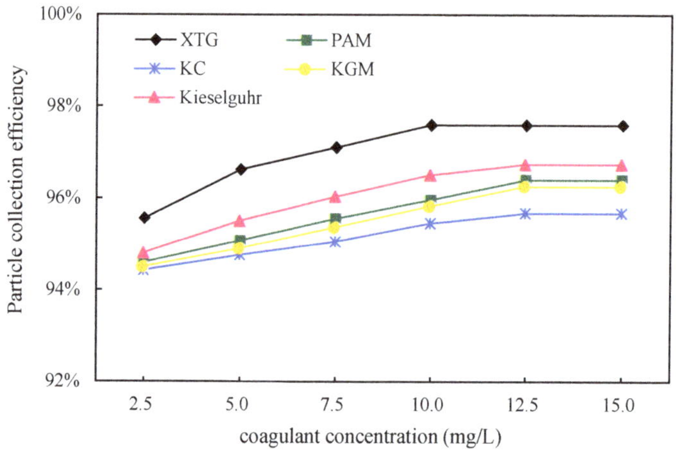

Figure 8. Effect of coagulant concentration on dust removal efficiency.

3.3. *Effect of Surfactants on Blast Furnace Dust Capture*

3.3.1. Influence of Surfactant Type

Coagulation Effect

In order to investigate the effect of surfactants on the efficiency of WESPs at trapping blast furnace dust, three different surfactants, dodecyl trimethyl ammonium chloride (DTAC), sodium dodecyl sulfate (SDS), and octylphenyl polyoxyethylene ether (TX-100), were selected. The surfactants, in line with the experimental method of chemical coagulants, were injected into the recirculation basin and entered the dust collector through the nozzle. After the experiment, the particle size distribution was measured by collecting the particles in the recirculation basin. The results are shown in Figure 9. The more obvious agglomeration effect was shown by DTAC, followed by SDS, and TX-100 basically had no agglomeration effect. There was no significant change in D_{10} after the addition of a surfactant compared with no agent. When no agent was added, the particle D_{50} and D_{90} were 5.8 and 14 μm, respectively. When the surfactant TX-100 was added, the D_{50} and D_{90} increased to 6.382 and 15.75 μm, respectively. When SDS was added, the D_{50} and D_{90} increased to 13.14 and 18.87 μm, respectively. When DTAC was added, the D_{50} continued to increase. Therefore, among the three surfactants, the best effect on the agglomeration of blast furnace dust was exhibited by DTAC. The surfactant molecules completed the adsorption of fine dust and water molecules through hydrophobic bonding and electrostatic interactions, forming an adsorption configuration in which hydrophobic groups were adsorbed on the surface of fine particles and hydrophilic groups were extended to water; thus, the wetting of the dust was achieved. At the particle/water interface, the surfactant molecules were adsorbed on the particle surface due to the influence of van der Waals forces. However, when the surfactant molecules were adsorbed on the surface of the particles, their oxygen-containing groups tended to form hydrogen bonds with water molecules, and such hydrogen bonds had a lifting effect on the surfactant molecules, causing the surfactant molecules to aggregate at the interface. Multiple such particles wrapped by water film collide with each other to achieve particle aggregation [35].

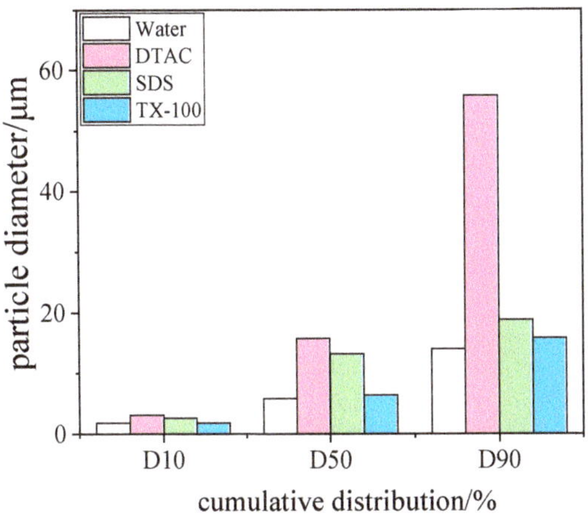

Figure 9. Histogram of particle size cumulative distribution of surfactant types.

Dust Removal Efficiency

We measured the inlet and outlet dust concentration after adding DTAC, SDS, and TX-100 and calculated the dust removal efficiency. As can be seen in Figure 10, the dust removal efficiencies were all improved with the addition of surfactants. When DTAC was added, the highest dust removal efficiency of 97.08% was achieved. When the surfactant was TX-100, the dust removal efficiency was the lowest at 96.17%. The dust removal efficiency was between that of DTAC and TX-100 when SDS was added (96.78%). Among the three surfactants, the addition of DTAC resulted in the best particle growth, the best corona discharge performance, and the highest dust removal efficiency. We find that the effects of the three selected surfactants on dust growth and corona discharge performance remained consistent, which was different from the effect of chemical coagulants, and this phenomenon is likely related to their different solubilities in water.

3.3.2. Effect of Surfactant Concentration

Coagulation Effect

In order to continue to investigate the effect of surfactants on the growth effects of particles, DTAC, which had the best agglomeration effect among the three surfactants, was selected to continue the blast furnace dust agglomeration experiment in this study with concentrations of 0, 5, 7, and 9 mg/L, and the results are shown in Figure 11. The content of large-size particles gradually increased with an increase in the surfactant concentration. At the DTAC concentration of 0 mg/L, 20.14% of $PM_{2.5}$, 79.77% of PM_{10}, and 20.23% of particles within 10–100 μm were found. After adding DTAC, the $PM_{2.5}$ content was significantly reduced. When the DTAC concentration was 5 mg/L, the $PM_{2.5}$ content was reduced to 8.64%, the PM_{10} content was reduced to 48.81%, and the content of particulate matter within 10–100 μm was significantly increased to 50.56%. There were particulate matter particle sizes larger than 100 μm, and the content of particulate matter particle sizes larger than 100 μm was 1.26%. When the DTAC concentration was 7 mg/L, the

$PM_{2.5}$ content remained basically unchanged, the PM_{10} content decreases slightly, and the particulate matter content within 10–100 μm was slightly higher than when DTAC was 5 mg/L. When the DTAC concentration was 9 mg/L, the PM_{10} content decreased to 41.76%, the dust content in the range of 10–100 μm was 54.59%, and the content of particulate matter with particle size larger than 100 μm increased to 3.65%. Therefore, with an increase in the surfactant concentration, the particulate matter grew better. DTAC, as a cationic surfactant, had the effect of reducing the surface charge and compressing the double electric layer for the adsorption of particulate matter, which was in a mosaic, chain-sequence fixation model on the surface of the particulate matter. As a result of this fixation, the surface charge of the particles was not uniformly distributed, and they easily attracted each other because they had opposite charges, and this caused flocculation.

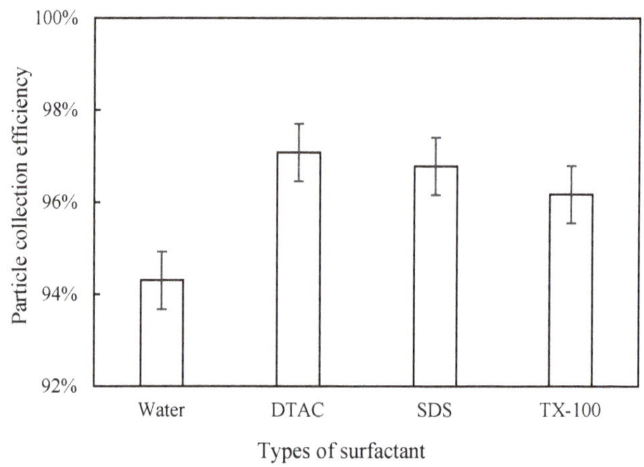

Figure 10. Effect of surfactant type on dust removal efficiency.

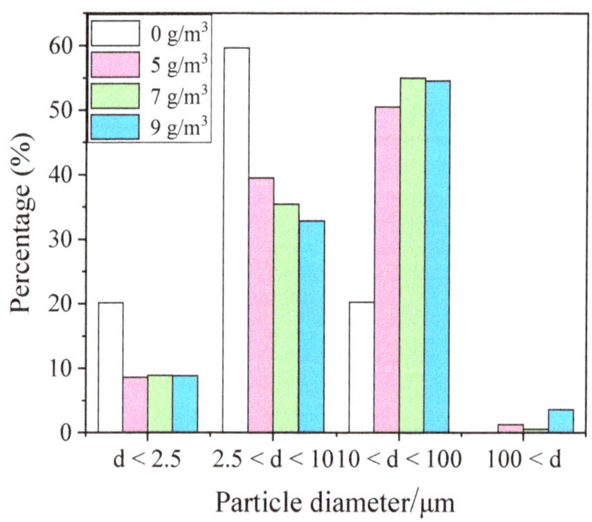

Figure 11. Histogram of particle size interval content of surfactant concentration.

Dust Removal Efficiency

To investigate the effect of surfactant concentration on the efficiency of WESPs at capturing blast furnace dust, the dust concentrations at the inlet and outlet of the dust collector were collected using sampling tubes, and the dust removal efficiency of the three surfactants at the different selected concentrations was calculated. The results are shown in Figure 12. The dust removal efficiency of the three surfactants at different concentrations was DTAC, SDS, and TX-100 in descending order, and it is obvious that the dust removal efficiency tends to gradually climb upward with an increase in the surfactant concentration in the figure. At low concentrations, the dust removal efficiency of DTAC and SDS remained the same, and the dust removal efficiency of SDS increased slowly as the concentration increased. Among several surfactants, DTAC has the best agglomeration effect and the highest dust removal efficiency. DTAC has a strong hydrophobic association effect, which can reduce the liquid film tension, improve the wettability of dust, and ensure the agglomeration of dust, thus improving the dust removal efficiency. Although higher concentrations of surfactants were not further investigated in this experiment, the growth rate of the dust removal efficiency gradually slows down as the concentration increases in Figure 13, especially for DTAC and SDS, and the curve tends to level off. Therefore, as with chemical coagulants, the amount of surfactant should be kept moderate and should be determined according to the concentration of dust to be removed. In addition, the dosage of both chemical coagulants and surfactants should be considered in terms of economic factors as well as dust removal efficiency, as excessive doses may lead to high costs.

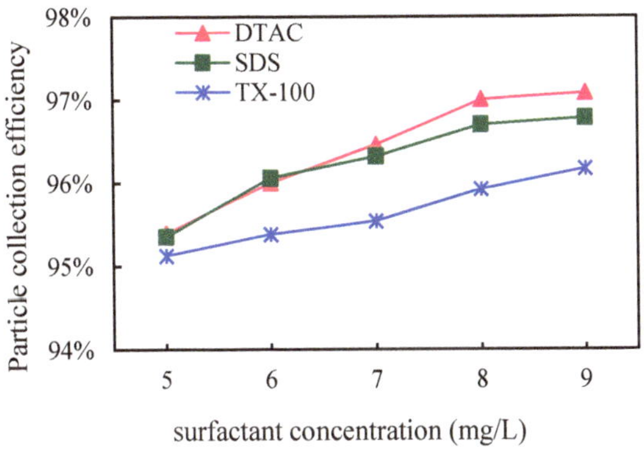

Figure 12. Effect of surfactant concentration on dust removal efficiency.

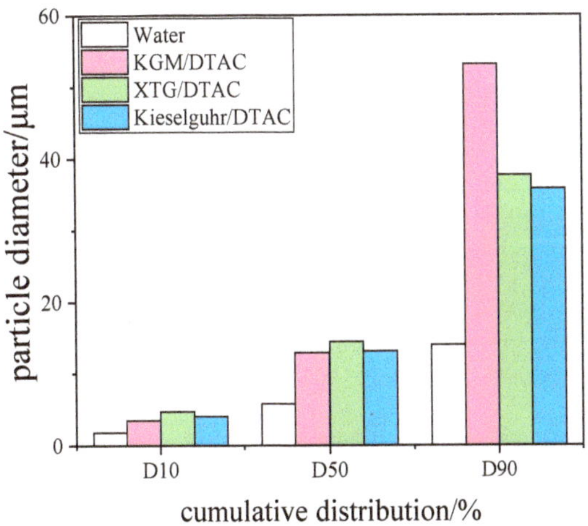

Figure 13. Histogram of cumulative particle size distribution of the synergistic effect of surfactant and coagulant.

3.4. *Effect of Synergy on Blast Furnace Dust Capture*

3.4.1. Impact of Synergistic Species

According to Nsengiyumva [36], when two different chemical coagulation solutions are mixed together, they may exhibit synergistic effects. DTAC at 9 mg/L was chosen as the surfactant to study the effect of synergistic interaction between surfactant and chemical coagulant on blast furnace dust capture. The chemical coagulants involved in the synergistic effect were changed to XTG, kieselguhr and KGM at 10 mg/L, and the effects of the synergistic effect of XTG/DTAC, kieselguhr/DTAC, and KGM/DTAC on the agglomeration effect of blast furnace dust were investigated. The results are shown in Figure 13. Compared with Figures 9 and 13, it was found that the D_{10} particle size increased slightly under the synergistic effect compared with the addition of DTAC only, but the D_{50} and D_{90} particle sizes decreased. When comparing Figures 5 and 13, it was found that the D_{10} particle size increased slightly under the synergistic effect of XTG/DTAC compared to the addition of XTG only, and the D_{10} particle size was 4.738 μm. However, the D_{50} and D_{90} particle sizes were smaller than that of XTG only. Similarly, the D_{10} particle size increased slightly with kieselguhr/DTAC synergy compared to that with kieselguhr only, and the D_{10} particle size was 4.031 μm. Unlike the XTG/DTAC synergy, the D_{50} and D_{90} particle sizes with kieselguhr/DTAC synergy were smaller than those with kieselguhr only. The particle size of D_{10} did not change under KGM/DTAC synergism compared with that of KGM only, but the particle sizes of D_{50} and D_{90} increased significantly, corresponding to 12.9 μm and 53.16 μm, respectively. At this time, the addition of polymeric chemical coagulants can make the small flocs become large flocs through adsorption and bridging effects. Therefore, the addition of a DTAC cationic surfactant increased the possibility of bridging with other polymeric chemical coagulants. In addition, polymer molecular chains tended to intertwine and form a mesh structure, thus increasing the possibility of particle capture and aggregation [37].

3.4.2. Dust Removal Efficiency

The inlet and outlet dust concentrations after adding KGM/DTAC, XTG/DTAC, and kieselguhr/DTAC were measured, and the dust removal efficiency was calculated separately. As can be seen in Figure 14, the highest dust removal efficiency of 97.82% was achieved with the synergistic effect of XTG/DTAC. Compared with the highest dust removal efficiency when only XTG or DTAC was added, both increased, but the increase was smaller. The dust removal efficiency reached 97.60% with the synergistic effect of kieselguhr/DTAC, which was a smaller increase than when only kieselguhr or DTAC was added. The dust removal efficiency with KGM/DTAC synergy was the lowest among the three, at 97.10%. The efficiency was significantly higher compared to when only KGM was added, but the dust removal efficiency was almost unchanged compared to when only DTAC was added. KGM is a gel with a stable cross-linking structure, which has hydrophobic and strong hydrogen bonding effects, and DTAC also has hydrophobic association effects. Under the synergistic effect of two chemical coagulants, the surface of blast furnace dust will generate a hydrophobic effect, and it will agglomerate into larger particles via the bridging effect of the chemical coagulant, thus improving the dust collection efficiency. Therefore, the addition of a surfactant to the chemical coagulant would improve the dust removal efficiency. The improvement of dust removal efficiency was not obvious, but this increases the economic cost. For a coagulant with a low dust removal efficiency, the addition of a surfactant can significantly improve the dust removal efficiency.

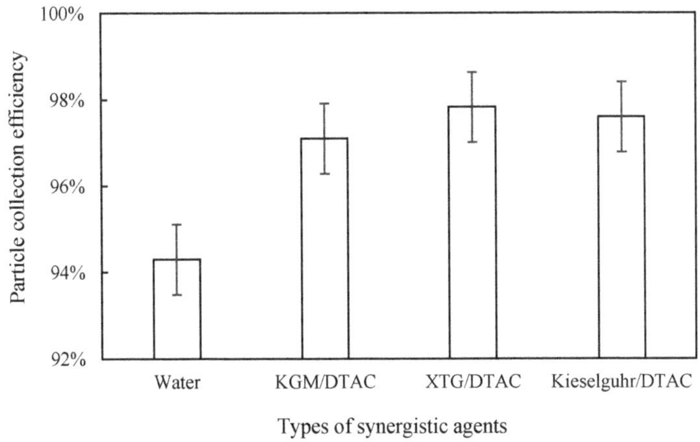

Figure 14. Effect of synergistic types on dust removal efficiency.

4. Conclusions

In this study, the effect of a chemical coagulant on blast furnace dust capture was investigated by coagulation experiments and dust removal efficiency experiments, and the synergistic effects of a chemical coagulant and a surfactant on the dust removal effect was found.

(1) All five chemical coagulants selected in this study can effectively promote the capture of blast furnace dust. The D_{50} of blast furnace dust and the proportion of large particles increased after the addition of coagulants. XTG had the best agglomeration effect on blast furnace dust, KC had the worst agglomeration effect, and their corresponding WESP removal efficiencies were 97.59% and 95.68%, respectively. With an increase in the XTG concentration, the agglomeration effect was better, and the dust removal efficiency reached its highest level when the XTG concentration was 10 mg/L.

(2) All three surfactants effectively improved the trapping efficiency of blast furnace dust in a WESP. As far as the agglomeration effect of the blast furnace dust was concerned, DTAC had the best agglomeration effect, and TX-100 had the worst agglomeration effect, corresponding to 97.08% and 96.17% removal efficiency of WESP, respectively; with an increase in the surfactant concentration, the blast furnace dust particles grew better.

(3) The synergistic effect of a chemical coagulant and a surfactant can improve the agglomeration effect and capture efficiency of furnace dust, but this effect was more obvious when the dust removal efficiency of the chemical coagulant was low. When XTG at 10 mg/L and DTAC at 9 mg/L acted synergistically, the number of respirable particles was significantly reduced, and the dust removal efficiency reached its maximum of 97.82%.

Author Contributions: Conceptualization, L.X. and Y.H.; methodology, L.X. and Y.H.; software, Y.H.; validation, L.X., Y.H. and H.C.; formal analysis, Y.H.; investigation, Y.H. and H.C.; resources, L.X.; data curation, L.X. and Y.H.; writing—original draft preparation, Y.H.; writing—review and editing, L.X.; visualization, L.X.; supervision, L.X.; project administration, L.X.; funding acquisition, L.X. All authors have read and agreed to the published version of the manuscript.

Funding: This research was supported by the Natural Science Foundation of Hebei province, China (Grant No. E2015203236). This research was supported by the Qinhuangdao science and technology research and development program (Grant No. 202003B033).

Institutional Review Board Statement: Not applicable.

Informed Consent Statement: Not applicable.

Data Availability Statement: Not applicable.

Conflicts of Interest: The authors declare no conflict of interest.

Nomenclature

d	Particle diameter
D_{10}	Diameter when the cumulative distribution of dust is 10%
D_{50}	Diameter when the cumulative distribution of dust is 50%
D_{90}	Diameter when the cumulative distribution of dust is 90%

References

1. BP. BP Statistical Review of China Energy 2015. Available online: http://www.bp.com/zh_cn/china/reports-and-publications/bp_20351.html (accessed on 30 October 2022).
2. Meij, R.; Te Winkel, B. The emissions and environmental impact of PM10 and trace elements from a modern coal-fired power plant equipped with ESP and wet FGD. *Fuel Process. Technol.* **2004**, *85*, 641–656. [CrossRef]
3. Senior, C.L.; Helble, J.J.; Sarofim, A.F. Emissions of mercury, trace elements, and fine particles from stationary combustion sources. *Fuel Process. Technol.* **2000**, *65*, 263–288. [CrossRef]
4. Pope Iii, C.A.; Burnett, R.T.; Thun, M.J.; Calle, E.E.; Krewski, D.; Ito, K.; Thurston, G.D. Lung cancer, cardiopulmonary mortality, and long-term exposure to fine particulate air pollution. *JAMA* **2002**, *287*, 1132–1141. [CrossRef] [PubMed]
5. Neas, L.M. Fine particulate matter and cardiovascular disease. *Fuel Process. Technol.* **2000**, *65*, 55–67. [CrossRef]
6. Tai, A.P.K.; Mickley, L.J.; Jacob, D.J. Correlations between fine particulate matter (PM 2.5) and meteorological variables in the United States: Implications for the sensitivity of PM 2.5 to climate change. *Atmos. Environ.* **2010**, *4*, 3976–3984. [CrossRef]
7. Avise, J.; Chen, J.; Lamb, B.; Wiedinmyer, C.; Guenther, A.; Salathe, E.; Mass, C. Attribution of projected changes in summertime US ozone and PM 2.5 concentrations to global changes. *Atmos. Chem. Phys.* **2009**, *9*, 1111–1124. [CrossRef]
8. Ylätalo, S.I.; Hautanen, J. Electrostatic precipitator penetration function for pulverized coal combustion. *Aerosol. Sci. Technol.* **1998**, *29*, 17–30. [CrossRef]
9. China Iron and Steel Industry Association: National Steel Production in June 2021. Available online: www.chinaisa.org.cn (accessed on 15 July 2021). (In Chinese)
10. China Iron and Steel Industry Association: Environmental Protection of Member Enterprises in June 2021. Available online: www.chinaisa.org.cn (accessed on 21 July 2021). (In Chinese)
11. Matino, I.; Dettori, S.; Colla, V.; Weber, V.; Salame, S. Forecasting blast furnace gas production and demand through echo state neural network-based models: Pave the way to off-gas optimized management. *Appl. Energy* **2019**, *253*, 113578. [CrossRef]
12. Zhao, Q.E. Trend analysis of sulfur in titanium slag from off-sheet titanium concentrate smelting. *Steel Vanadium Titan.* **2018**, *39*, 97–101. (In Chinese)
13. Guo, Y.H. Current status and future development trend of blast furnace gas purification and quality utilization technology. *J. Iron Steel Res.* **2020**, *32*, 525–531. (In Chinese)
14. Durham, M.D.; Schlager, R.J.; Ebner, T.G.; Stewart, R.M.; Bustard, C.J. Method and Apparatus for Decreased Undesired Particle Emissions in Gas Streams. U.S. Patent 5,893,943, 13 April 1999.
15. Johansen, A.; Schæfer, T. Effects of physical properties of powder particles on binder liquid requirement and agglomerate growth mechanisms in a high shear mixer. *Eur. J. Pharm. Sci.* **2001**, *14*, 135–147. [CrossRef] [PubMed]
16. Wei, F.; Zhang, J.Y.; Wang, C.M.; Zheng, C.G. Research progress of coal combustion ultrafine particle agglomeration promotion technology. *Coal Convers.* **2003**, *26*, 27–31. (In Chinese)
17. Zhao, Y.C.; Zhang, J.Y.; Wei, F.; Chen, J.; Zheng, C.G. Experimental study on the agglomeration promotion mechanism of ultrafine particulate matter from coal combustion. *J. Chem. Eng.* **2007**, *58*, 2876–2881. (In Chinese)
18. Rajniak, P.; Mancinelli, C.; Chern, R.T.; Stepanek, F.; Farber, L.; Hill, B.T. Experimental study of wet granulation in fluidized bed: Impact of the binder properties on the granule morphology-ScienceDirect. *Int. J. Pharm.* **2007**, *334*, 92–102. [CrossRef] [PubMed]
19. Li, H.L.; Zhang, J.Y.; Zhao, Y.C.; Ding, F.; Zhao, C.G. Experimental study on solid-liquid agglomeration of fine coal combustion particles. *Chin. J. Electr. Eng.* **2009**, *29*, 62–66. (In Chinese)
20. Campo, V.L.; Kawano, D.F.; Campo, V.L.; Kawano, D.F.; Da Silva, D.B., Jr.; Carvalho, I. Carrageenans: Biological properties, chemical modifications, and structural analysis–A review. *Carbohydr. Polym.* **2009**, *77*, 167–180. [CrossRef]
21. Forbes, E. Shear, selective and temperature responsive flocculation: A comparison of fine particle flotation techniques. *Int. J. Miner. Process.* **2011**, *99*, 1–10. [CrossRef]
22. Zhao, W.; Liu, Y.; Bao, J.J.; Geng, J.F.; Yang, L.J. Experimental study of chemical agglomeration to promote the removal of fine particulate matter from coal combustion. *Chin. J. Electr. Eng.* **2013**, *33*, 52–58. (In Chinese)
23. Thonglek, V.; Kiatsiriroat, T. Agglomeration of sub-micron particles by a non-thermal plasma electrostatic precipitator. *J. Electrost.* **2013**, *72*, 33–38. [CrossRef]
24. Liu, Y.; Yang, L.J.; Pan, D.P.; Huang, R.T. Experiments on the Removal of Fine Particles in Existing Air Pollution Control Devices by Chemical Agglomeration. *Adv. Mater. Res.* **2014**, *3384*, 756–760.

25. Balakin, B.V.; Kutsenko, K.V.; Lavrukhin, A.A.; Kosinski, P. The collision efficiency of liquid bridge agglomeration. *Chem. Eng. Sci.* **2015**, *137*, 590–600.
26. Liu, Y.; Hu, B.; Zhou, L.; Jiang, Y.; Yang, L. Improving the Removal of Fine Particles with an Electrostatic Precipitator by Chemical Agglomeration. *Energy Fuels* **2016**, *30*, 8441–8447. [CrossRef]
27. Guo, Y.; Zhang, J.; Zhao, Y.; Wang, S.; Jiang, C.; Zheng, C. Chemical agglomeration of fine particles in coal combustion flue gas: Experimental evaluation. *Fuel* **2017**, *203*, 557–569. [CrossRef]
28. Hu, B.; Yang, Y.; Zhou, L.; Ao, S.; Cai, L.; Linjun, Y.; Roszak, S. Experimental and DFT studies of PM2.5 removal by chemical agglomeration. *Fuel* **2018**, *212*, 27–33.
29. Sun, Z.; Yang, L.; Shen, A.; Zhou, L.; Wu, H. Combined effect of chemical and turbulent agglomeration on improving the removal of fine particles by different coupling mode. *Powder Technol.* **2019**, *344*, 242–250. [CrossRef]
30. Zhou, L.; Chen, W.; Wu, H.; Shen, A.; Yuan, Z.; Yang, L. Investigation on the relationship of droplet atomization performance and fine particle abatement during the chemical agglomeration process. *Fuel* **2019**, *245*, 65–77. [CrossRef]
31. Li, R.; Li, C.; Zhuang, J.; Zhu, H.; Fang, L.; Sun, D. Mechanistic Influence of Chemical Agglomeration Agents on Removal of Inhalable Particles from Coal Combustion. *ACS Omega* **2020**, *5*, 25906–25912. [CrossRef]
32. Gao, S.; Xiao, L.C. Study on Dust Turbulence-Chemical Agglomeration for Electrostatic Precipitation Technology. *E3S Web Conf.* **2021**, *245*, 03011.
33. Yang, G.Z.; Zhao, Y.C.; Xiong, Z.; Gong, B.G.; Gao, T.; Zhang, J.Y. Study on chemical agglomeration enhanced dedusting and co-desulfurization of 300MW coal-fired power plant with zero wastewater discharge. *Chin. J. Electr. Eng.* **2021**, *41*, 5274–5282. (In Chinese)
34. Lyklema, J. Modern Trends of Colloid Science in Chemistry and Biology. In *Adsorption of Polyelectrolytes and Their Effect on the Interaction of Colloid Particles*; Eicke, H.F., Ed.; Birkhäuser: Basel, Switzerland, 1985; pp. 55–73.
35. Zhang, J.G.; Liu, Y.T.; Wang, M.; Wang, Y.; Li, H.; Xie, J.; Zhao, W.; Zhou, W.; Ye, S.; Li, L.; et al. Mechanism of non-ionic surfactant effect on coal wettability based on molecular dynamics simulation. *Eng. Sci. Technol.* **2022**, *54*, 191–202. (In Chinese)
36. Nsengiyumva, E.M.; Alexandridis, P. Xanthan gum in aqueous solutions: Fundamentals and applications. *Int. J. Biol. Macromol.* **2022**, *216*, 583–604. [CrossRef] [PubMed]
37. Xu, X.J. (Ed.) *Principles of Chemical Flocculant Action*; Science Press: Beijing, China, 2005; pp. 86–92. (In Chinese)

coatings

MDPI

Article

Eco-Friendly Electrowinning for Metals Recovery from Waste Electrical and Electronic Equipment (WEEE)

Henrikas Cesiulis [1,2,*] and Natalia Tsyntsaru [1,3,*]

1 Department of Physical Chemistry, Vilnius University, LT-03225 Vilnius, Lithuania
2 JSC "Elektronikos Perdirbimo Technologijos", LT-06140 Vilnius, Lithuania
3 Institute of Applied Physics, Moldova State University, MD-2028 Chisinau, Moldova
* Correspondence: henrikas.cesiulis@chf.vu.lt (H.C.); ashra_nt@yahoo.com (N.T.)

Abstract: The amount of waste electrical and electronic equipment (WEEE) has been intensely increasing over the recent decades. In this view, the efficient recovery of metals from WEEE will allow a secure supply of raw materials and will contribute to a circular economy. Among many factors currently affecting the contribution of recycling, is the lack of suitable technologies for WEEE treatment in an environmentally friendly way. Current trends in eco-friendly technologies applied for gold, silver, copper, and tin recovery by electrowinning are reviewed in this paper. In addition, a case study on the perspectives of tin electrowinning has been evaluated. Tin can be present in rather high quantities in WEEE; moreover, its price is about three times higher than that for copper. The electrorecovery of tin has been carried out in cooperation with JSC "Elektronikos perdirbimo technologijos". The eco-friendly process based on electrowinning from citric acid-containing leachates is elaborated. The citrate-based solutions have been chosen because citric acid is considered to be an environmentally friendly component. A high deposition rate and current efficiency have been achieved at a deposition potential −0.85 V at 60 °C. However, additional steps would be beneficial to diminish the interference of metals present in the scraps, such as Pb(II) and Cu(II), on tin electrorecovery.

Keywords: electrowinning; recycling; electrical and electronic waste; leaching; tin; noble metals

Citation: Cesiulis, H.; Tsyntsaru, N. Eco-Friendly Electrowinning for Metals Recovery from Waste Electrical and Electronic Equipment (WEEE). *Coatings* **2023**, *13*, 574. https://doi.org/10.3390/coatings13030574

Academic Editors: Emerson Coy and Deling Yuan

Received: 11 January 2023
Revised: 24 February 2023
Accepted: 3 March 2023
Published: 7 March 2023

1. Introduction

The amount generated by waste electrical and electronic equipment (WEEE) is increasing dramatically, especially in economically developed countries where markets are overwhelmed by huge quantities of new electronic goods. In order to control the environmental problems, a number of EU directives has been issued [1]. In this view, the efficient recovery of metals from WEEE will allow a secure supply of raw materials and will contribute to a circular economy. Advanced circular economy systems and sophisticated recycling technologies build the backbone for the development of a resource efficient and sustainable society [2]. Recycling is also regarded as a tool that improves sustainability due to the potentially lower environmental impacts of secondary material provision in comparison with production from primary raw materials. Sustainable recycling has many advantages that include energy efficiency, less influence on air, soil, and water, etc. The full spectra of environmental parameters are included in life cycle assessments [2]. Furthermore, the diverse range of materials found in WEEE makes it difficult to give a generalized material composition for the entire waste stream in order to design different recycling chains for the same group of waste.

The key factors that currently limit the input of recycling to meet demands for raw materials can be summarized as: (1) recycling of many materials from end-of-life products and waste streams is currently not economically feasible; (2) there is a lack of suitable environmentally friendly technologies for collection and separation of valuable metals; (3) some materials are embedded in products that are in use for long time periods (e.g., buildings or wind turbines); and (4) the demand for many metals is growing [3].

The presence of precious metals in WEEE such as Au, Pt, and Pd and trends in the market prices of metals, makes it economically attractive to recycle metals from WEEE, i.e., it is wiser for e-waste to be treated to recover these precious metals. In addition, evaluating the results obtained in [4], it is evident that the percentage of precious metals in WEEE has decreased a few times during last 20 years. This fact suggests that the elaboration of new technologies for metal recovery from WEEE with relatively small percentages of valuable metals is required.

There are a number of WEEE treatment routes comprising mechanical shredding following the various chemical leaching processes to obtain the metals via reduction [5]. Recovery of leached metal via the electrowinning technique is one of the alternative ways for metal recovery. In most cases, this method does not require additional chemicals ("chemical-free") and can be considered an eco-friendly method for metal recovery.

The set-up used for electrowinning is simple in nature. Namely, a bath resistant to chemicals with anodes and cathodes submersed in a solution with current passing through the electrodes as a fundamental process unit. Besides ores, electrowinning is also used to recover valuable metals from electronic and galvanic industrial waste. The most common metals recovered using electrowinning are gold, silver, copper, cadmium, and zinc due to their relative value.

Actually, tin is also a valuable metal and its price is approx. three times higher than that for copper. The electrowinning of Sn is an attractive refining process because its electrochemical equivalent is relatively higher (6.2×10^{-4} g/C) than that for Cu (3.3×10^{-4} g/C) and Zn (3.5×10^{-4} g/C), and it is comparable with the electrochemical equivalent for Au (6.8×10^{-4} g/C). Therefore, in this paper a brief review on electrowinning of metals is provided. In addition, the case study of tin electrowinning from citrate-based baths has been carried out, and the perspectives of its industrial application are discussed here. The citrate-based solutions have been chosen since citric acid is considered an environmentally friendly component. It is food-safe, easily biodegradable, requires lower acid concentrations, and does not generate toxic fumes or hazardous waste [6]. The elaborated process of electrowinning of tin might be implemented by small and medium enterprises because of the relatively simple and affordable technologies for electrolysis and the usage of non-hazardous chemicals.

2. Brief Review on Experimental Set-Up for Electrowinning

2.1. Cathode Configurations

Electrowinning is a heterogeneous process. Therefore, the rate of mass gain on the electrodes depends on the total area of the electrodes. Hence, a number of attempts have been made to increase the specific area of the cathodes or to increase the "area of cathode per volume of solution" ratio. The simplest and conventional way to increase the mentioned ratio is the usage of a series of anodes and cathodes arranged in a "sandwich" configuration, where cathodes and anodes are typically arranged in alternating order on either side of the cathode compartments (see Figure 1). Increasing the metal concentration and solution flow rate has the largest effect on the metal deposition rate (particularly for copper) [7].

Another option for cathode configuration is spiral-wound electrodes. Theoretically, for any scale-up, the spiral electrode maintains a high specific surface area and low internal resistance [8]. In a cell with the spiral cathodes, a sufficient amount of metal ions is supplied to the cathode surface at very low concentrations of the metal. This high mass transport of ions to the cathode surface enables the cell to operate outside of the mass transport limited regime of conventional electrowinning cells [9].

The most interesting material for cathodes applied to electrowinning processes might be metal foams. The characteristics of Cu foams (see Figure 2) for electrowinning have been comprehensively studied in [10].

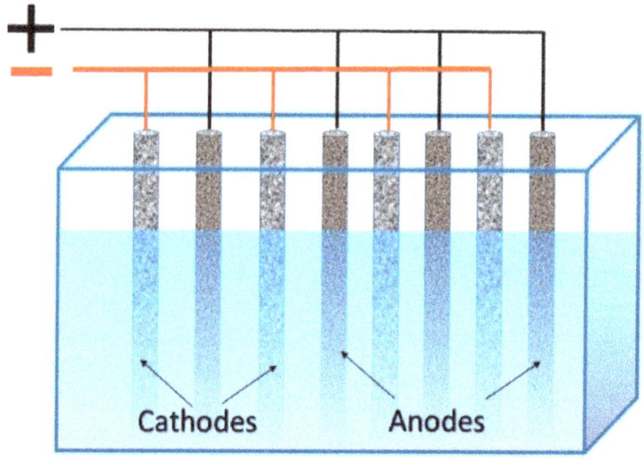

Figure 1. Arrangement of cathodes and anodes in "sandwich" configuration.

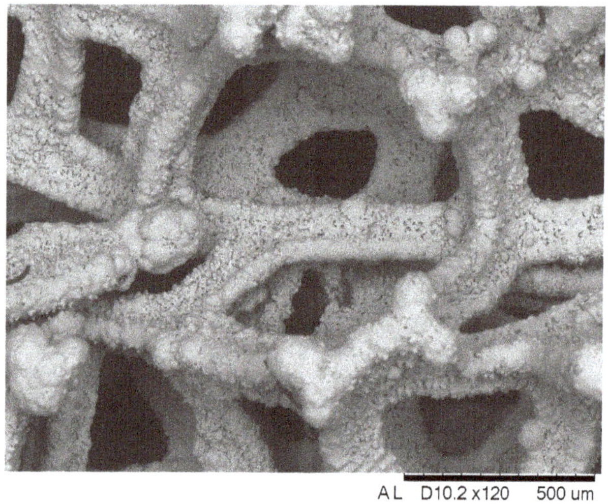

Figure 2. SEM image of 3D copper foam.

Using various electrochemical methods, it was determined that the rate-limiting step in copper deposition is the diffusion. The main processes occurring on the electrode are the charging-up of the double electric layer, charge transfer in the electrochemical reaction, and diffusion. The specific electrochemically active area of Cu foam has been estimated from electrochemical impedance spectroscopy (EIS) data, and based on the values of the double electric layer. It was shown that this area could be 7–14 times higher than that for a plane electrode. In addition, based on the EIS data, it was determined that the charge transfer resistance of the Cu foam electrode is 1.5–1.7 times lower than that of a Cu plane electrode, which results in an increase in the charge transfer rate by approximately 2 times. Based on the analysis of the diffusion impedance and chronopotentiometry data, it was found that Cu^{2+} mass transfer and the copper deposition rate was up to 3 times faster on the foam surface in comparison with a flat one having the same geometric area under the same deposition potential range.

Other types of cathodes can be used for the treatment of wastewater containing copper ions and various surfactants. In the case of anionic surfactants, the negatively charged hydrophilic parts can electrostatically attract positively charged metal ions, leading to the interference of electrolytic metal recovery [11]. Additionally, if nonionic surfactants coexist with the anionic surfactant in the wastewater, they will attract each other with their hydrophobic tails. The mixed anionic and nonionic surfactants may inhibit the mass transfer of metal ions and reduce the metal deposition rate per electrode area unit. Therefore, the use of steel wool as cathodes with enhanced active surface areas is attractive for electrowinning applications [12]. On the steel wool cathodes (Figure 3) metals could be rapidly recovered from dilute solutions with an acceptable current efficiency in comparison to a parallel-plate reactor, e.g., the time of copper recovery was shortened up to 40 min [13].

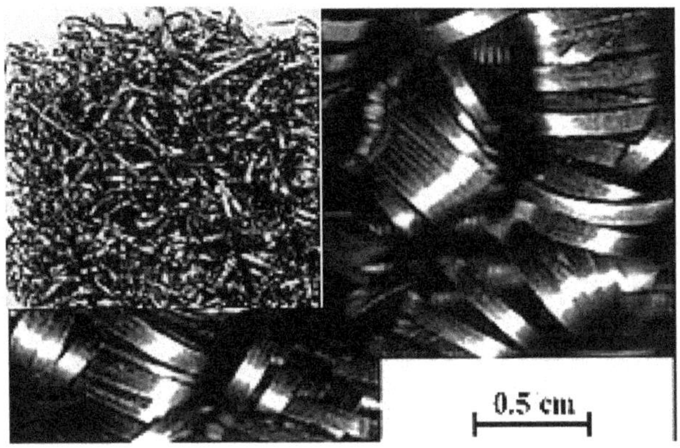

Figure 3. An example of stainless steel wool cathodes.

Another type of electrodes for electrowinning are powder cathodes. The use of powder materials makes it possible to combine the advantages of fibrous electrodes with their enhanced surface with the advantages of compact cathodes including low cost and reprocessing simplicity. Powder cathodes may be made not only of iron or titanium, but also of more electro-negative metals such as aluminum or magnesium. In turn, the use of these powders in the Au-containing solutions makes it possible to extract precious metal by electrowinning and cementation simultaneously [14].

2.2. *Solutions for Electrowinning*

Electrowinning is often combined with leaching of WEEE [15–17]. Different solutions for leaching and for electrowinning have been tested and briefly discussed in [18]. An evolving interest has grown in organic acids as leaching agents due to their biodegradability and stability under leaching conditions. The examples of organic acids used in previous studies to substitute the strong toxic leaching agents and reduce the environmental pollution include: ascorbic acid, citric acid, succinic acid, oxalic acid, formic acid, phosphoric acid, acetic acid, DL-malic acid, and L-tartaric acid [19,20].

The electrowinning of gold and silver in many cases is an important way to obtain pure metals using cyanide-based solutions of Au, and in some cases of Ag. This process had been used safely for many years, but currently the application of the cyanide-based methods is restricted in many countries. In this paper, we review some alternative solutions for gold and silver electrowinning.

Attempts have been made to implement thiosulfate-based leaching technology for industrial Au production, as it may reduce the use of cyanide. The electrowinning of leachate was carried out at rather high constant current density of ~12 A/m^2. However, the drawback of this approach is the low recovery efficiency of Au, because some amount of Au is cemented on the anodes when the electrolytic bath is off [21].

In addition, the obtained results revealed that, when an excess of thiosulfate is present in the electrolyte, cathodic side reactions, such as hydrogen evolution or thiosulfate decomposition, occur. This leads to the electrodeposition of gold layers having smaller grains and rougher surfaces in comparison with sulfite-based electrolytes. Consequently, sulfite can represent a better-controlled electroreduction reaction, resulting in the formation of finer grains, and purer and more even layers of gold [22]. A selective recovery by electrowinning of Au and Cu was described in [23]. The process was carried out from the solutions obtained from the leaching of printed circuit boards of mobile phones in an ammonia solution containing thiosulphate ions.

Another alternative method for Au recovery is electrowinning of leachates obtained during leaching of gold-containing raw materials in acidic solutions of thiourea. However, if leachate contains gold at low concentrations, the electrowinning process is uneconomical. Instead, the process in which Au is specifically adsorbed in the form of gold–thiourea complexes onto activated carbon and eluted by dilute acidic thiourea solutions containing alcohol can be adopted. Such eluates contain 5–50 mg dm^{-3} of Au or more, which can then be recovered by electrowinning [24].

Gold and silver recovery by electrowinning from various electronic connector waste was studied in [25]. Thus, from the solution containing ~100 mg/L of Au and Ag, the metals were successfully recovered by electrowinning with high yields (89.95% for Au and 87.98% for Ag) after 1.5 h of electrolysis. Such an effective process could be achieved due to a copper cathode, which favors reduction processes for both noble metals. The only issue when using a copper electrode is its proneness to react with thiourea and decompose it. This will lead to a 20% decrease in thiourea concentration at the end of the process, whereas it was possible to achieve 95% recovery on a graphite cathode within 1.5 h at room temperature.

The recovery of Ag by electrowinning is also considered in the case of treatment of spent batteries. Thus, an environmentally friendly process for recycling of such batteries was developed in [26]. A high electrowinning rate and recovery efficiency (98.5%) of ultra-pure Ag (≥99.9%) was achieved under optimized operating potentials. The elaborated process can be considered as a sustainable technology for precious metal recovery from scraps containing silver oxide batteries.

One of the effective way to recover Cd from metallic mixtures is the hydrometallurgical process described in [27]. Sulfuric acid is often used as the most common agent to dissolve base metals, and in this case, it is possible to regulate different levels of impurities. Namely, it is possible to selectively leach Cd and Zn from a mixture of Cd, Zn, Cu, and Pb. After the leaching, Cd and Zn dissolve and remain in the filtrate, while Cu and Pb are precipitated and can be separated by filtration. Then, the metallic Cd will be obtained by cementation with zinc dust. In order to remove any co-precipitated Zn, electrowinning can be performed.

Currently, high purity tin is generally produced by an electrolytic method using an acidic tin solution such as from leachates based on sulfuric acid [28], alkaline solutions [29], or a mixture of fluorosilicic acid and sulfuric acid [30]. Notably, Sn can be separated from sulfuric acid leachate containing Cu by its precipitation in the form of SnO_2 at a pH of 3, and the pure Cu can be recovered by electrowinning [31]. Citric acid-based solutions are very promising for metal leaching and electrowinning due to its biodegradability. Citrate ions form stable complexes with many metals and it is widely used for metal and alloy electrodeposition, including electrowinning of metals after treatment of WEEE (e.g., Cu [32], Ni [33], Sn [34], Pb and Zn [35], Sn-Zn [36], Cu-Sn [37], FeCoNiCr [38]). The agents facilitating metals leaching are EDTA [39] and H_2O_2 [40], which can also be used for electrowinning.

3. Results

3.1. Methodology of Experiments

Two leachates of base metals ("L1" and "L2") were prepared on the premises of JSC *Elektronikos Perdirbimo Technologijos* (EPT) throughout WEEE treatment. Namely, leachates "L1" and "L2" were obtained after treatment of telecommunication boards and random access memory (RAM) cards, respectively. It should be noted that the scrap had been shredded up into 1–5 mm size pieces. Then, the leaching was carried out in the rotating reactor RRM 7 (producer EMAK, Turkey). The load of 7 kg of scrap was treated for 2 h at

60 °C in the leaching solution with a total volume 12 L, which contained 200 g/L citric acid and 10 g/L ammonium persulfate at pH 3.2–6.5. The pH of the solutions had been adjusted using sodium hydroxide.

In addition, for comparison, a reference solution ("Ref") containing citric acid (200 g/L) and Sn (II) (20 g/L) at pH 3.5 was prepared and evaluated (see Table 1).

Table 1. The composition of investigated solutions. Leachates "L1" and "L2" are obtained after treatment of WEEE.

		Composition of g/L					
Solution	**Tag**	**Pb**	**Sn**	**Cu**	**Al**	**Zn**	**Ni**
Sn–citric acid	"Ref"	-	20	-	-	-	-
leachate	"L1"	1.63	20.38	0.02	0.06	0.10	0.09
leachate	"L2"	1.36	4.80	2.29	0.02	0.031	0.10

The composition of the leachates was determined using inductively coupled plasma optical emission spectroscopy, ICP-OES, using a Perkin Elmer Optima 7000 DV spectrometer (Waltham, MA, USA). The composition of leachates is provided in Table 1. The main difference between "L1" and "L2" arise from the Cu(II) concentration, which is comparable with Sn(II) for the last case.

The electrochemical studies at a lab scale were carried out in a standard 3-electrode cell using a programmable potentiostat/galvanostat AUTOLAB PGSTAT 302N (Metrohm, Utrecht, the Netherlands).

All potentials were recorded and presented vs. a saturated Ag/AgCl/KCl electrode. The stainless steel (type 304) plates served both as the anode and cathode. In the last case, the working area was 4 cm^2. After electrolysis, the deposits were evaluated by SEM equipped with EDS for structural and compositional analyses.

3.2. Perspectives of Tin Electrowinning from Citrate-Based Solutions

The cathodic current densities for Sn electroreduction from the "Ref" solution were rather high at the potentials more negative than −0.7 V at 60 °C (see Figure 4).

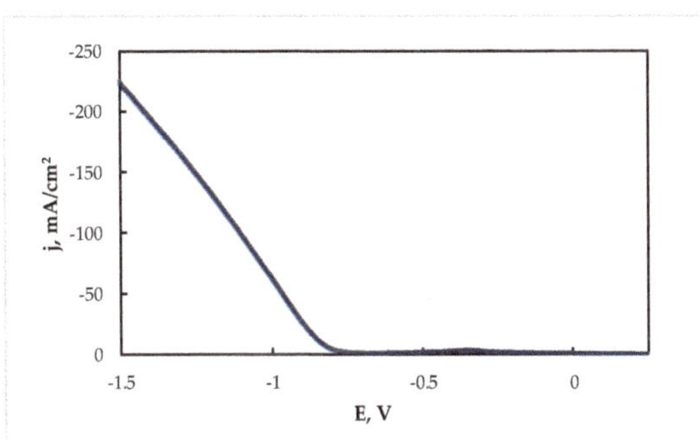

Figure 4. Cathodic polarization curve of Sn electroreduction at 60 °C from the "Ref" solution (g/L): Sn(II)—20; citric acid—200.

Furthermore, the electrowinning of tin under potentiostatic conditions at −0.85 V can be considered as optimal, because a high deposition rate (0.0035 g cm^{-2} min^{-1}) and current efficiency (86%) were achieved. Under these conditions, the current increased sufficiently with deposition time (Figure 5).

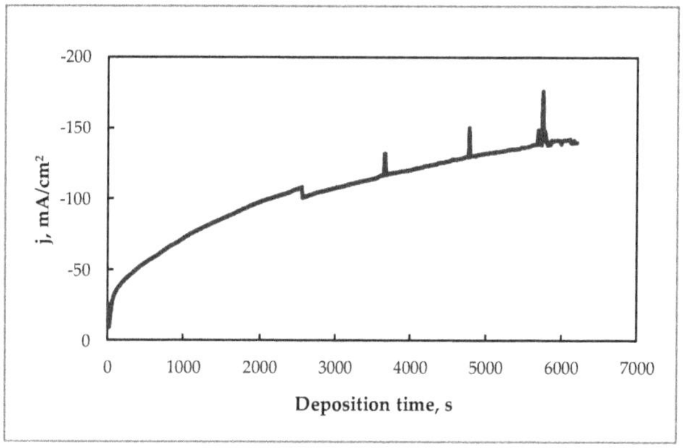

Figure 5. Chronoamperogram of Sn electrodeposition at E = −0.85 V and 60 °C from the citrate solution. Content of the solution as indicated in Figure 4, substrate—Cu electrode.

The rise of apparent current density during electrodeposition was triggered by the formation of a coarse crystalline deposit on the cathode, as shown in Figure 6. The growing of such crystallites increases the active area of the cathode, therefore the current increased under potentiostatic deposition conditions.

Figure 6. SEM image of Sn electrodeposit obtained at E = −0.85 V and 60 °C from the solution "Ref".

The solution "L1" obtained after treatment of WEEE contained a similar concentration of Sn(II) as the "Ref" solution, namely ~20 g/L (see Table 1). However, irrespectively of the pH, the obtained cathodic current densities were essentially lower, almost 10 times, than in the "Ref" solution, as is shown in Figure 7.

The metal deposition rates from leachate "Ll" throughout the studied range of pH was also ~10 times lower. The elemental analysis of the deposits obtained at −0.85 V showed that the main component in the deposits was lead (see Table 2).

This means that Pb inhibited the electroreduction of Sn(II) in the citrate solutions, although at pH 3.2 the content of Sn was essentially increased up to 30.8%, compared to higher pHs. Further studies are needed in order to find the optimal conditions for electrowinning from the given leachate, i.e., when the electrodeposits ratio of Sn to Cu and Pb content would be closer to their concentration ratio in the solution.

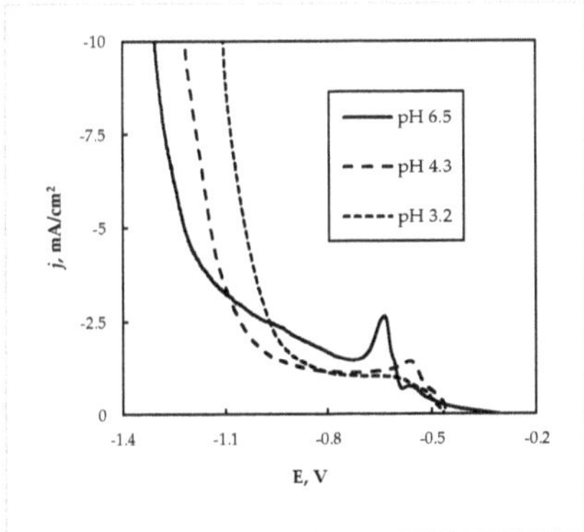

Figure 7. Cathodic polarization curve on Sn electrode at 60 °C and scan rate 5 mV/s from the leachate "L1".

Table 2. The composition of deposits obtained at −0.85 V and 60 °C from leachate "L1"at various pHs.

Metal	Content of Elements in the Deposit at Various pHs, wt.%		
	6.5	4.3	3.2
Al	0.26	0.18	0.31
Cu	0.18	0.46	0.52
Pb	88.3	92.0	68.4
Sn	11.23	7.40	30.8

The leachate "L2" was obtained from the selected leaching of WEEE, namely from the waste of printed circuit boards, and therefore it contains essential contents of Cu(II) and Sn(II), and the content of Sn(II) is higher than Cu(II) (see Table 1). As shown in Figure 8, in this case, the cathodic current strongly depends on temperature, and at 60 °C the currents were 2–3 times higher than those obtained at 20 °C. Therefore, performing electrowinning at elevated temperatures will be beneficial for the process. The potentiostatic conditions were also applied in this case. The deposition at higher cathodic potentials (−0.85 V and higher) led to the formation of crumble powder from the electrode, which is an unacceptable result from a practical point of view.

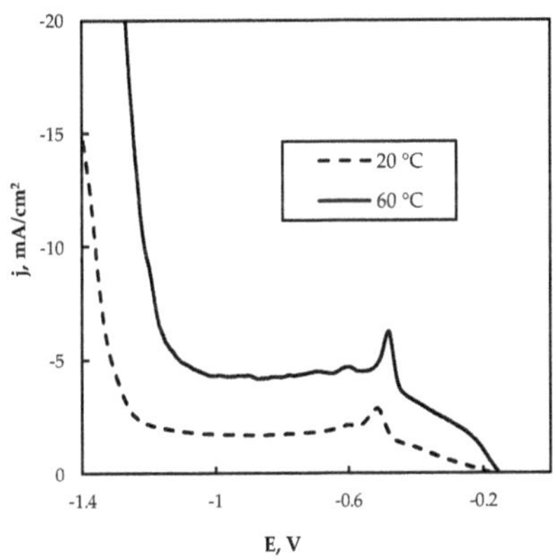

Figure 8. Cathodic polarization curve on Cu electrode at 20 °C and 60 °C from the leachate "L2"; potential scan rate 5 mV/s, pH 6.5.

23. Kasper, A.C.; Carrillo, A.J.; García Gabaldón, M.; Veit, H.M.; Pérez, H.V. Determination of the potential gold electrowinning from an ammoniacal thiosulphate solution applied to recycling of printed circuit board scraps. *Waste Manag. Res.* **2016**, *34*, 47–57. [CrossRef]
24. Urbanski, T.S.; Fornari, P.; Abbruzzese, C. Gold electrowinning from aqueous–alcoholic thiourea solutions. *Hydrometallurgy* **2000**, *55*, 137–152. [CrossRef]
25. Ippolito, N.M.; Birloaga, I.; Ferella, F.; Centofanti, M.; Vegliò, F. Preliminary study on gold recovery from high grade e-waste by thiourea leaching and electrowinning. *Minerals* **2021**, *11*, 235. [CrossRef]
26. Wang, Z.; Peng, C.; Yliniemi, K.; Lundström, M. Recovery of high-purity silver from spent silver oxide batteries by sulfuric acid leaching and electrowinning. *ACS Sustain. Chem. Eng.* **2020**, *8*, 15573–15583. [CrossRef]
27. Safarzadeh, M.S.; Bafghi, M.S.; Moradkhani, D.; Ilkhchi, M.O. A review on hydrometallurgical extraction and recovery of cadmium from various resources. *Miner. Eng.* **2007**, *20*, 211–220. [CrossRef]
28. Takemoto, K.; Imori, T.; Ouchi, T.; Takahashi, H. Method For Manufacturing High Purity Tin, Electrowinning Apparatus for High Purity Tin and High Purity Tin. U.S. Patent 20160097139, 2016. Available online: https://www.freepatentsonline.com/y2016/0097139.html (accessed on 27 December 2022).
29. Kanazawa, T.; Morimoto, K.; Yamaguchi, I. Tin Electrowinning Method. Japan Patent JP2020117798A, 6 August 2020.
30. Ishiwatari, M. Method and apparatus for electrolytic purification of high purity tin. Japan Patent JP3882608B2, 21 February 2020.
31. Guo, X.; Qin, H.; Tian, Q.; Li, D. Recovery of metals from waste printed circuit boards by selective leaching combined with cy-clone electrowinning process. *J. Hazard. Mater.* **2020**, *384*, 121355. [CrossRef] [PubMed]
32. Jin, L.; Li, W.Q.; Wang, Z.Y.; Yang, Y.J.-Q.; Zheng, A.-N.; Yang, F.-Z.; Zhan, D.; Wu, D.-Y.; Tian, Z.-Q. Novel, simple, and green citrate-based copper electronic electroplating bath in microvia void-free filling for printed circuit board application. *ACS Sustain. Chem. Eng.* **2022**, *10*, 14204–14211. [CrossRef]
33. Bigos, A.; Wolowicz, M.; Janusz-Skuza, M.; Starowicz, Z.; Szczerba, M.J.; Bogucki, R.; Beltowska-Lehman, E. Citrate-based baths for electrodeposition of nanocrystalline nickel coatings with enhanced hardness. *J. Alloys Compd.* **2021**, *850*, 156857. [CrossRef]
34. Sharma, A.; Min Park, Y.; Lee, S.; Ahn, B. Morphology, resistivity and corrosion behavior of tin coatings plated from citric acid bath. *Mater. Res. Express* **2019**, *6*, 116589. [CrossRef]
35. Halli, P.; Agarwal, V.; Partinen, J.; Lundström, M. Recovery of Pb and Zn from a citrate leach liquor of a roasted EAF dust using precipitation and solvent extraction. *Sep. Purif. Technol.* **2020**, *236*, 116264. [CrossRef]
36. Tsurusaki, T.; Ohgai, T. Mechanical properties of solder-jointed copper rods with electrodeposited Sn-Zn alloy films. *Materials* **2020**, *13*, 1330. [CrossRef]
37. Ding, L.; Chen, C.; Li, Q.; Yuan, J.; Li, H.; Xue, Y.; Dong, H.; Li, B.; Niu, Y. Process and theoretical research on electroplating Cu–Sn alloys of low Sn. *J. Appl. Electrochem.* **2021**, *51*, 1287–1299. [CrossRef]
38. Guo, F.Y.; Yu, J.K.; Xiao, J.J.; Qiao, Q.; Yang, H.B.; Guo, Y.Q. Preparation of FeCoNiCr high entropy alloy coatings and optimization of process parameters. *Rare Metal. Mat. Eng.* **2021**, *50*, 2337–2342.
39. Deng, D.; Deng, C.; Liu, T.; Xue, D.; Gong, J.; Tan, R.; Mi, X.; Gong, B.; Wang, Z.; Liu, C.; et al. Selective recovery of copper from electroplating sludge by integrated EDTA mixed with citric acid leaching and electrodeposition. *Sep. Purif. Technol.* **2022**, *301*, 121917. [CrossRef]
40. Jadhav, U.; Su, C.; Hocheng, H. Leaching of metals from large pieces of printed circuit boards using citric acid and hydrogen peroxide. *Environ. Sci. Pollut. Res.* **2016**, *23*, 24384–24392. [CrossRef] [PubMed]

coatings

MDPI

Article

Lignin and Starch Derivatives with Selenium Nanoparticles for the Efficient Reduction of Dyes and as Polymer Fillers

Anna Modrzejewska-Sikorska [1,*], Mariola Robakowska [2,*], Emilia Konował [1], Hubert Gojzewski [3], Łukasz Gierz [4,*], Bartosz Wieczorek [4], Łukasz Warguła [4] and Wiktor Łykowski [4]

1 Faculty of Chemical Technology, Institute of Chemistry and Technical Electrochemistry, Poznan University of Technology, Berdychowo 4, 60-965 Poznan, Poland; emilia.konowal@put.poznan.pl
2 Faculty of Chemical Technology, Institute of Chemical Technology and Engineering, Poznan University of Technology, Berdychowo 4, 60-965 Poznan, Poland
3 Sustainable Polymer Chemistry, Department of Molecules & Materials, Faculty of Science and Technology, University of Twente, 7522 NB Enschede, The Netherlands; h.gojzewski@utwente.nl
4 Faculty of Mechanical Engineering, Institute of Machine Design, Poznan University of Technology, Piotrowo 3, 60-965 Poznan, Poland; bartosz.wieczorek@put.poznan.pl (B.W.); lukasz.wargula@put.poznan.pl (Ł.W.); wiktor_lykowski@o2.pl (W.Ł.)
* Correspondence: anna.modrzejewska-sikorska@put.poznan.pl (A.M.-S.); mariola.robakowska@put.poznan.pl (M.R.); lukasz.gierz@put.poznan.pl (Ł.G.); Tel.: +48-61-6652311 (A.M.-S.); +48-61-6653683 (M.R.); +48-61-2244516 (Ł.G.)

Abstract: Selenium nanoparticles (SeNPs) were synthesized and stabilized by biopolymers, namely, sodium lignosulfonate (LS) and starch sodium octenyl succinate (OSA). The obtained selenium nanoparticles were studied for their catalytic activity in the reduction of a dye (C.I. Basic Blue 9, methylene blue) by sodium borohydride. The SeNPs-OSA and SeNPs-LS nanoparticles were also dispersed in a photosensitive matrix and studied as polymer composites. The research confirmed the catalytic abilities of the prepared SeNPs in the reduction of the organic dye. Mechanical tests on the polymers and their composites showed an improvement in the composites' strength in all tested cases. An increase in hardness and Young's modulus values of the filled materials compared to the pure matrix was found as well.

Keywords: selenium nanoparticles; lignosulfonate; starch sodium octenyl succinate; biopolymer; organic dye

Citation: Modrzejewska-Sikorska, A.; Robakowska, M.; Konował, E.; Gojzewski, H.; Gierz, Ł.; Wieczorek, B.; Warguła, Ł.; Łykowski, W. Lignin and Starch Derivatives with Selenium Nanoparticles for the Efficient Reduction of Dyes and as Polymer Fillers. *Coatings* **2023**, *13*, 1185. https://doi.org/10.3390/coatings13071185

Academic Editor: Shoufeng Tang

Received: 24 May 2023
Revised: 23 June 2023
Accepted: 28 June 2023
Published: 30 June 2023

1. Introduction

Selenium nanostructures (SeNPs) are particles of elemental selenium or its aggregates with the size in the order of nanometers. SeNPs can take a spherical form, but also appear as nano-ribbons or nano-rods [1,2]. SeNPs find a wide range of applications in many fields of science, industry, and technology. They unveil, similar as other nanoparticles, interesting chemical and physical properties and characteristics, i.e., catalytic, thermoelectric, optical, and piezoelectric properties, as well as very good photoconductivity [3–7]. For this reason, they are used in electronics and xerography, in the utility glass industry, biomedicine, pharmacy, as well as biotechnology and, due to their high antioxidant capacity, in the packaging industry, as additives to materials whose main component is plastics.

The desired properties of SeNPs, such as size, shape, and surface charge, are driven by the choice of the processing routes—physical, chemical, and biological [8–12]. The physical routes can employ, for instance, solvothermal synthesis, microwave treatment, electrodeposition, or pulsed laser ablation using an excimer laser [8,9]. The chemical routes can employ reduction, i.e., reducing selenium salts to stable colloidal suspensions. This reduction is most often carried out with the participation of surfactants or ascorbic acid [11]. This method requires the appropriate selection of a stabilizer, a reducing agent, and a solvent. However, some chemical reactions accompanying the production of selenium nanostructures may adversely affect the environment. The biological route usually involves an alternative processing, which includes biological and green synthesis; both are simple and cheap and lack a negative environmental impact. The biological synthesis uses bacteria that reduce a selenium ion to a nanoparticle, while the green synthesis uses metabolites from plant extracts that can act as potential selenium-reducing agents. In addition, fungi can also be used in the production of SeNPs [2,10–12].

Lignin and starch are crucial renewable feedstocks. Especially, lignin is seen as a candidate for replacing non-biodegradable fossil-derived plastics, as it is the second abundant biopolymer in the world [13,14]. The use of lignin or starch derivates, such as sodium lignosulfonate and starch octenyl succinate, in the preparation of SeNPs can require a specific combination of a chemical method and green chemistry. Both types of these bio-based compounds can act as stabilizers for the selenium nanostructures produced [15,16]. The use of sodium lignosulfonate as a stabilizer allows obtaining SeNPs in liquid form, while the use of a starch octenyl succinate results in selenium nanostructures on a solid matrix.

Lignosulfonates (LSs) are anionic polyelectrolytes containing a significant number of sulfonic groups. They are produced in a large amount—approx. 1 million tons of dry matter per year worldwide [17]. The structural and chemical properties of lignosulfonates include their relatively high molecular weight (15,000–50,000 g/mol), solubility in water, a high degree of polydispersity, and a high ash content [18]. Due to the presence of hydrophobic chains and numerous hydrophilic functional groups in their structure, in aqueous solutions they have surface-active properties [19]. Many potential industrial applications [20,21] are linked to the presence of a multitude of functional groups such as carboxyl, phenolic, and sulfonic groups in the structure of LSs [22–24]. According to the literature, this is the reason why LSs have already been used, among others, as reducers and stabilizers in the manufacturing of gold and silver nanostructures [25–27]. Water-soluble lignosulfonates (LSs) were used for a simple and inexpensive synthesis of gold [25], while silver [26] nanoparticles in aqueous solutions at room temperature during the reaction with gold or silver salts. As a possible application, the obtained nanoparticles were studied for their catalytic activity in the reduction of organic dyes by sodium borohydride [26].

Starch $(C_6H_{10}O_5)_n$ is a glucose polymer, specifically, a biopolymers of a plant origin. Due to the presence of free hydroxyl groups in its structure, it can be relatively easily modified. One of the modified starches is sodium starch octenyl succinate (OSA). The synthesis of OSA starch was patented by Caldwell and Wurzburg [28,29]. The synthesis requires a cyclic octenyl succinic acid anhydride and starch under basic conditions. The obtained starch derivate has favorable emulsifying and stabilizing properties due to the presence of both hydrophobic and hydrophilic groups on its surface.

Synthetic dye pollutants in water are hazardous for plants, animals, and humans and cause environmental problems [30]. The removal of the synthetic dye contaminants pose both scientific and industrial challenges [31]. However, it was suggested that SeNPs are among the most promising agents for their efficient catalytic activity in degrading hazardous dyes such as methylene blue and bromothymol blue and others organic compounds in water. SeNPs are also attracting increasing attention due to their low toxicity, as well as their potential use in disease prevention and treatment [32].

Using SeNPs as fillers in polymer (nano)composites to fine-tune the characteristics of polymer matrices is of interest as well [33,34]. Different polymer composites with selenium are reported, e.g., in the form of polymeric films as well as of polymers containing polyvinyl alcohol-stabilized selenium nanoparticles [33], polymer nanocomposites of selenium biofabricated using fungi [34], or composite materials for prosthetic dentures based on acrylic reinforced with nano-selenium [34]. In the study by Shah [33], it was found that the introduction of selenium nanoparticles caused a reduction in the mechanical properties (transverse, impact, and hardness) of an acrylic polymer, but the SeNPs were found to be an interesting material for anti-pollution processing (catalysis) and to design/control the physicochemical properties of polymer matrices.

This work aimed to use SeNPs as a catalyst and a filler of polymer matrices. For this purpose, SeNPs were firstly prepared and stabilized with starch and lignin derivatives. Second, SeNPs were used as a catalyst in the reduction process of a selected organic dye and as a filler for polymer matrices. The materials were tested mechanically (strength properties), thermally (thermal stability), and microscopically (surface morphology was examined to determine the dispersion of the filler). Our results provide new insights for the application of selenium nanoparticles.

2. Materials and Methods

2.1. Materials

Selenium (IV) oxide and ascorbic acid were received from Sigma Aldrich (Saint Louis, MO, USA), sodium lignosulfonate (LS) from softwood, molecular weight 6400 Da, was received from Borregaard LignoTech (Sarpsborg, Norway), and Sylobloc K500 silica was received from Grace GmbH & Co. (Columbia, MD, USA) KG. The monomers, 2-hydroxyethyl methacrylate (HEMA) and polyethylene glycol diacrylate, with MW = 575 (PEGDA), were purchased from Sigma Aldrich (Saint Louis, MO, USA) and were purified by column chromatography before use. The photoinitiator, 2,2-dimethoxy-2-phenylacetophenone (Irgacure 651) was also purchased from Sigma-Aldrich (Saint Louis, MO, USA).

2.2. Methods

2.2.1. Preparation of Sodium Starch Octenyl Succinate

Sodium starch octenyl succinate (OSA) containing 2.5% of octenyl succinate groups was synthesized according to the procedure described in [35].

2.2.2. Synthesis of SeNPs

The synthesis of colloidal selenium was carried out by chemical reduction using ascorbic acid as a reducer of the selenium compounds, selenium (IV) oxide and OSA, with LS as a stabilizers of the obtained SeNPs. The use of two different stabilizers made it possible to obtain selenium nanoparticles SeNPs in two forms, i.e., starch octenyl succinate provided selenium nanoparticles (SeNPs-OSA) on a solid matrix (=starch derivate), while the use of sodium lignosulfonate allowed obtaining SeNPs-LS in liquid form, which were used to impregnate the commercial silicon dioxide Sylobloc K500 in the next step.

A selenium solution with a concentration of 5.5 g/L was used for the study. Details about the preparation of the selenium nanoparticles: the concentration and the volume of SeO_2 were, respectively, 5.5 g/L and 40 mL, the concentration and the volume of the reducer $C_6H_8O_6$ were, respectively 42.2 g/L and 20 mL, the amount of the first stabilizer, sodium starch octenyl succinate (OSA), was 3 g, the concentration and the volume of the second stabilizer, a sodium salt of lignosulfonic acid (LS), were, respectively 18.75 g/L and 13.33 mL.

The resulting mixtures were stirred for 24 h at room temperature and then allowed to stand quiescent for a few days to complete the reaction. After this time, the resulting selenium nanoparticles were subjected to a physicochemical analysis. The obtained liquid selenium nanoparticles were also deposited on the commercial silica Sylobloc K500. The details of the prepared selenium nanoparticles are shown in Table 1.

2.2.3. Preparation of SeNPs–Matrix/Polymer

To obtain the SeNPs–carrier/polymer composite, a mixture of two monomers, i.e., 2-hydroxyethyl methacrylate (HEMA) and polyethyl glycol diacrylate (PEGDA) in a 2:3 ratio (by weight), was prepared. PEGDA was used as a crosslinking agent for the photopolymerization of HEMA in order to obtain HEMA/PEGDA-based polymers [36]. Following this preparation, the 2,2-dimethoxy-2-phenylacetophenone photoinitiator was added (1 wt% relative to the monomer mixture). SeNPs-OSA and SeNPs-LS842-Sylobloc K500 in the amounts of 1%, 3%, and 5% wt.% were added to mixture of the monomers. The homogenization process was carried out using a mechanical shaker (15 h, 3000 1/min) and an ultrasonic bath (24 h, 50 kHz, 310 W). Samples of the polymer composite (dimensions: diameter 12 mm, thickness 1 mm) were obtained by the photopolymerization of the mixtures in a mold. The samples were irradiated for 3 min on both sides through PET foils with the full spectrum (320–395 nm) of a DYMAX-Blue Wave 50 lamp (Torrington, CT, USA).

Table 1. The details of the selenium nanoparticles.

Sample Name	Synthesis and Composition	
SeNPs-LS842 (liquid sample)	40 mL (5.5 g/L) SeO_2 20 mL (42.2 g/L) $C_6H_8O_6$—reducer 13.33 mL (18.75 g/L) sodium salt of lignosulfonic acid (LS)—liquid stabilizer	The components were stirred for 24 h at room temperature and then allowed to stand quiescent for a few days to complete the reaction.
SeNPs-OSA (solid sample)	40 mL (5.5 g/L) SeO_2 20 mL (42.2 g/L) $C_6H_8O_6$—reducer 3 g sodium starch octenyl succinate (OSA)—solid stabilizer	
SeNPs-LS842-Sylobloc K500 (solid sample)	40 mL (5.5 g/L) SeO_2 20 mL (42.2 g/L) $C_6H_8O_6$—reducer 13.33 mL (18.75 g/L) sodium salt of lignosulfonic acid (LS)—liquid stabilizer Commercial silica Sylobloc K500 used as a solid matrix	

2.2.4. SEM—Surface Composition and Imaging

A Scanning Electron Microscope (S-3400N, Hitachi, Tokyo, Japan) equipped with the EDS Ultradry Analysis Attachment (Thermo-Scientific, Waltham, MA, USA) was used. Based on energy-dispersion microanalysis, the selenium content was determined on individual matrices, and additionally, SEM micrographs of the solidified samples were acquired.

2.2.5. Catalytic Properties

The catalytic properties of the selenium colloids were checked using the USB 4000 UV–VIS spectrophotometer (Ocean Optics, Dunedin, FL, USA). In our study, methylene blue (C.I. Basic Blue 9) was used as a dye, and sodium borohydride ($NaBH_4$) was used as a reducer.

These properties were investigated for the SeNPs-LS842 colloidal system as well as for the SeNPs-OSA and SeNPs-LS842-Sylobloc K500 systems. The analysis was performed using methylene blue and sodium borohydride. In the case of a liquid sample—SeNPs-LS842 (see sample coding later on)—20 μL of selenium, 2 mL of the dye (4 mg/L), and 50 μL of sodium borohydride (10 g/L) were added to a measuring cuvette. Then. the absorption bands in the UV spectrum were recorded every 30 s. The study of nanoparticles in a solid form—SeNPs-OSA and SeNPs-LS842-Sylobloc K500—involved placing them in vials together with the dye and sodium borohydride (in an amount analogous to that of the liquid colloids) and then observing the color change of the organic dye.

2.2.6. Thermogravimetric Analysis

The thermogravimetric analysis (TGA) was performed using a PerkinElmer Pyris 1 TGA instrument (NETZSCH–Geratebau GmbH, Germany). The sample holder made of Pt was filled with approximately 10 mg of material. Mass loss measurements as a function of the temperature were performed from 25.0 to 590.0 °C, at a heating rate of 10.0 °C/min under a nitrogen atmosphere.

2.2.7. Mechanical Properties

The mechanical properties of the composite materials were studied at room temperature by testing the Young's modulus, tensile strength, elongation at break, and hardness according to PN-EN ISO 527-1:1998 (crosshead speed of 5 mm/min) with a Zwick/Roell universal testing machine, model Z020 (Zwick GmbH & Co. KG, Ulm, Germany). The

Shore A hardness was measured according to DIN 53 505. The results were averaged (at least 6 individual measurements for each sample).

2.2.8. Optical Microscopy

Optical microscopy images were collected using an Olympus BX60 microscope (Tokyo, Japan). The samples were fractured at room temperature.

2.2.9. Atomic Force Microscopy (AFM)

Following the optical microscopy analysis, the samples' surface was studied by AFM in air and at room temperature, using a MultiMode 8 instrument with a NanoScope V controller (Bruker, Billerica, MA, USA). The AFM was operated in the PeakForce Quantitative Nanomechanical Mapping mode (PF-QNM); thus, force–distance curves were recorded (512×512 per map) and further processed (NanoScope Analysis software, version 1.9). During scanning, the ScanAsyst control for the feedback loop gain and applied load was set to "off" to apply the chosen constant scanning parameters. The force–distance curves were collected following a sine-wave sample-tip trajectory with a frequency of 2 kHz and utilizing a peak-force amplitude value of 100–150 nm. Medium-soft cantilevers were chosen to conduct the imaging (nominal spring constant of 5 N/m, nominal radius of 10 nm, Budgetsensors, Tap150-G, Sofia, Bulgaria). The AFM optical sensitivity (deflection sensitivity) was calculated based on the thermal tune method [37].

3. Results and Discussion

In order to analyze and characterize the obtained biocomposites, i.e., the starch–selenium nanostructures (SeNPs-OSA) and the silica–lignosulfonate–selenium nanostructures (SeNPs-LS842-Sylobloc K500), tests of their surface structure and elemental composition (SEM, EDS, AFM analyses), stability (TGA), and selected mechanical properties, such as Young's modulus, tensile strength, elongation at break, and hardness, were carried out. The ability of the biocomposites to catalyze reduction reactions of selected dyes in model aqueous solutions was also studied.

3.1. Energy Dispersion Microanalysis (EDS) and SEM Imaging (Filler)

On the one hand, a significant amount of selenium (5.09%) was found by energy dispersion microanalysis (EDS) on the starch derivative (SeNPs-OSA, stabilizer OSA) (Figure 1). The EDS mapping of the sample area showed that the distribution of selenium on the surface of both starch and silica grains was uniform.

On the other hand, the impregnation of silicon dioxide with the selenium nanoparticles, where the stabilizer of the produced nanoparticles was a water-soluble lignin derivative (sodium lignosulfonate LS842), allowed for much smaller amounts of SeNPs (0.23%) on the solid matrix Sylobloc K500. Therefore, when the goal is to obtain solid selenium, it is much more advantageous to use the stabilizer in the correct form. The SEM micrographs indicated the shape and size of the obtained composites containing modified starch or silica and SeNPs. They showed small particles with diameters in the range of 10–50 µm for SeNPs-OSA and smaller particles for the SeNPs-LS842-Sylobloc K500 systems, with a diameter below 5 µm (Figure 2). Selenium nanoparticles were deposited on the surface of starch or silica grains (which was confirmed by the EDS analysis). The recorded starch grain size was characteristic of potato starch grains. Visible damage to the middle part of the grain (cup-shaped grain) was caused by the chemical modification of the native starch.

3.2. Selenium as a Catalyst for Reduction Reactions

In chemistry, one of the strongest and most effective reducing agents is sodium borohydride, but many reduction reactions are kinetically inhibited. These kinds of reactions require a longer time to be completed or a larger amount of a reducer. However, sodium borohydride has low stability and auto-decomposes in aqueous media. The problem can be solved by introducing nanostructures of elements such as gold, silver, or selenium into the system [26].

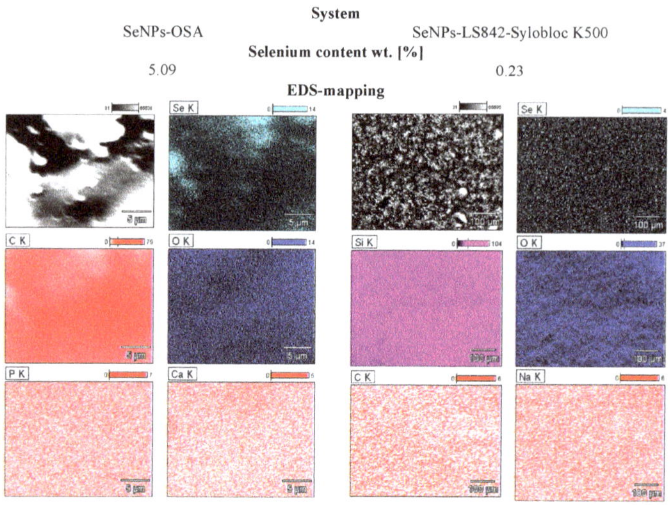

Figure 1. Selenium content in individual matrices—EDS analysis.

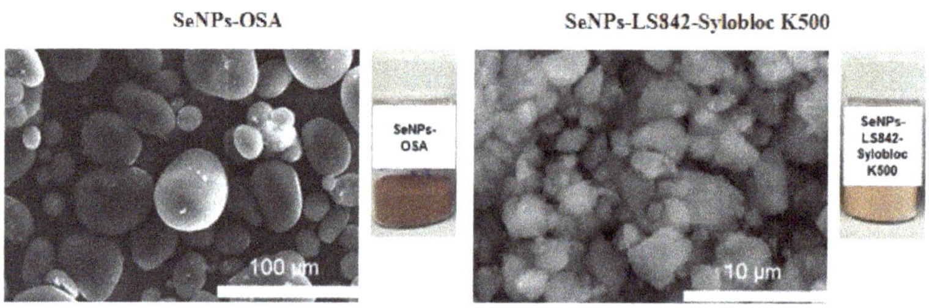

Figure 2. Selenium content in individual matrices—SEM pictures and photographs of the obtained samples.

The measured UV–Vis absorbance spectra of the colloid in liquid form (SeNPs-LS842) confirmed the catalytic ability of the produced selenium nanoparticles (Figure 3).

After adding the selenium colloid to the dye and reducing agent mixture ($NaBH_4$), a gradual decrease in the UV–Vis absorbance in the dye spectrum was observed. This proves the catalytic ability of the nanostructures used. During the first 3 min from the initiation of the reduction process, the spectrum of methylene blue rapidly decreased, so that, after approx. 25 min, the entire dye was reduced. The catalytic efficiency of SeNPs in solid form, i.e., SeNPs-OSA and the SeNPs-LS842-Sylobloc K500 systems, was also studied (Figure 4).

In the vial containing the organic dye and sodium borohydride (Figure 4A), no color change was observed for a long time. However, the systems enriched with SeNPs on a starch derivative—SeNPs-OSA (Figure 4B)—or silica—SeNPs-LS842-Sylobloc K500 (Figure 4C)—discolored; hence, one can conclude that the SeNPs also catalyzed the dye reduction reaction.

3.3. Thermogravimetric Analysis

A thermogravimetric analysis of the obtained copolymers filled with SeNPs-OSA or SeNPs-LS842-Sylobloc K500 was performed (Figure 5). This analysis confirmed the stability of the obtained copolymers and composites. At temperatures below 140 °C, the weight loss for all samples was similar. at around 3%. In the higher temperature range, 140–330 °C, the weight loss was still negligible and oscillated around 5%. It was only above 330 °C that thermal decomposition of the samples occurred. The mass of the dry residue corresponded to the filler content in the composite material.

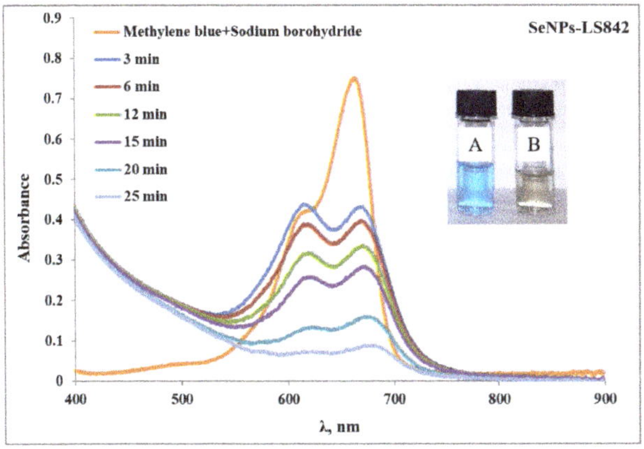

Figure 3. UV–Vis absorbance spectra for SeNPs in liquid form; inset photos: A—methylene blue and sodium borohydride, B—methylene blue, sodium borohydride, and SeNPs-LS842 25 min after the addition of the selenium nanoparticles.

Figure 4. Catalytic properties of SeNPs. (**A**)—methylene blue and sodium borohydride, (**B**)—methylene blue, sodium borohydride, and SeNPs-OSA 90 s after the addition of the selenium nanoparticles, (**C**)—methylene blue, sodium borohydride, and SeNPs-LS842-Sylobloc K500 6 min after the addition selenium nanoparticles.

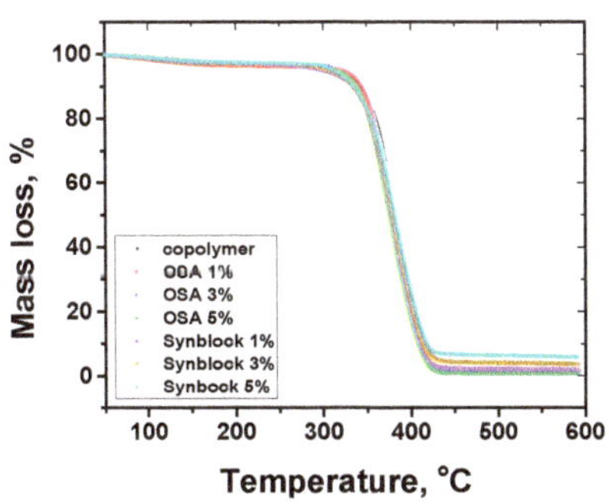

Figure 5. Thermogravimetric curves of the neat copolymer filled with SeNPs-OSA or SeNPs-LS842-Sylobloc K500 at the indicated contents.

3.4. *Influence of the SeNPs–Carrier System on the Mechanical Parameters of the Composite Materials*

The applications of polymers are always limited due to their poor mechanical properties. The mechanical properties of a polymer composite are affected by the filler used, as well as by the continuous phase.

The studied composite polymers varied in strength properties depending on their composition, as well as on the filler content (Figure 6). Depending on the selenium content and the filler type, various correlations were noted. The Young's modulus increased with an increase in the filler content, indicating an increase in the material stiffness caused by the fact that chain mobility and intermolecular motion became more difficult in the composite, and the added inorganic particles had a higher stiffness than the polymer matrices (Figure 6a). With a 5% content of the SeNPs-LS842-Sylobloc K500 filler, the Young's modulus value increased by 57% compared to the Young's modulus obtained for the unfilled copolymer. Since the Young's modulus of a polymer composite depend on the properties of its ingredients (matrix and filler) and only to a slight extent on their interfacial adhesion, both types of fillers contribute to the modulus increase in a similar way. However, the strength strongly depends on the stress transfer between particles and the polymer matrix. When there is a good interaction between the matrix and the particles, the applied stresses can be effectively transferred to the particles from the matrix, thus enhancing the strength. As in the case of the Young's modulus, the strength of the composite materials containing SeNPs-LS842-Sylobloc K500 also increased with the increasing filler content (Figure 6b). In contrast to the Young's modulus, the tensile strength is very sensitive to interfacial adhesion and can therefore be used even as an indicator of the copolymer/filler interactions. The filler with SeNPs-OSA (Figure 6b) did not ensure such large interactions between the filler and the polymer matrix, and as a result, it did not affect the strength properties of the obtained composite materials effectively; in addition, agglomerates of large sizes were responsible for creating an additional stress concentration that facilitated the samples' cracking.

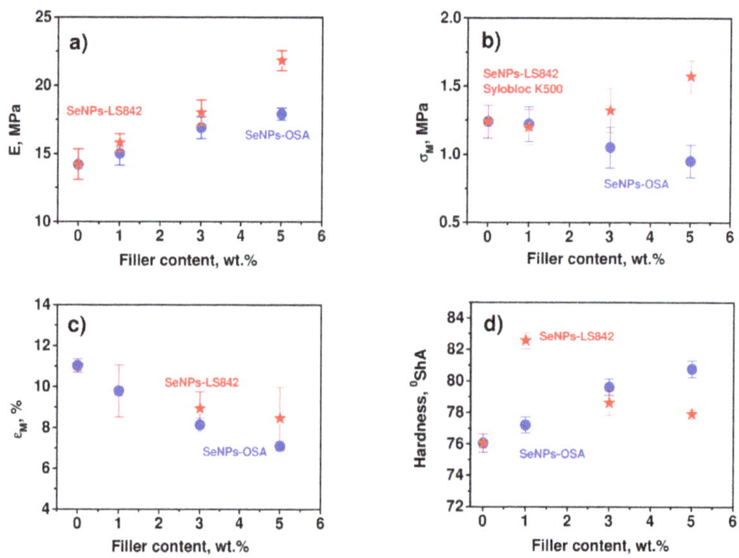

Figure 6. Young's modulus E (**a**), tensile strength δ_M (**b**), elongation at break ε_M (**c**), and hardness (**d**) as a function of the filler content.

According to the literature data, the value of elongation at break usually decreases after adding the filler, which was also visible in our case [38]. The introduction of an inorganic filler strengthened the polymer matrix at the expense of deteriorating its plasticity, which resulted in a decrease in the value of ε_M (Figure 6c). Elongation at break reflects the ductility of an investigated sample. A decrease in ductility occurred and indicated a reduction in the matrix deformation due to the mechanical restraints imposed by the almost non-deformable filler particles, which made the material more brittle and relatively weak.

The hardness of both types of composites increased with the filler content (Figure 6d). This resulted from the increased moduli and the presence of hard filler domains. Somewhat surprising was the hardness of the SeNPs-LS842-Sylobloc K500-filled materials, although this parameter was expected to be higher for this filler. One can speculate that this resulted from the lower crosslinking density of SeNPs-LS842-Sylobloc K500 with the polymer matrix.

3.5. AFM and Optical Microscopy Analysis of the Composites

The optical microscopy images (Figure 7) of the fractured copolymer and its composites showed typical cracking features (crack propagation) across the bulk surface. As the SeNPs-LS842-Sylobloc K500 sample employed the smallest—in terms of size—filler, the sample surface showed many crack propagation lines, but with a modest bulk penetration (white arrows). Since a crack typically starts and ends at the fillers, one can appreciate the good filler dispersity in the matrix in the images shown. Locally, the copolymer sample surface was flat enough for AFM imaging to be performed. The surface of the copolymer was found to be homogeneous, with a room mean square roughness of only 8.3 nm. The (surface) deformation and adhesion force maps showed a slight change in value (see cross-section plots) that indicated a good mixability of the monomers used in the study (monophase surface morphology). Due to the size of the fillers used, AFM imaging of the composites was not possible within the AFM scanner range.

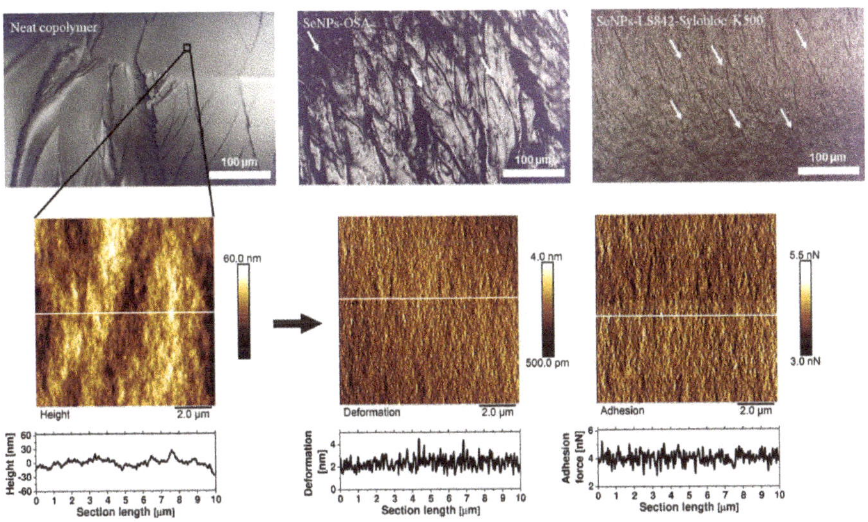

Figure 7. Optical microscopy of the neat copolymer and of the copolymer filled with SeNPs–OSA or SeNPs–LS842–Sylobloc K500 at the highest content (5% wt.) White arrows indicate examples of crack propagation and their directions. AFM images with the cross-section plots represent the height, deformation, and adhesion force mapping of the area indicated by the black square.

To sum up, the filler SeNPs-LS842-Sylobloc K500 turned out to be the most optimal additive. Even its small admixture (1% by weight) improved the stiffness and hardness of the copolymer material.

4. Conclusions

Selenium nanoparticles, both in liquid form (SeNPs-LS842) and in combination with a suitable matrix (SeNPs-OSA or SeNPs-LS842-Sylobloc K500), have catalytic properties that can be used in the reduction of organic dyes. Furthermore, natural biopolymers, such as sodium lignosulfonate and starch octenyl succinate, can be used as stabilizers for the obtained nanoparticles.

In this article, the process and properties of methacrylate and acrylate-based compounds containing SeNPs-OSA were investigated and compared with those of compounds containing SeNPs-LS842-Sylobloc K500. As a result, it can be stated that tensile strength and elongation at break of the SeNPs-LS842-Sylobloc K500-filled materials increased monotonically within the whole range of the filler content, but elongation at break of the SeNPs-OSA filled materials decreased due to mechanical restraints imposed by the almost non-deformable filler, which made the material more brittle. Furthermore, studies have shown that even a small admixture of SeNPs-LS842-Sylobloc K500 significantly improved the stiffness and hardness of polymer materials. Last but not least, the copolymers obtained appeared to be thermostable at a temperature below 330 °C.

Author Contributions: Conceptualization, A.M.-S., M.R. and E.K.; methodology, A.M.-S., M.R. and E.K.; software, M.R. and Ł.G.; validation, A.M.-S., M.R. and Ł.G.; formal analysis, A.M.-S., M.R. and H.G.; investigation, A.M.-S., M.R., E.K. and H.G.; resources, A.M.-S., M.R. and E.K.; data curation, A.M.-S. and M.R.; writing—original draft preparation, A.M.-S., M.R., E.K., H.G. and Ł.G.; writing—review and editing, A.M.-S., M.R., H.G., Ł.G., B.W., Ł.W. and W.Ł.; visualization, A.M.-S., M.R. and Ł.G.; supervision, A.M.-S. and Ł.G.; project administration, A.M.-S. and M.R.; funding acquisition, Ł.G. All authors have read and agreed to the published version of the manuscript.

Funding: This work was supported by the Polish Ministry of Education and Science.

Institutional Review Board Statement: Not applicable.

Informed Consent Statement: Not applicable.

Data Availability Statement: Not applicable.

Conflicts of Interest: The authors declare no conflict of interest.

References

1. Presentato, A.; Piacenza, E.; Anikovskiy, M.; Cappelletti, M.; Zannoni, D.; Turner, R.J. Biosynthesis of selenium-nanoparticles and -nanorods as a product of selenite bioconversion by the aerobic bacterium *Rhodococcus aetherivorans* BCP1. *New Biotechnol.* **2018**, *41*, 1–8. [CrossRef]
2. Diko, C.S.; Zhang, H.; Lian, S.; Fan, S.; Li, Z.; Qu, Y. Optimal synthesis conditions and characterization of selenium nanoparticles in *Trichoderma* sp. WL-Go culture broth. *Mater. Chem. Phys.* **2020**, *246*, 122583. [CrossRef]
3. Vahdati, M.; Moghadam, T.T. Synthesis and Characterization of Selenium Nanoparticles-Lysozyme Nanohybrid System with Synergistic Antibacterial Properties. *Sci. Rep.* **2020**, *10*, 510. [CrossRef] [PubMed]
4. Tripathi, R.M.; Hameed, P.; Rao, R.P.; Shrivastava, N.; Mittal, J.; Mohapatra, S. Biosynthesis of Highly Stable Fluorescent Selenium Nanoparticles and the Evaluation of Their Photocatalytic Degradation of Dye. *Bionanoscience* **2020**, *10*, 389–396. [CrossRef]
5. Jamróz, E.; Kulawik, P.; Kopel, P.; Balková, R.; Hynek, D.; Bytesnikova, Z.; Gagic, M.; Milosavljevic, V.; Adam, V. Intelligent and active composite films based on furcellaran: Structural characterization, antioxidant and antimicrobial activities. *Food Packag. Shelf Life* **2019**, *22*, 100405. [CrossRef]
6. Menon, S.; Ks, S.D.; Santhiya, R.; Rajeshkumar, S.; Kumar, V. Selenium nanoparticles: A potent chemotherapeutic agent and an elucidation of its mechanism. *Colloids Surf. B Biointerfaces* **2018**, *170*, 280–292. [CrossRef]
7. Umar, A.; Khan, M.S.; Alam, S.; Zekker, I.; Burlakovs, J.; Rubin, S.S.; Bhowmick, G.D.; Kallistova, A.; Pimenov, N.; Zahoor, M. Synthesis and Characterization of Pd-Ni Bimetallic Nanoparticles as Efficient Adsorbent for the Removal of Acid Orange 8 Present in Wastewater. *Water* **2021**, *13*, 1095. [CrossRef]
8. Lai, S.K.M.; Tang, H.-W.; Lau, K.C.; Ng, K.M. Nanosecond UV Laser Ablation of Gold Nanoparticles: Enhancement of Ion Desorption by Thermal-Driven Desorption, Vaporization, or Phase Explosion. *J. Phys. Chem. C* **2016**, *120*, 20368–20377. [CrossRef]
9. Panahi-Kalamuei, M.; Salavati-Niasari, M.; Zarghami, Z.; Mousavi-Kamazani, M.; Taqriri, H.; Mohsenikia, A. Synthesis and characterization of Se nanostructures via co-precipitation, hydrothermal, microwave and sonochemical routes using novel starting reagents for solar cells. *J. Mater. Sci. Mater. Electron.* **2015**, *26*, 2851–2860. [CrossRef]

10. Qiao, L.; Dou, X.; Yan, S.; Zhang, B.; Xu, C. Biogenic selenium nanoparticles synthesized by *Lactobacillus casei* ATCC 393 alleviate diquat-induced intestinal barrier dysfunction in C57BL/6 mice through their antioxidant activity. *Food Funct.* **2020**, *11*, 3020–3031. [CrossRef]
11. Mates, I.; Antoniac, I.; Laslo, V.; Vicas, S.; Brocks, M.; Luminita, F.; Milea, C.; Mohan, A.; Cavalu, S. Selenium nanoparticles: Production, characterization and possible applications in biomedicine and food science. *Sci. Bull. B Chem. Mater. Sci. UPB* **2019**, *81*, 205–216.
12. Alam, H.; Khatoon, N.; Raza, M.; Ghosh, P.C.; Sardar, M. Synthesis and Characterization of Nano Selenium Using Plant Biomolecules and Their Potential Applications. *BioNanoScience* **2019**, *9*, 96–104. [CrossRef]
13. Sohouli, E.; Irannejad, N.; Ziarati, A.; Ehrlich, H.; Rahimi-Nasrabadi, M.; Ahmadi, F.; Luque, R. Application of polysaccharide-based biopolymers as supports in photocatalytic treatment of water and wastewater: A review. *Environ. Chem. Lett.* **2022**, *20*, 3789–3809. [CrossRef]
14. Mukherjee, S.; Mukhopadhyay, S. Recent progress in sulfur-containing technical lignin-based polymer composites. *Express Polym. Lett.* **2023**, *17*, 120–151. [CrossRef]
15. Ndwandwe, B.K.; Malinga, S.P.; Kayitesi, E.; Dlamini, B.C. Selenium nanoparticles–Enhanced potato starch film for active food packaging application. *Int. J. Food Sci. Technol.* **2022**, *57*, 6512–6521. [CrossRef]
16. Modrzejewska-Sikorska, A.; Konował, E.; Klapiszewski, Ł.; Nowaczyk, G.; Jurga, S.; Jesionowski, T.; Milczarek, G. Lignosulfonate-stabilized selenium nanoparticles and their deposition on spherical silica. *Int. J. Biol. Macromol.* **2017**, *103*, 403–408. [CrossRef]
17. Hu, T.Q. *Chemical Modification, Properties, and Usage of Lignin*; Springer: New York, NY, USA, 2002.
18. Belgacem, A.G.M.N. *Monomers, Polymers and Composites from Renewable Resources*, 1st ed.; Elsevier: Amsterdam, The Netherlands, 2008.
19. Qiu, X.; Kong, Q.; Zhou, M.; Yang, D. Aggregation behavior of sodium lignosulfonate in water solution. *J. Phys. Chem. B* **2010**, *114*, 15857–15861. [CrossRef] [PubMed]
20. Vishtal, A.; Kraslawski, A. Challenges in industrial applications of technical lignins. *BioResources* **2011**, *6*, 3547–3568. [CrossRef]
21. Vanholme, R.; Morreel, K.; Ralph, J.; Boerjan, W. Lignin engineering. *Curr. Opin. Plant Biol.* **2008**, *11*, 278–285. [CrossRef] [PubMed]
22. Pang, Y.X.; Qiu, X.Q.; Yang, D.J.; Lou, H.M. Influence of oxidation, hydroxymethylation and sulfomethylation on the physico-chemical properties of calcium lignosulfonate. *Colloids Surf. A Physicochem. Eng. Asp.* **2008**, *312*, 154–159. [CrossRef]
23. Yan, M.; Yang, D.; Deng, Y.; Chen, P.; Zhou, H.; Qiu, X. Influence of pH on the behavior of lignosulfonate macromolecules in aqueous solution. *Colloids Surf. A Physicochem. Eng. Asp.* **2010**, *371*, 50–58. [CrossRef]
24. Deng, Y.; Wu, Y.; Qian, Y.; Ouyang, X.; Yang, D.; Qiu, X. Adsorption and desorption behaviors of lignosulfonate during the self-assembly of multilayers. *BioResources* **2010**, *2010*, 1178–1196.
25. Konował, E.; Modrzejewska-Sikorska, A.; Milczarek, G. Synthesis and multifunctional properties of lignosulfonate-stabilized gold nanoparticles. *Mater. Lett.* **2015**, *159*, 451–454. [CrossRef]
26. Modrzejewska-Sikorska, A.; Konował, E.; Cichy, A.; Nowicki, M.; Jesionowski, T.; Milczarek, G. The effect of silver salts and lignosulfonates in the synthesis of lignosulfonate-stabilized silver nanoparticles. *J. Mol. Liq.* **2017**, *240*, 80–86. [CrossRef]
27. Modrzejewska-Sikorska, A.; Konował, E. Silver and gold nanoparticles as chemical probes of the presence of heavy metal ions. *J. Mol. Liq.* **2020**, *302*, 112559. [CrossRef]
28. Sweedman, M.C.; Tizzotti, M.J.; Schäfer, C.; Gilbert, R.G. Structure and physicochemical properties of octenyl succinic anhydride modified starches: A review. *Carbohydr. Polym.* **2013**, *92*, 905–920. [CrossRef]
29. Caldwell, C.G.; Wurzburg, O.B. Polysaccharide Derivatives of Substituted Dicarboxylic Acids. U.S. Patent 2,661,349, 1 December 1953.
30. Khatami, M.; Iravani, S. Green and Eco-Friendly Synthesis of Nanophotocatalysts: An Overview. *Comments Inorg. Chem.* **2021**, *41*, 133–187. [CrossRef]
31. Nguyen, N.T.T.; Nguyen, L.M.; Nguyen, T.T.T.; Liew, R.K.; Nguyen, D.T.C.; van Tran, T. Recent advances on botanical biosynthesis of nanoparticles for catalytic, water treatment and agricultural applications: A review. *Sci. Total Environ.* **2022**, *827*, 154160. [CrossRef]
32. Asl, F.D.; Mousazadeh, M.; Azimzadeh, M.; Ghaani, M. Mesoporous selenium nanoparticles for therapeutic goals: A review. *J. Nanoparticle Res.* **2022**, *24*, 210. [CrossRef]
33. Shah, C.P.; Singh, K.K.; Kumar, M.; Bajaj, P.N. Vinyl monomers-induced synthesis of polyvinyl alcohol-stabilized selenium nanoparticles. *Mater. Res. Bull.* **2010**, *45*, 56–62. [CrossRef]
34. Tsivileva, O.; Pozdnyakov, A.; Ivanova, A. Polymer Nanocomposites of Selenium Biofabricated Using Fungi. *Molecules* **2021**, *26*, 3657. [CrossRef] [PubMed]
35. Prochaska, K.; Kędziora, P.; Le Thanh, J.; Lewandowicz, G. Surface properties of enzymatic hydrolysis products of octenylsuccinate starch derivatives. *Food Hydrocoll.* **2007**, *21*, 654–659. [CrossRef]
36. González-Henríquez, C.M.; Del Pizarro, G.C.; Sarabia-Vallejos, M.A.; Terraza, C.A.; López-Cabaña, Z.E. In situ-preparation and characterization of silver-HEMA/PEGDA hydrogel matrix nanocomposites: Silver inclusion studies into hydrogel matrix. *Arab. J. Chem.* **2019**, *12*, 1413–1423. [CrossRef]

37. Hutter, J.L.; Bechhoefer, J. Calibration of atomic-force microscope tips. *Rev. Sci. Instrum.* **1993**, *64*, 1868–1873. [CrossRef]
38. Aso, O.; Eguiazábal, J.I.; Nazábal, J. The influence of surface modification on the structure and properties of a nanosilica filled thermoplastic elastomer. *Compos. Sci. Technol.* **2007**, *67*, 2854–2863. [CrossRef]

coatings

Article

Enhanced Adsorption of Methylene Blue Using Phosphoric Acid-Activated Hydrothermal Carbon Microspheres Synthesized from a Variety of Palm-Based Biowastes

Saeed Alhawtali [1], Mohanad El-Harbawi [1,*], Abdulrhman S. Al-Awadi [2], Lahssen El Blidi [1], Maher M. Alrashed [1] and Chun-Yang Yin [3]

[1] Department of Chemical Engineering, King Saud University, Riyadh 11421, Saudi Arabia; saeerr2014@gmail.com (S.A.); lelblidi@ksu.edu.sa (L.E.B.); mabdulaziz@ksu.edu.sa (M.M.A.)
[2] K.A. CARE Energy Research and Innovation Center in Riyadh, King Saud University, Riyadh 11421, Saudi Arabia; alawadi@ksu.edu.sa
[3] Newcastle University in Singapore, 537 Clementi Road #06-01, SIT Building @ Ngee Ann Polytechnic, Singapore 599493, Singapore; chunyang.yin@newcastle.ac.uk
* Correspondence: melharbawi@ksu.edu.sa

Abstract: In the present study, the ability for novel carbon microspheres (CMs) derived from date palm (*Phoenix dactylifera*) biomass using a hydrothermal carbonization (HTC) process and activated using phosphoric acid to remove methylene blue dye was investigated. Three types of palm-based wastes (seeds, leaflet, and inedible crystallized date palm molasses) were used and converted to CMs via the HTC process. The prepared samples were then activated using phosphoric acid via the incipient wetness impregnation method. The CMs samples before and after activation were analyzed using scanning electron microscopy (SEM), elemental analysis and scanning (CHNS), and the Fourier transform infrared (FTIR) and Brunauer–Emmet–Teller (BET) methods. The samples exhibited high BET surface areas after activation (1584 m^2/g). The methylene blue adsorption results showed good fitting to the Langmuir, Fruendlich, and Temkin isotherm models for all activated samples. The maximum adsorption capacity achieved was 409.84 mg/g for activated CM obtained from the palm date molasses, indicating its high potential for application as a dye-based adsorption material.

Keywords: date palm biomass; carbon microspheres; phosphoric acid activation; methylene blue; adsorption

Citation: Alhawtali, S.; El-Harbawi, M.; Al-Awadi, A.S.; El Blidi, L.; Alrashed, M.M.; Yin, C.-Y. Enhanced Adsorption of Methylene Blue Using Phosphoric Acid-Activated Hydrothermal Carbon Microspheres Synthesized from a Variety of Palm-Based Biowastes. *Coatings* **2023**, *13*, 1287. https://doi.org/10.3390/coatings13071287

Academic Editor: Deling Yuan

Received: 30 June 2023
Revised: 16 July 2023
Accepted: 20 July 2023
Published: 22 July 2023

1. Introduction

Hydrothermal carbonization (HTC) is an environmentally friendly, cost-effective, and efficient process for converting agricultural waste into valuable porous carbon (hydrochar). In recent years, hydrochars have received growing attention due to their low production costs and potential applications in many industrial fields. The synthesized hydrochars are generally manifested in the form of carbon microspheres (CMs). Carbon microspheres/hydrochars can be synthesized via a two-step HTC process: (1) the dehydration of carbohydrates to form a furan-like molecule (either 5-hydroxymethyl-2-furaldehyde or furfural) and (2) the subsequent condensation and polymerization of the furan compound(s) to form carbon materials [1,2]. In general, CMs can be produced by digesting various organic feedstocks, such as biomass or saccharides, at moderate temperatures (160–250 °C) and pressures. This approach is simple, straightforward, environmentally friendly, and cost-effective. Many researchers have prepared CMs/hydrochar using different materials, such as agricultural biomass [3,4], sewage sludge [5,6], food waste [7,8], saccharides [9,10], algae [11], and date palm molasses [12]. Hydrochar has been proven to be very effective in various industrial applications, especially in the adsorption of toxic substances from water, such as dyes, pigments, and heavy metals [3,13]. Various wastewater treatment processes, including adsorption, membrane filtration, chemical precipitation, coagulation, ion exchange, electrochemistry, and electrochemical removal, can remove dyes and heavy metals from wastewater [14,15]. Some of these methods have substantial downsides, including high capital expenditure, high energy requirements, high maintenance and operation costs, partial removal, time-consuming regeneration, and the formation of toxic sludge [16,17]. Adsorption is one of the most effective and economical techniques for removing water contaminants [18].

In line with the global awareness of sustainable development, many researchers have worked to sustainably convert waste from various sources into useful materials. Therefore, it is favorable to repurpose these agricultural wastes, convert them into carbon-concentrated materials, and apply them in wastewater treatment. The use of date palm-based waste in wastewater treatment is, therefore, a two-pronged solution—it solves the issue of the abundance of palm wastes without burning them indiscriminately, and it converts them into useful adsorbent materials for the removal of toxic pollutants, such as dye contaminants, from wastewater.

In the Kingdom of Saudi Arabia, there are more than 31 million date palms [19], and each palm produces approximately 20 kg of waste per year [20]. The total waste of date palms is about 620,000 MT. Several attempts have been made to convert palm date biomass into valuable materials, such as activated carbon [21], hydrogel [22], biodiesel [23], and syngas [24]. However, most date palm wastes, such as leaves, twigs, and fibers, are not fully utilized in most cases and are usually burned, leading to numerous environmental problems (e.g., the unrestricted release of hazardous flue gas consisting of soot, CO_2, NO_x, SO_x, etc.).

The annual production of synthetic dyes is increasing worldwide as the demand for consumer goods increases. It has been estimated that textile mills currently produce more than 10,000 different types of dyes and pigments, with an approximated annual production of about 7×10^5 tons [25]. About 10–15% of this amount is discharged directly into the environment in the form of wastewater [26,27]. As the consumption of dyes is rapidly increasing due to high demand in various sectors, especially in the textile industry, the production of dyes is expected to increase, which will certainly lead to an increase in waste generated from the textile industry. Dye concentrations in textile effluents have been reported in a wide range of values (10–800 mg/L) [28].

Hydrochar can be synthesized at mild reaction conditions [29], which makes its production cost-effective. In addition, the presence of oxygen functional groups, such as hydroxyl, phenol, carbonyl, and carboxyl, on the surface of hydrochar enables the material to have an enhanced pollutant sorption ability [30]. On the other hand, the main disadvantages of hydrochar are its poor surface area and porosity, which hinder its ability to adsorb contaminants. Therefore, increasing their surface area and porosity can increase the adsorptive effectiveness of CMs. In previous studies, researchers concluded that activated hydrochar prepared from biomass waste had a much better adsorption capacity than non-activated hydrochar. Islam and co-researchers [31] synthesized activated hydrochar from coconut shells and sodium hydroxide and obtained a methylene blue (MB) adsorption capacity of 200.01 mg/g. Tran and co-researchers [32] prepared activated hydrochar using potassium hydroxide from coffee husk waste and obtained a maximum adsorption capacity of 418.8 mg/g for MB. Zhou and co-researchers [33] activated hydrochar made from sugarcane bagasse using phosphoric acid and sodium hydroxide. A maximum adsorption capacity of MB, 357.14 mg/g was obtained.

As far as we know, there is no known investigation focused on the use of a variety of date palm-based wastes for the synthesis of phosphoric acid-activated CMs in applications for dye-based wastewater treatment. It should be noted that these different types of date palm-based wastes are not homogenous, and their hydrothermal carbonization processes need to be properly optimized—this represents a novel aspect of the field of HTC. The findings from such a study would bode well for the sustainable use and repurpose of date palm-based wastes, which reflects our underlying motivation. We prepared three samples of CMs from several types of biomass precursors of date palm (*Phoenix dactylifera*), namely, seeds, leaflets, and inedible crystallized date palm molasses. The prepared samples were subsequently activated with phosphoric acid to develop enhanced adsorption characteristics for enhanced methylene blue removal.

2. Materials and Methods

2.1. Raw Materials and Chemicals

Date palm seeds and leaflets were obtained from a local farm located on the outskirts of Riyadh, Saudi Arabia, while inedible crystallized palm molasses were obtained from a date factory located in Riyadh. The seeds and leaflets were washed several times with tap water to remove dust and dirt. The material was then dried in a drying oven at 80 °C for 48 h. The material was then ground into a powder using a Fritsch Pulverisette 15 cutting mill, Germany, equipped with a 0.25-mm sieve. Finally, the powdered material was sent to undergo the HTC process described in our previous studies [3,12]. Deionized (DI) water was utilized to clean the solid samples and for the HTC and dye adsorption processes. MB (purity 95%) was obtained from LOBA Chemie, India. Absolute ethanol was obtained from VWR, Spain, and used only for washing the solid materials prepared via the HTC process.

2.2. Hydrothermal Carbonization

An amount of 4 g of seeds, 2.5 g of leaflets and 4 g of inedible crystallized date palm molasses were placed in Erlenmeyer flasks with deionized water (25 mL) and magnetically stirred for 5 h. The samples were then placed in 45 mL Teflon-lined Paar reactors for HTC. The Paar vessels were hermetically sealed and heated in a muffle oven at 230 °C for 4 h with seeds, 3 h with inedible crystallized molasses, and 8 h with leaflets. The reason for the different reaction times is attributed to the chemical compositions of the parts, and the different times reflect the difficulty in biomass digestion processes associated with different types of wastes. It was found that molasses that contain about 47% glaucous [34] and seeds were relatively easier to convert to CMs than leaflets because they require lower temperatures and reaction times [35,36]. This could be due to the high cellulose content in leaflets (47.14% cellulose, 36.73% lignin, and 16.13% hemicellulose) [37]. In contrast, the seeds and date molasses contain (32.77% cellulose, 30.20% hemicellulose, and 37.03% lignin) [37] and (47.1% glucose, 28% fructose, and 4.7% sucrose), respectively [34]. The dark-colored liquids were filtered to produce solid products of HTC. They were then washed a few times with DI water and absolute ethanol. The wet products were dried at a temperature of 110 °C for 12 h in an oven. Subsequently, the dry final products were ground with a laboratory mortar and then stored in sealed bottles in a desiccator prior to the activation process.

2.3. Activation after HTC

The samples were activated with phosphoric acid according to the method of incipient wetness impregnation [38,39]. The effects of the phosphoric acid/sample ratio and activation temperature were studied. One gram of the CM sample prepared during the HTC process was impregnated via the dropwise addition of phosphoric acid solution. The sample was vigorously stirred until a thin liquid film of phosphoric acid formed on the sample surface. The effects of phosphoric acid concentrations and activation temperatures were investigated. The following ratios of phosphoric acid/CM sample were investigated: 0.5, 0.75, 1, 1.3, 1.7, 2.3, 3, 4, and 5. The impregnated sample was then dried overnight at 80 °C. Then, the impregnated sample was carbonized in a horizontal tube furnace (OTF-1200X-S, MTI, CA, USA) at three different temperatures (450, 500, and 650 °C) and an N_2 flow of 50 CCM for 1.5 h. The obtained products were washed several times with deionized water to remove any residual phosphoric acid, filtered off, and dried at 80 °C for 24 h. The samples are referred to as "NAS" (non-activated seed), "AS" (activated seed), and "AAS" (activated seed after adsorption). Finally, the prepared samples were stored in a dry place before being used for dye removal.

2.4. Characterizations

The absorbance rates before and after activation were characterized via scanning electron microscopy (SEM) using the Tuscan VEGA II LSU (Tuscan Inc., Addison, IL, USA) to study the surface morphology of the prepared materials. The ImageJ software (version 1.53t) was used to determine the diameter of CMs. The adsorbents were further characterized using the Micromeritics TriStar II PLUS (Norcross, GA, USA) surface analyzer to evaluate their surface area, pore diameter, and pore volume. The samples were degassed for 90 min at 250 °C under a vacuum. The data obtained from this analysis, which included pore size and surface area, were further analyzed and compared to determine changes in the surface characteristics of the CMs due to various chemical treatments and their effects on dye removal. A Fourier transform infrared (FTIR) analysis was performed on the three samples before and after activation. The FT-IR spectra are important as they indicate the presence of bonds on the microspheres, such as C–H, C=O, etc. The CMS samples were mixed with KBr to form a pallet for the FT-IR analysis. Wavelength ranges in the infrared region adsorbed by the adsorbents were measured using an FTIR spectrophotometer (model: Shimadzu IRPrestige-21, Tokyo, Japan). An elemental analysis of the samples was conducted via a PerkinElmer series II CHNS/O 2400 analyzer (VELP, Wisconsin, NY, USA). Oxygen content was established via the calculation of the remaining mass. Boehm titration [40,41] was performed on the produced samples to quantify the acidic (oxygenated) functional groups on the surface of inactivated and activated samples.

2.5. Adsorption Studies

The adsorptive abilities of the inactivated and activated prepared CMs were tested for MB removal using a batch adsorption series. The effects of the following parameters on the adsorption process were evaluated: pH, contact time, adsorbent dosage, and initial dye concentration. A stock solution (1000 mg/L of MB) (95% purity) was produced by dissolving MB in deionized water. Sodium hydroxide or sulfuric acid (0.1 mol/L) was used to adjust the pH and maintain it at 6. The pH of the MB solutions was analyzed with a pH meter. The batch adsorption experiment was conducted in 300 mL Erlenmeyer flasks with 250 mL solutions of MB added to each flask at an initial concentration range of 25 to 500 mg/L. Samples (0.125 g) with a particle size of <0.25 mm were put into the MB solution (250 mL) and agitated in a shaker (150 rpm) at 25 °C. For the equilibrium studies, the batch was run for 540 min to ensure that equilibrium conditions were achieved.

For the batch adsorption experiment, after certain times (e.g., 15, 30 min, 60 min, . . ., 540 min), 10 mL was taken with a pipette from the supernatant of each flask and centrifuged at 5000 rpm for 6 min in a Hettich EBA 20 centrifuge to separate suspended particles from the prepared material. A small amount (2 mL) of supernatant solution was obtained from each flask, diluted with suitable amounts of deionized water, and then stored in a sealed bottle for analysis using a UV–Vis spectrophotometer (Jasco V-770, Oklahoma City, OK, USA). The remaining 8 mL was returned to the flask to avoid loss of adsorbance. The concentration of MB was established by comparing the adsorbance with a calibration curve. Equation (1) was used to determine the total MB adsorbed, q_t (mg/g), onto the CMs at time t:

$$q_t = \frac{(C_o - C_t)\,V}{M} \tag{1}$$

where q_t (mg/g) is the amount of adsorption at time t, C_o represents the initial concentration of the solution (mg/L), C_t represents the concentration of the solution at time t (mg/L), V represents the volume of the solution (L), and M represents the mass of the dry adsorbent (g).

The amount of MB adsorbed at equilibrium, q_e (mg/g), was calculated using Equation (2):

$$q_e = \frac{(C_o - C_e)\,V}{M} \quad (2)$$

where q_e (mg/g) represents the amount of adsorption at equilibrium while C_e represents the solution equilibrium concentration (mg/L).

The removal efficiency can be calculated using Equation (3):

$$R(\%) = \frac{(C_o - C_e)}{C_o} \times 100 \quad (3)$$

2.6. Reusability and Recovery of Adsorbent

A desorption study, or regeneration study, was conducted to assess the ability of the adsorbate to be removed from the mobile phase. This analysis is important to explain the transfer process of the adsorbate between the aqueous phase and the solid phase. Five cycles of adsorption—desorption were studied in this experimental work. Ethanol was used as a regenerative solution, as MB is soluble in ethanol. The adsorption and desorption process was conducted using a method elucidated by Genli et al. [42] and Al-Awadi et al. [3]. Briefly, three activated samples were separately mixed with 100 mL of ethanol solution in a series of conical flasks, and the mixtures were agitated at 150 rpm for 2 h using a digital orbital shaker (Jeio tech rotary shaker, SI -600, Seoul, Korea). The conical flasks were closed with rubber stoppers to avoid vaporization of the regenerative solution. After the mixing process was completed, the solutions were then separated from the solid particles via centrifuge (Bioevopeak CFG-18.5B, Jinan, China) at 6000 rpm speed for 6 min, followed by the measuring of adsorbance values using a UV–Vis spectrophotometer (UV–Vis, UV5, Mettler Toledo, Greifensee, Switzerland). The adsorbance values (nm) of the filtrates were then recorded. The adsorbent was then dried at 80 °C overnight. The adsorption and desorption cycles of MB on activated materials were carried out by washing with water and ethanol followed by regeneration and then subsequent reuse. The activated samples were regenerated into their original form and reutilized for five adsorption–desorption cycles. The percentage of removal (%) for each cycle in the batch adsorption process was then calculated. The AM sample was tested for the reusability study and multistage adsorption. Experiments were performed with 200 mL of the MB solution and 100 mg of the activated material at room temperature. The operating conditions were 240 min of contact time and an initial concentration of 300 mg/L MB.

3. Results and Discussions

3.1. Formation and Morphologies of Carbon Microspheres

The yields of the conversions of the three biomass samples—seeds, molasses, and leaflets—were 43%, 32%, and 52%, respectively. The micrographs of CMs for the three samples are presented in Figure 1. Figure 1a shows that CMs were dispersed on the surface of the seed sample at 230 °C 4 h before activation with phosphoric acid. The CMs appear to possess high degrees of sphericity and are well-developed and reasonably homogeneous. As shown in Figure 1a, the largest and smallest CM diameters were 16.6 and 4.7 µm, respectively. The average diameter of the CMs is 8.6 µm. Figure 1a′ shows that the CMs were slightly deformed from their original morphology after the activation process. The largest and smallest CM diameters were 15.7 and 4.4 µm, respectively. However, the average size of the particles remained in the same range at about 8.2 µm. Figure 1b,b′ shows the CMs on the leaflet sample before (NAL) and after activation (AL), respectively. It should be noted that CMs on the leaflets were not very well-developed or homogeneous as compared to those on the seeds and palm date molasses. In addition, the spheres were slightly smaller in size than those of the seeds or palm date molasses. From Figure 1b, it can be seen that the diameter of the smallest CM was 3.1 µm, while the diameter of the largest CM was 11.7 µm. In contrast, for the activated sample (Figure 1b′), the smallest and largest diameters were 3.1 µm and 9.6 µm, respectively. The average diameter was 5.8 µm. Figure 1c,c′ illustrates the CMs developed for the palm date crystallized molasses before (NAM) and after activation (AM), respectively. The CMs appear to be fully formed and very regular. The average sizes of the particles before and after activation were about 7.5 and 7.2 µm respectively. It is important to note that obtaining CMs from cellulosic materials is more complex than obtaining them from saccharides, such as glucose, sucrose, and fructose, due to the compact structure of highly crystallized cellulose. Consequently, CMs formed from leaflet samples required an 8 h residence time, which was longer than the time for CMs obtained from the seed sample and crystallized molasses. Figure 1a′,c′ shows that the CMs retained their original spherical morphology without any noticeable change in the diameter of the particles. The average sizes of the particles before and after activation were about 8.5 and 8.2 µm, respectively.

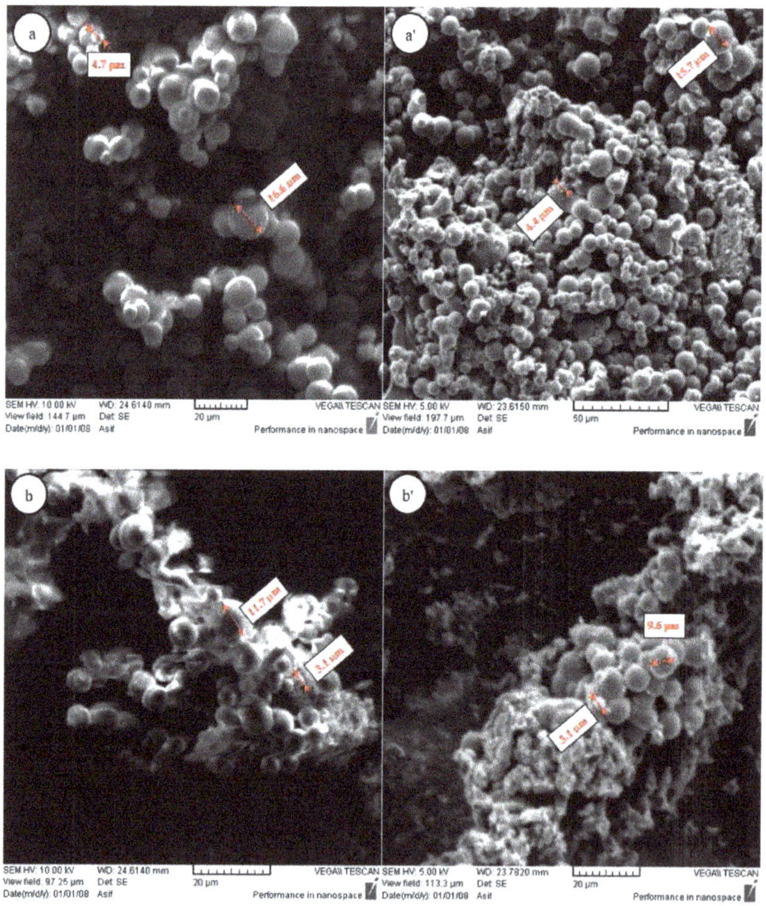

Figure 1. *Cont.*

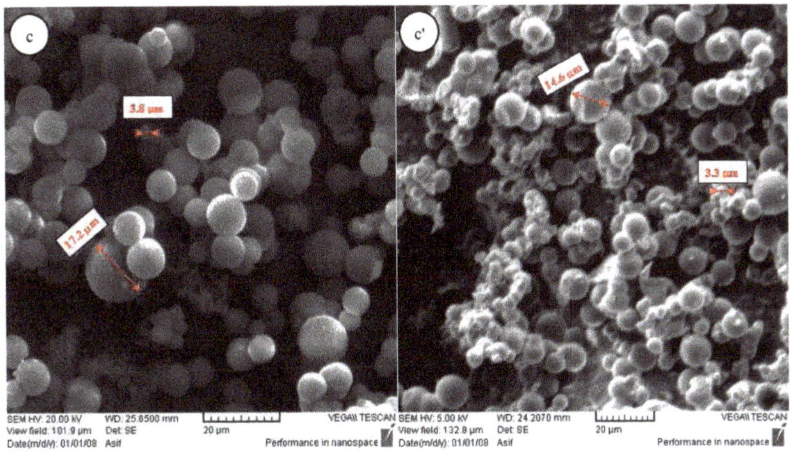

Figure 1. Micrographs for (**a**) non-activated seeds (NAS) at 230 °C and 4 h, (**a'**) activated seeds (AS) at 230 °C and 4 h, (**b**) non-activated leaflets (NAL) at 230 °C and 8 h, (**b'**) activated leaflets (AL) at 230 °C and 8 h, (**c**) non-activated molasses (NAM) at 230 °C and 3 h, and (**c'**) activated molasses (AM) at 230 °C and 3 h.

3.2. Surface Textural Properties of Carbon Microspheres

Figure 2 shows the N_2 adsorption–desorption isotherms of activated and non-activated CM samples. The isotherms for all non-activated samples can be generically classified as type 1 in the IUPAC classification. N_2 adsorption on these materials was observed only at a very low relative pressure ($P/P°$), indicative of the predominant microporosity of the samples, even though there was a slight uptake (deviation) at $P/P°$ higher than 0.95. On the other hand, the activated CMs (AS and AL) show a slight presence of mesoporosity in their textures through marginal hysteresis loops—giving them a quasi type IV IUPAC classification. The palpable absence of a plateau at the end of hysteresis implies that it is not solely type IV. The figure on pore size distribution suggests that the increases in pore volume are attributed to the addition of mesoporosity within the 20–50 Angstrom range. The existence of marginal hysteresis loops in the AS and AL samples indicates the existence of peripheral mesopores, which may be promising for adsorption in a solution environment.

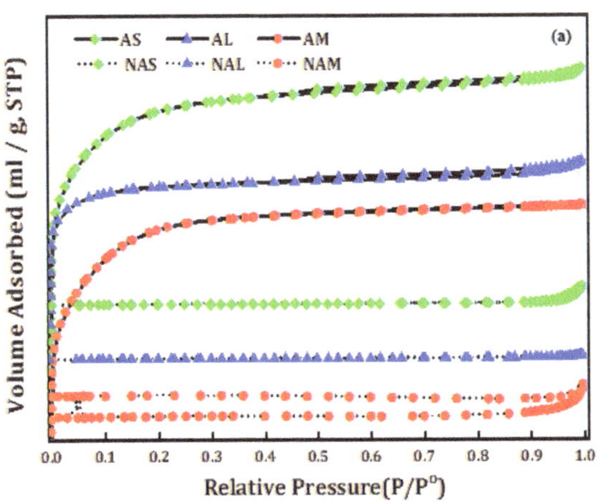

Figure 2. *Cont.*

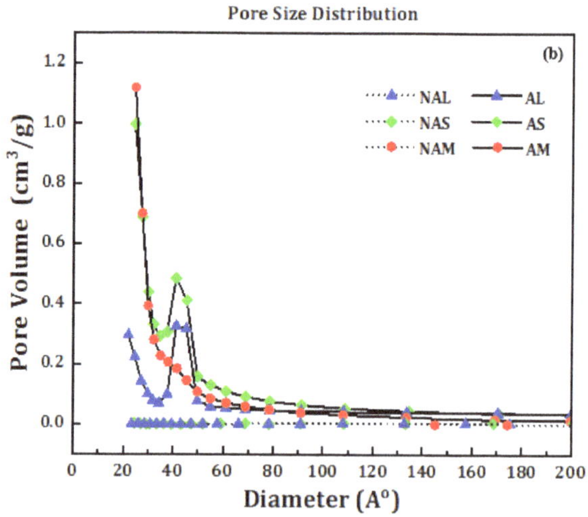

Figure 2. (**a**) Nitrogen physisorption isotherms and (**b**) pore size distribution for the inactivated and activated samples.

Table 1 shows the BET surface areas of the three synthesized samples before and after activation. The three samples possessed small surface areas before activation (<60 m^2/g), indicating that the internal pores on the CM particles were not fully developed. However, upon chemical treatment with phosphoric acid, pores developed on the surface of the microspheres, greatly increasing their surface areas (808–1584 m^2/g). Table 1 reveals that the pore sizes of the activated samples increased substantially after activation. In reference to the International Union of Pure and Applied Chemistry (IUPAC) categorization [43], all samples could be classified as mesoporous. The molecular size of MB is 0.95 nm in width and length (1.382 nm–1.447 nm) [44,45]. This indicates that the size of MB molecules is smaller compared to the pore size of all developed adsorbents. Thus, the MB molecules could be easily attached and fitted into the pores of the inactivated and activated materials, which facilitated MB removal from the aqueous solution.

Table 1. BET surface area for the inactivated and activated samples.

CM Samples	Before Activation		
	BET (m^2/g)	Pore Volume (cm^3/g)	Pore Size (nm)
Non-activated seeds (NAS)	5.02	0.041	30.44
Non-activated leaflets (NAL)	2.21	0.0086	18.04
Non-activated molasses (NAM)	0.72	0.0033	36.66
Activated seeds (AS)	1584	0.47	2.52
Activated leaflets (AL)	808	0.156	3.33
Activated molasses (AM)	1543	0.48	2.22

Boehm titration was performed for all prepared samples. The results are shown in Figure 3, indicating that the AM samples contained more phenolic groups that the other samples. Indeed, the phenolic groups dominated the surfaces of both the activated and non-activated samples. The total oxygenated (acidic) functional groups on all samples ranged from 0.1 to 0.4 mmol/g. The elemental analysis results are summarized in Table 2 for all samples. In general, the activated samples possessed higher carbon and lower hydrogen contents than non-activated samples, with the exception of AS. The results also show that AS had a higher carbon content than the other samples. A high oxygen concentration in the AM sample could indicate that many oxygen-containing groups were present in addition to the carboxyl groups [46].

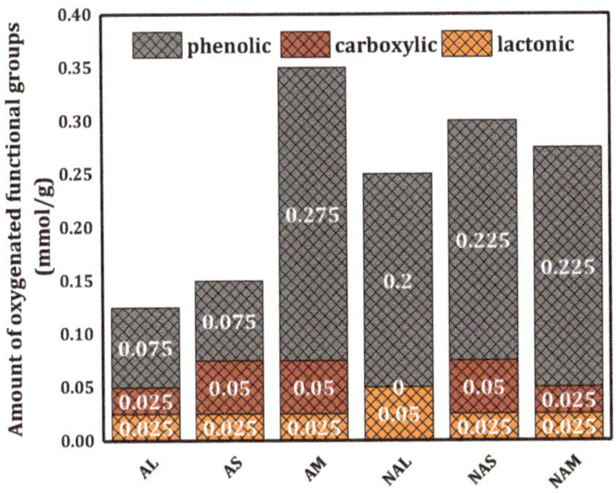

Figure 3. The total acidic (oxygenated) functional groups on the surface of the inactivated and activated CM samples.

Table 2. Elemental analysis results of the prepared CM materials.

CM Sample	Chemical Composition					O/C (Atomic)	H/C (Atomic)
	C (wt.%)	H (wt.%)	N (wt.%)	S (wt.%)	O (wt.%)		
NAS	65.53 ± 0.58	4.86 ± 0.09	12.65 ± 0.14	0.69 ± 0.005	16.27 ± 0.16	0.19	0.89
NAL	58.99 ± 0.53	6.58 ± 0.13	21.65 ± 0.24	0.92 ± 0.007	11.86 ± 0.13	0.15	1.34
NAM	67.80 ± 0.60	4.81 ± 0.09	14.78 ± 0.16	0.62 ± 0.005	11.99 ± 0.14	0.13	0.85
AS	77.48 ± 0.69	2.96 ± 0.06	11.42 ± 0.12	0.36 ± 0.003	7.78 ± 0.12	0.08	0.46
AL	70.12 ± 0.63	2.57 ± 0.05	25.35 ± 0.27	0.72 ± 0.006	1.24 ± 0.01	0.01	0.44
AM	49.41 ± 0.44	3.57 ± 0.07	8.29 ± 0.09	0.49 ± 0.004	38.24 ± 0.31	0.58	0.87

Variations related to the elemental compositions of the biomass and hydrothermal-based carbon microspheres can be studied using the conventional Van Krevelen diagram (not shown for brevity). From this plot, it appears that the H/C and O/C atomic ratios adhere to the same trend as the dehydration process, with a slight deviation in the direction of the H/C atomic ratio, indicating a low occurrence of the decarboxylation process [47]. The FTIR spectra for the CM samples are shown in Figure 4. Stretching vibrations attributed to aliphatic C–H groups are exhibited at around 2920 cm^{-1} [10]. The bands at approximately 1700 and 1610 cm^{-1} can be assigned to C=O and C=C, respectively, implying the existence of aromatic rings (*ibid*).

3.3. *Adsorption Studies*

3.3.1. Effect of pH

Adsorption efficiency is closely related to pH, since a change in pH can cause the surface charge of an adsorbent to be affected by protonation and deprotonation. The effect of pH on the adsorption process of MB on the three activated samples was examined by varying the pH values from 3 to 8 under the following conditions: an adsorbent dosage of 25 mg, an MB concentration of 300 ppm, a 50 mL MB solution, and a stirring speed of 150 rpm for 2 h. The amount of adsorbed MB, q, against pH for the activated samples was plotted, as shown in Figure 5. The uptake of MB enhances with the increase in pH from 3 to 6. The optimum pH of the solution was found to be 6 for all adsorbents, and equilibrium was reached after about 120 min of contact time. The maximum adsorption capacity of MB on the activated CM samples (AS, AL, AM) was 306 ± 8, 223 ± 7, and 393.4 ± 4.7 mg/g, respectively. After a pH of 6, the maximum amounts of MB gradually decrease.

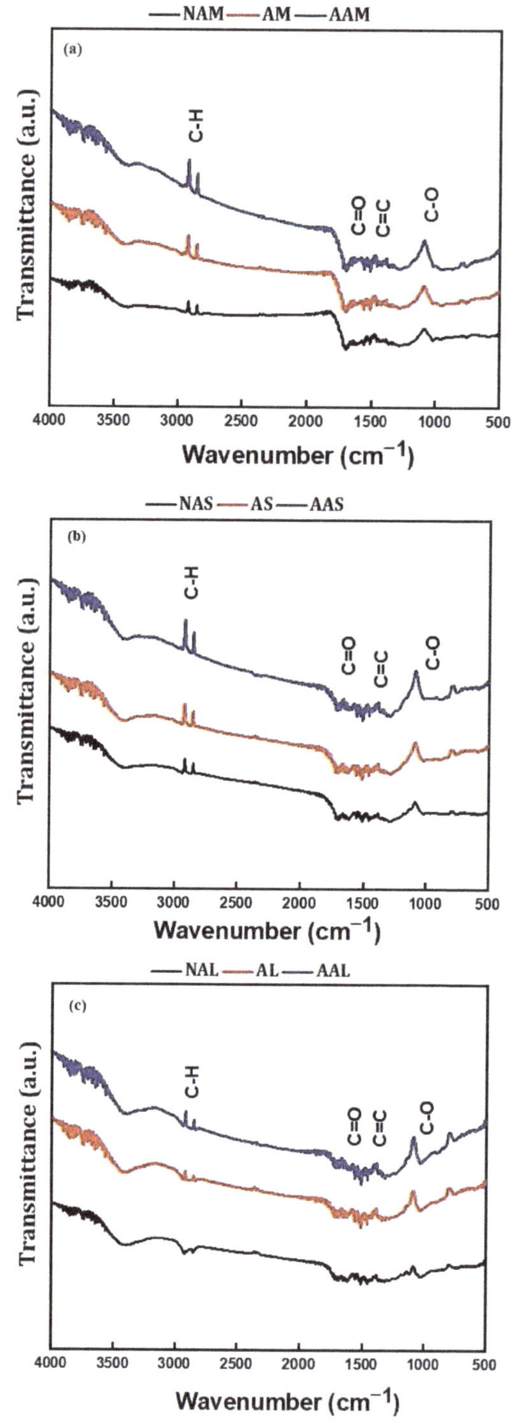

Figure 4. FT–IR spectra for (**a**) molasses, (**b**) seeds, and (**c**) leaflets.

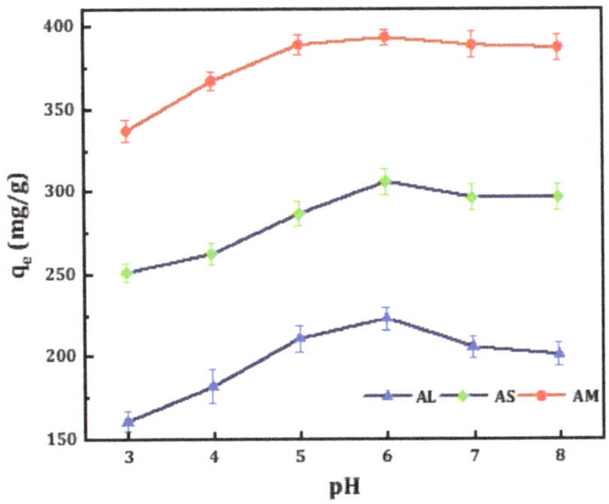

Figure 5. The influence of pH on the adsorption of MB on activated samples.

These results are in agreement with the literature, where it has been found that the increase in the adsorption amount of MB with an increase in pH can be explained by the high electrostatic interaction between the negatively charged adsorbent surface and the cationic MB [48–50]. Bharathi et al. [51] found that the removal of cationic dye increases at a high pH due to the high concentration of H+ competing with the cationic groups in MB for adsorption.

3.3.2. The Influence of Initial Concentration and Contact Time

The effect of the initial concentration of MB solution on the adsorption capacity of the activated CM samples was investigated at dye concentrations from 25 to 500 mg/L and a fixed contact time of 540 min. The adsorption capacities of the adsorbents were determined. Equilibrium was achieved after 120 min. The removal of MB was faster in the initial stage, then gradually decreased, and became constant after reaching equilibrium. Figure 6 shows the adsorption capacities for MB at 25 °C in activated and activated samples. It should be noted that the inactivated samples (NAS, NAL, and NAM) adsorbed MB only at low concentrations (e.g., 25 and 50 ppm). This can be attributed to differences in the surface area and pore size of the adsorbents, which affect their adsorption capacity. The maximum amounts of MB adsorbed, q_e, by NAS, NAL, and NAM at 25 ppm were 2.25, 5.77, and 12.06 mg/g, respectively. Whereas, for the activated samples, the adsorption capacity of the prepared samples increased with an increase in the concentration of MB. The maximum adsorption capacities were observed for AM, 410.4 ± 11.1 mg/g, followed by AS, 327.8 ± 6.7 mg/g, and AL, 249.2 ± 5.4 mg/g. The results showed that the three activated samples' adsorption capacities enhanced with an increase in the initial dye concentration. The error values for the results of inactivated samples are less than 0.5%.

3.3.3. Adsorption Isotherm

The adsorption isotherm explains how adsorbent particles interact with adsorbent material, which is vital for optimizing adsorption mechanisms, elucidating surface properties and adsorption capacities, and designing effective adsorption systems. Therefore, various adsorption models were used in their non-linear forms to analyze the obtained data. The program OriginPro—Graphing and Data Analysis was used for non-linear regression and the parameters of the models were determined and compared to establish the adsorption behavior of the studied system.

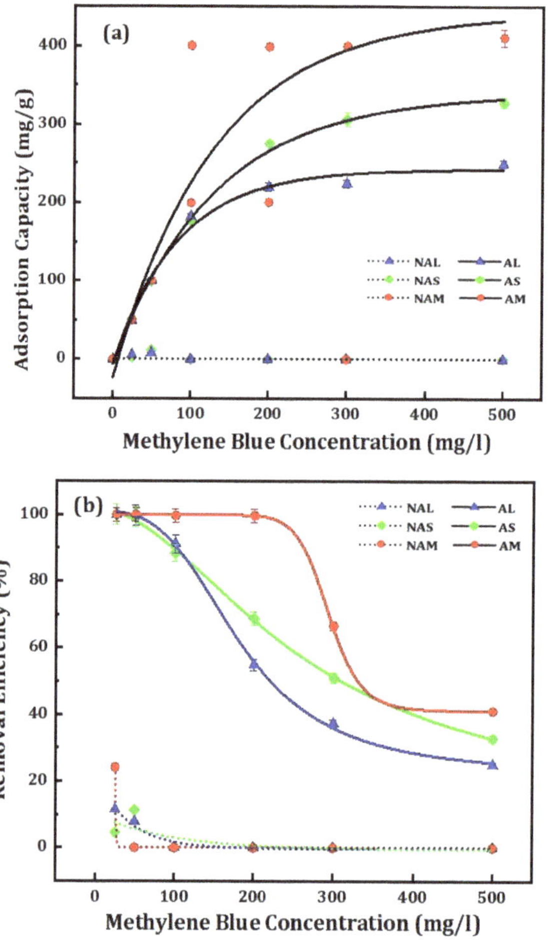

Figure 6. (**a**) Adsorption capacities of inactivated and activated samples. (**b**) Removal efficiency of inactivated and activated samples.

Langmuir Isotherm

The Langmuir adsorption model [52] was applied to one-layer adsorption on active locations of the adsorbent (Equation (4)):

$$q_e = \frac{q_m K_L C_e}{1 + K_L C_e} \tag{4}$$

in which q_m (mg/g) represents the amount of adsorbed MB per unit mass of CMs relative to the complete monolayer coverage, and K denotes the Langmuir constant. A non-linear form of this model was conducted and its parameters are listed in Table 3. It can be seen that the experimental data fit the Langmuir isotherm model quite well, with determination coefficients, R^2, close to 0.90 for all prepared samples. The maximal monolayer adsorption capacity obtained for the samples was in the order of AM > AS > AL, with AM having the highest q_m (414.6 mg/g).

Table 3. Langmuir, Freundlich, and Temkin isotherm constants for the activated samples.

CM Samples	Langmuir Constants			Freundlich Constants			Temkin Constants		
	Q_{max}	K_L	R^2	1/n	K_F	R^2	B	A	R^2
AS	322.54	0.124	0.903	0.178	122.10	0.967	35.76	27.81	0.964
AL	219.71	11.55	0.919	0.098	138.90	0.953	16.69	57.18×10^2	0.955
AM	414.63	4.54	0.880	0.054	306.20	0.842	19.59	62.16×10^5	0.844

In addition, the Langmuir isotherm is linked to a dimensionless constant—the separation factor (R_L) (Equation (5)). The values of R_L can be obtained from the results of the experiment, and a plot of the separation factor against the initial concentration (C_o) can be generated. The dimensionless constant in the mathematical expression can be written as follows:

$$R_L = \frac{1}{(1 + K_L C_o)} \tag{5}$$

in which C_o represents the highest initial MB concentration (mg/L). The value of R_L indicates the type of adsorption process: $R_L < 1$, the adsorption is favorable; $R_L = 1$, the adsorption is linear; $R_L > 1$, the adsorption is unfavorable; and $R_L = 0$, the adsorption is irreversible. Values of R_L have been obtained for AS, AL, and AM and are 0.993, 0.980, and 0.0033, respectively, suggesting that adsorption is a favorable process for all ranges of dye concentrations in this experimental work.

Freundlich Isotherm

The Freundlich adsorption model [53] is given by Equation (6):

$$q_e = K_F C_e^{1/n} \tag{6}$$

where K_F and n are constants, estimated from the slope and intercept of the linear plot of $\log q_e$ versus $\log C_e$ (Equation (6)). The $1/n$ value is observed to be between the range of 0–1, and adsorption becomes more heterogeneous when the value of $1/n$ approaches 0. A steep slope is seen when $1/n$ is closer to 1, and it reflects a higher adsorption capacity at higher equilibrium concentrations that decreases rapidly at a low equilibrium concentration [54]. Fitting the experimental data to a non-linear form of Freundlich adsorption resulted in R^2 values of 0.967, 0.953, and 0.842 for AS, AL, and AM, respectively.

Temkin Isotherm

The Temkin isotherm model [55] is expressed by Equation (7):

$$q_e = B \ln(A_T C_e) \tag{7}$$

where $B = \frac{RT}{b_T}$ R gas constant and T is the study temperature °C, while b_T and A_T are model constants that depend on the system and adsorbent. The Temkin isotherm observes that the heat of adsorption of all molecules in the layer reduces linearly with coverage due to interactions between adsorbate molecules [56]. Unlike the Freundlich isotherm, the Temkin isotherm assumes an unvarying dispersal of binding energies on the surface. Therefore, the Temkin isotherm is often used to model adsorption on homogeneous surfaces, whereas the Freundlich isotherm is more appropriate for heterogeneous surfaces [2]. The constants of the Temkin model were determined and are listed in Table 3. The coefficients of determination were 0.964, 0.955, and 0.844 for AS, AL, and AM, respectively.

In comparison to all R^2 values, it was seen that the experimental data fit well with the investigated models for activated seeds and leaflet samples and, to a lesser extent, with the molasses sample. Hence, it can be deduced that monolayer adsorption took place during the batch adsorption process, utilizing the activated samples, as the experimental data fit the Langmuir adsorption isotherm quite well. The data also revealed that dye molecules in the solution were not being deposited onto the other dye molecules that had been adsorbed onto the adsorptive sites of the activated samples. Table 4 lists a comparison of the maximum adsorption capacities of MB on different adsorbents. It shows that the activated molasses (AM) in this work possessed outsized and superior adsorption capacities as compared with those of other previously reported biomass, which bodes well for its application in dye contaminant removal from wastewater.

Table 4. Comparison of the maximum adsorption capacities of MB on different hydrochar-based adsorbents synthesized via the HTC process.

Adsorbent	q_{max}, (mg/g)	Reference
Coconut shell	200.01	[31]
Peanut shell	1368	[56]
Canola stalk	93.4	[57]
Bamboo	655.76	[30]
Sewage sludge	52.56	[58]
Pine wood	86.7	[59]
Bamboo	1155.57	[60]
Rice husk	53.21	[61]
Coffee husk	418.78	[32]
Filtrasorb 400	295 ± 3	[62]
Norit	276 ± 3	[62]
Picacarb	248 ± 2	[62]
Palm date molasses-activated hydrochar (AM)	414.63	This work
Palm date seed-activated hydrochar (AS)	322.54	This work
Palm date leaflet-activated hydrochar (AL)	219.71	This work

The generally good fit of our adsorption data with all three models, together with the existence of oxygenated functional groups, implies the prominence of adsorption mechanisms pertaining to electrostatic forces and hydrogen bonding. The π-π interactions between the aromatic rings of MB and the surface of the microspheres may also play a significant role [49].

3.3.4. Adsorption Kinetics

An adsorption kinetic analysis was performed to assess the adsorptive capability of the adsorbent as well as to investigate the mechanism of mass transfer and the reaction rate. In addition, this analysis can afford insight into the adsorption pathway and probable mechanisms involved. The pseudo-second-order kinetic model [63] was adopted in this work in order to outline the dynamic of the adsorption process.

The pseudo-second-order model is expressed by Equation (8):

$$q_t = k_2 q_e^2 t / (1 + k_2 q_e t) \tag{8}$$

in which k_2 is the rate constant of the second-order reaction (g/mg. min), and q_e and q_t are the amounts of adsorbate (mg/g) adsorbed at equilibrium and time t (min), respectively. The pseudo-second-order model is centered on the notion that the rate-determining step of adsorption is chemisorption relating to valence forces via the sharing or swap of electrons between the adsorbent and adsorbate. Pseudo-second-order constants (k_2) were established for the three activated samples and are listed in Table 5. It was observed that the values of the linear regression for AS, AL, and AM were 0.994, 0.985, and 0.9999, respectively. This agrees with Ho and McKay [64], who concluded that, of all adsorption systems that they analyzed, pseudo-second-order (PSO) kinetics provided the best correlation with experimental data. The theoretical, q_{theo}, and experimental, q_{exp}, adsorption values were compared and are presented in Table 5.

Table 5. Parameters of pseudo-second-order kinetics model established from experimental data for the synthesized CM samples after activation.

CM Samples	q_{exp}	$q_{theo.}$	K	R^2
AS	275.1 ± 4.7	270.78 ± 2.3	0.0012	0.994
AL	219.6 ± 5.6	223.77 ± 3.7	2.5901	0.985
AM	398.5 ± 0.065	399.62 ± 0.8	0.0014	0.999

3.4. Multistage Extraction, Reusability, and Recovery Studies

The MB was not completely removed in a single cycle of adsorption. Thus, it is worthwhile to consider a multistage extraction approach. The evolution of the amount of MB along the extraction cycles is depicted in Figure 7A. As expected, after three successive extraction steps, less than 3% of MB the amount was detected. On the other hand, the reusability of AM without any treatment was limited where the surface of sample was saturated by MB, and less than 1% removal was recorded for the second and third time, as shown in Figure 7B. The regenerated adsorbent samples were studied for five adsorption–desorption cycles, as shown in Figure 7C. According to the results, the AM, AS, and AL retained reuse efficiencies of 46.6, 36.2, and 22.01% respectively, from the original concentration in MB removal subsequent to five successive adsorption–desorption cycles.

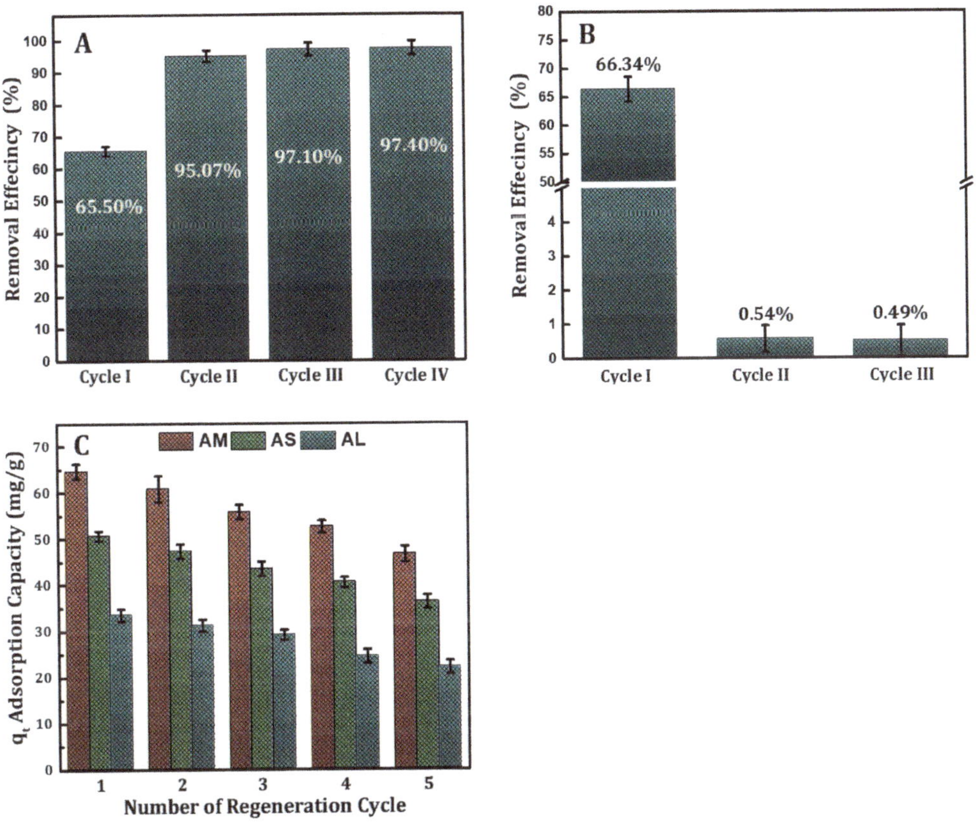

Figure 7. Number of cycles and reusability of activated molasses sample (**A**,**B**) and regeneration study of prepared samples (**C**).

4. Conclusions

By utilizing the method of hydrothermal carbonization integrated with incipient wetness impregnation, we have synthesized distinctive and comparatively well-defined CMs with outsized BET surface area values from a variety of palm-based biowastes. In terms of surface functionality, phenolic groups dominated the surfaces of both the activated and non-activated samples. Our physisorption analysis suggests that the phosphoric acid-activated process induced higher pore volumes while generating more mesoporosities on the surface of the CMs—such mesoporosities are favorable for liquid-based adsorption processes. The highest MB adsorption capacity, in excess of 400 mg/g and recorded in palm date molasses-activated hydrochar, represents the suitability of our CMs for the adsorption of MB in a fluidic environment. Results from our study are vital because they afford a three-pronged outcome in the form of repurposing and reusing bio-waste resources, the synthesis of valuable microparticles, and applications in dye-based wastewater treatment. The model fitting findings from our work can be employed to scale-up the bench-scale adsorption system for wastewater treatment applications.

Author Contributions: Conceptualization, M.E.-H., A.S.A.-A., L.E.B., M.M.A. and C.-Y.Y.; methodology, M.E.-H., A.S.A.-A. and S.A.; validation, L.E.B. and C.-Y.Y.; formal analysis, M.E.-H., A.S.A.-A. and C.-Y.Y.; investigation, L.E.B., A.S.A.-A. and C.-Y.Y.; resources, L.E.B. and S.A.; data curation, C.-Y.Y.; writing—original draft, M.E.-H.; writing—review and editing, C.-Y.Y.; supervision, M.E.-H., L.E.B. and C.-Y.Y.; dunding acquisition, M.E.-H. and M.M.A. All authors have read and agreed to the published version of the manuscript.

Funding: This research was funded by the Deputyship for Research and Innovation, "Ministry of Education", in Saudi Arabia for funding this research work through project number IFKSUDR_F158.

Institutional Review Board Statement: Not applicable.

Informed Consent Statement: Not applicable.

Data Availability Statement: Not applicable.

Acknowledgments: The authors would like to thank the Deputyship for Research and Innovation, "Ministry of Education", in Saudi Arabia for funding this research work through project number IFKSUDR_F158.

Conflicts of Interest: The authors declare no conflict of interest.

References

1. Ryu, J.; Suh, Y.W.; Suh, D.J.; Ahn, D.J. Hydrothermal preparation of carbon microspheres from mono-saccharides and phenolic compounds. *Carbon* **2010**, *48*, 1990–1998. [CrossRef]
2. Demir-Cakan, R.; Baccile, N.; Antonietti, M.; Titirici, M.-M. Carboxylate-rich carbonaceous materials via one-step hydrothermal carbonization of glucose in the presence of acrylic acid. *Chem. Mater.* **2009**, *21*, 484–490. [CrossRef]
3. Al-Awadi, A.S.; El-Harbawi, M.; Algarawi, A.; Alalawi, A.; Alrashed, M.M.; Yin, C.Y. Synthesis of carbon microspheres via hydrothermal carbonization of Sabal palms (*Sabal palmetto*) biomass for adsorption of methylene blue. *Biomass Conv. Bioref.* **2022**, 1–11. [CrossRef]
4. Yek, P.N.Y.; Liew, R.K.; Mahari, W.A.W.; Peng, W.; Sonne, C.; Kong, S.H.; Tabatabaei, M.; Aghbashlo, M.; Park, Y.K.; Lam, S.S. Production of value-added hydrochar from single-mode microwave hydrothermal carbonization of oil palm waste for dechlorination of domestic water. *Sci. Total Environ.* **2022**, *833*, 154968. [CrossRef] [PubMed]
5. Tasca, A.L.; Puccini, M.; Gori, R.; Corsi, I.; Galletti, A.M.R.; Vitolo, S. Hydrothermal carbonization of sewage sludge: A critical analysis of process severity, hydrochar properties and environmental implications. *Waste Manag.* **2019**, *93*, 1–13. [CrossRef]
6. Zhang, J.H.; Lin, Q.M.; Zhao, X.R. The hydrochar characters of municipal sewage sludge under different hydrothermal temperatures and durations. *J. Integr. Agric.* **2014**, *13*, 471–482. [CrossRef]
7. Liu, Y.; Sun, Y.; Wan, Z.; Jing, F.; Li, Z.; Chen, J.; Tsang, D.C. Tailored design of food waste hydrochar for efficient adsorption and catalytic degradation of refractory organic contaminant. *J. Clean. Prod.* **2021**, *310*, 127482. [CrossRef]
8. Saqib, N.U.; Baroutian, S.; Sarmah, A.K. Physicochemical, structural and combustion characterization of food waste hydrochar obtained by hydrothermal carbonization. *Bioresour. Technol.* **2018**, *266*, 357–363. [CrossRef]
9. Pak, S.; Ahn, J.; Kim, H. Synthesis of Saccharide-based Hydrochar with Macroporous Structure for Effective Organic Pollutant Removal. *Fibers Polym.* **2022**, *23*, 1789–1796. [CrossRef]

10. Sevilla, M.; Fuertes, A.B. Chemical and structural properties of carbonaceous products obtained by hydrothermal carbonization of saccharides. *Chem. Eur. J.* **2009**, *15*, 4195–4203. [CrossRef]
11. Jabeen, S.; Gao, X.; Hayashi, J.I.; Altarawneh, M.; Dlugogorski, B.Z. Effects of product recovery methods on the yields and properties of hydrochars from hydrothermal carbonization of algal biomass. *Fuel* **2023**, *332*, 126029. [CrossRef]
12. El-Harbawi, M.; Alhawtali, S.; Al-Awadi, A.S.; El Blidi, L.; Alrashed, M.M.; Alzobidi, A.; Yin, C.Y. Synthesis of Carbon Microspheres from Inedible Crystallized Date Palm Molasses: Influence of Temperature and Reaction Time. *Materials* **2023**, *16*, 1672. [CrossRef] [PubMed]
13. Zhang, Y.N.; Guo, J.Z.; Wu, C.; Huan, W.W.; Chen, L.; Li, B. Enhanced removal of Cr (VI) by cation functionalized bamboo hydrochar. *Bioresour. Technol.* **2022**, *347*, 126703. [CrossRef]
14. Kamarudzaman, A.N.; Adan, S.N.A.C.; Hassan, Z.; Wahab, M.A.; Makhtar, S.M.Z.; Seman, N.A.A.; Jalil, M.F.A.; Handayani, D.; Syafiuddin, A. Biosorption of Copper (II) and Iron (II) using Spent Mushroom Compost as Biosorbent. *Biointerface Res. Appl. Chem.* **2022**, *12*, 7775–7786.
15. Salami, B.A.; Oyehan, T.A.; Gambo, Y.; Badmus, S.O.; Tanimu, G.; Adamu, S.; Lateef, S.A.; Saleh, T.A. Technological trends in nanosilica synthesis and utilization in advanced treatment of water and wastewater. *Environ. Sci. Pollut. Res.* **2022**, *29*, 42560–42600. [CrossRef]
16. Crini, G.; Lichtfouse, E. Advantages and disadvantages of techniques used for wastewater treatment. *Environ. Chem. Lett.* **2019**, *17*, 145–155. [CrossRef]
17. Eccles, H. Treatment of metal-contaminated wastes: Why select a biological process? *Trends Biotechnol.* **1999**, *17*, 462–465. [CrossRef]
18. Da'ana, D.A.; Zouari, N.; Ashfaq, M.Y.; Abu-Dieyeh, M.; Khraisheh, M.; Hijji, Y.M.; Al-Ghouti, M.A. Removal of toxic elements and microbial contaminants from groundwater using low-cost treatment options. *Curr. Pollut. Rep.* **2021**, *7*, 300–324. [CrossRef]
19. Ali, A.M.; Shahbaz, M.; Shahzad, K.; Inayat, M.; Naqvi, S.; Al-Zahrani, A.A.; Rashid, M.I.; Rehan, M.; Mahpudz, A.B. Polygeneration syngas and power from date palm waste steam gasification through an Aspen Plus process modeling. *Fuel* **2023**, *332*, 126120. [CrossRef]
20. Faiad, A.; Alsmari, M.; Ahmed, M.M.; Bouazizi, M.L.; Alzahrani, B.; Alrobei, H. Date palm tree waste recycling: Treatment and processing for potential engineering applications. *Sustainability* **2022**, *14*, 1134. [CrossRef]
21. Danish, M.; Hashim, R.; Ibrahim, M.M.; Sulaiman, O. Optimized preparation for large surface area activated carbon from date (*Phoenix dactylifera* L.) stone biomass. *Biomass Bioenergy* **2014**, *61*, 167–178. [CrossRef]
22. Sabzevari, A.; Kabiri, K. Converting date seed biomass into highly absorbing hydrogel. *Iran. Polym. J.* **2016**, *25*, 597–606. [CrossRef]
23. Abu-Jrai, A.M.; Jamil, F.; Ala'a, H.; Baawain, M.; Al-Haj, L.; Al-Hinai, M.; Al Abri, M.; Rafiq, S. Valorization of waste Date pits biomass for biodiesel production in presence of green carbon catalyst. *Energy Convers. Manag.* **2017**, *135*, 236–243. [CrossRef]
24. Elnajjar, E.; Syam, M.M.; Al-Omari, S.A.B. Experimental investigations of bio-syngas production using microwave pyrolysis of UAE'S palm date seed pits. *Fuel* **2021**, *303*, 121348. [CrossRef]
25. Bhatia, D.; Sharma, N.R.; Singh, J.; Kanwar, R.S. Biological methods for textile dye removal from wastewater: A review. *Crit. Rev. Environ. Sci. Technol.* **2017**, *47*, 1836–1876. [CrossRef]
26. Bayomie, O.S.; Kandeel, H.; Shoeib, T.; Yang, H.; Youssef, N.; El-Sayed, M.M. Novel approach for effective removal of methylene blue dye from water using fava bean peel waste. *Sci. Rep.* **2020**, *10*, 7824. [CrossRef] [PubMed]
27. Góralczyk-Bińkowska, A.; Długoński, A.; Bernat, P.; Długoński, J.; Jasińska, A. Environmental and molecular approach to dye industry waste degradation by the ascomycete fungus *Nectriella pironii*. *Sci. Rep.* **2021**, *11*, 23829. [CrossRef] [PubMed]
28. Yaseen, D.A.; Scholz, M. Textile dye wastewater characteristics and constituents of synthetic effluents: A critical review. *Int. J. Environ. Sci. Technol.* **2019**, *16*, 1193–1226. [CrossRef]
29. Li, Y.; Meas, A.; Shan, S.; Yang, R.; Gai, X.; Wang, H.; Tsend, N. Hydrochars from bamboo sawdust through acid assisted and two-stage hydrothermal carbonization for removal of two organics from aqueous solution. *Bioresour. Technol.* **2018**, *261*, 257–264. [CrossRef]
30. Qian, W.C.; Luo, X.P.; Wang, X.; Guo, M.; Li, B. Removal of methylene blue from aqueous solution by modified bamboo hydrochar. *Ecotoxicol. Environ. Saf.* **2018**, *157*, 300–306. [CrossRef]
31. Islam, M.A.; Ahmed, M.J.; Khanday, W.A.; Asif, M.; Hameed, B.H. Mesoporous activated coconut shell-derived hydrochar prepared via hydrothermal carbonization-NaOH activation for methylene blue adsorption. *J. Environ. Manag.* **2017**, *203*, 237–244. [CrossRef]
32. Tran, T.H.; Le, A.H.; Pham, T.H.; Nguyen, D.T.; Chang, S.W.; Chung, W.J.; Nguyen, D.D. Adsorption isotherms and kinetic modeling of methylene blue dye onto a carbonaceous hydrochar adsorbent derived from coffee husk waste. *Sci. Total Environ.* **2020**, *725*, 138325. [CrossRef] [PubMed]
33. Zhou, F.; Li, K.; Hang, F.; Zhang, Z.; Chen, P.; Wei, L.; Xie, C. Efficient removal of methylene blue by activated hydrochar prepared by hydrothermal carbonization and NaOH activation of sugarcane bagasse and phosphoric acid. *RSC Adv.* **2022**, *12*, 1885–1896. [CrossRef]
34. Lotfiman, S.; Awang Biak, D.R.; Ti, T.B.; Kamarudin, S.; Nikbin, S. Influence of date syrup as a carbon source on bacterial cellulose production by *Acetobacter xylinum* 0416. *Adv. Polym. Technol.* **2018**, *37*, 1085–1091. [CrossRef]

35. Sevilla, M.; Fuertes, A.B. The production of carbon materials by hydrothermal carbonization of cellulose. *Carbon* **2009**, *47*, 2281–2289. [CrossRef]
36. Sun, X.; Li, Y. Colloidal carbon spheres and their core/shell structures with noble-metal nanoparticles. *Angew. Chem. Int. Ed. Engl.* **2004**, *116*, 607–611. [CrossRef]
37. Nasser, R.A.; Salem, M.Z.; Hiziroglu, S.; Al-Mefarrej, H.A.; Mohareb, A.S.; Alam, M.; Aref, I.M. Chemical analysis of different parts of date palm (*Phoenix dactylifera* L.) using ultimate, proximate and thermo-gravimetric techniques for energy production. *Energies* **2016**, *9*, 374. [CrossRef]
38. Romero-Anaya, A.J.; Lillo-Ro'denas, M.A.; Salinas-Martı'nez de Lecea, C.; Linares-Solano, A. Hydrothermal and conventional H3PO4 activation of two natural bio-fibers. *Carbon* **2012**, *50*, 3158–3169. [CrossRef]
39. Romero-Anaya, A.J.; Ouzzine, M.; Lillo-Rodenas, M.A.; Linares-Solano, A. Spherical carbons: Synthesis, characterization and activation processes. *Carbon* **2014**, *68*, 296–307. [CrossRef]
40. Boehm, H.-P. Chemical identification of surface groups. In *Advances in Catalysis*; Eley, D.D., Pines, H., Weisz, P.B., Eds.; Academic Press: Cambridge, UK, 1966; pp. 179–274.
41. Boehm, H.-P. Surface chemical characterization of carbons from adsorption studies. In *Adsorption by Carbons*; Elsevier: Amsterdam, The Netherlands, 2008; pp. 301–327.
42. Genli, N.; Kutluay, S.; Baytar, O.; Şahin, Ö. Preparation and characterization of activated carbon from hydrochar by hydrothermal carbonization of chickpea stem: An application in methylene blue removal by RSM optimization. *Int. J. Phytoremediat.* **2021**, *24*, 1–13. [CrossRef]
43. Everett, D.H. Manual of symbols and terminology for physicochemical quantities and units, appendix II: Definitions, terminology and symbols in colloid and surface chemistry. *Pure Appl. Chem.* **1971**, *31*, 577–638. [CrossRef]
44. De Souza Macedo, J.; da Costa Júnior, N.B.; Almeida, L.E.; da Silva Vieira, E.F.; Cestari, A.R.; de Fátima Gimenez, I.; Villarreal Carreno, N.L.; Barreto, L.S. Kinetic and calorimetric study of the adsorption of dyes on mesoporous activated carbon prepared from coconut coir dust. *J. Colloid Interface Sci.* **2006**, *298*, 515–522. [CrossRef]
45. Jia, P.; Tan, H.; Liu, K.; Gao, W. Removal of Methylene Blue from Aqueous Solution by Bone Char. *Appl. Sci.* **2018**, *8*, 1903. [CrossRef]
46. Fernandez, M.E.; Ledesma, B.; Román, S.; Bonelli, P.R.; Cukierman, A.L. Development and characterization of activated hydrochars from orange peels as potential adsorbents for emerging organic contaminants. *Bioresour. Technol.* **2015**, *183*, 221–228. [CrossRef] [PubMed]
47. Zheng, Q.; Morimoto, M.; Takanohashi, T. Production of carbonaceous microspheres from wood sawdust by a novel hydrothermal carbonization and extraction method. *RSC Adv.* **2017**, *7*, 42123–42128. [CrossRef]
48. Eltaweil, A.S.; Abd El-Monaem, E.M.; El-Subruiti, G.M.; Ali, B.M.; Abd El-Latif, M.M.; Omer, A.M. Graphene oxide incorporated cellulose acetate beads for efficient removal of methylene blue dye; isotherms, kinetic, mechanism and co-existing ions studies. *J. Porous Mater.* **2023**, *30*, 607–618. [CrossRef]
49. Basha, K.; El-Monaem, A.; Eman, M.; Khalifa, R.E.; Omer, A.M.; Eltaweil, A.S. Sulfonated graphene oxide impregnated cellulose acetate floated beads for adsorption of methylene blue dye: Optimization using response surface methodology. *Sci. Rep.* **2022**, *12*, 1–17. [CrossRef]
50. Hurairah, S.N.; Lajis, N.M.; Halim, A.A. Methylene blue removal from aqueous solution by adsorption on *Archidendron jiringa* seed shells. *J. Geosci. Environ. Prot.* **2020**, *8*, 128–143. [CrossRef]
51. Bharathi, K.S.; Ramesh, S.T. Removal of dyes using agricultural waste as low-cost adsorbents: A review. *Appl. Water Sci.* **2013**, *3*, 773–790. [CrossRef]
52. Langmuir, I. The constitution and fundamental properties of solids and liquids. Part I. Solids. *J. Am. Chem. Soc.* **1916**, *38*, 2221–2295. [CrossRef]
53. Freundlich, H. Über die adsorption in lösungen. *Z. Für Phys. Chem.* **1907**, *57*, 385–470. [CrossRef]
54. Fytianos, K.; Voudrias, E.; Kokkalis, E. Sorption-desorption Behaviour of 2, 4-dichlorophenol by Marine Sediments. *Chemosphere* **2000**, *40*, 3–6. [CrossRef]
55. Temkin, M.I. Adsorption equilibrium and kinetics of processes on nonhomogeneous surfaces and at interaction between adsorbed molecules. *Zh. Fiz. Khim.* **1941**, *15*, 296–332.
56. Sun, Z.; Liu, Y.; Srinivasakannan, C. Green Preparation and Environmental Application of Porous Carbon Microspheres. *ChemistrySelect* **2020**, *5*, 9308–9312. [CrossRef]
57. Salimi, M.; Balou, S.; Kohansal, K.; Babaei, K.; Tavasoli, A.; Andache, M. Optimizing the preparation of meso-and microporous canola stalk-derived hydrothermal carbon via response surface methodology for methylene blue removal. *Energy Fuels* **2017**, *31*, 12327–12338. [CrossRef]
58. Ferrentino, R.; Ceccato, R.; Marchetti, V.; Andreottola, G.; Fiori, L. Sewage sludge hydrochar: An option for removal of methylene blue from wastewater. *Appl. Sci.* **2020**, *10*, 3445. [CrossRef]
59. Madduri, S.; Elsayed, I. Novel oxone treated hydrochar for the removal of Pb (II) and methylene blue (MB) dye from aqueous solutions. *Chemosphere* **2020**, *260*, 127683. [CrossRef]
60. Li, B.; Guo, J.; Lv, K.; Fan, J. Adsorption of methylene blue and Cd (II) onto maleylated modified hydrochar from water. *Environ. Pollut.* **2019**, *254*, 113014. [CrossRef]

61. Zou, W.; Li, K.; Bai, H.; Shi, X.; Han, R. Enhanced cationic dyes removal from aqueous solution by oxalic acid modified rice husk. *J. Chem. Eng. Data* **2011**, *56*, 1882–1891. [CrossRef]
62. Raposo, F.; De La Rubia, M.A.; Borja, R. Methylene blue number as useful indicator to evaluate the adsorptive capacity of granular activated carbon in batch mode: Influence of adsorbate/adsorbent mass ratio and particle size. *J. Hazard. Mater.* **2009**, *165*, 291–299. [CrossRef]
63. Ho, Y.S.; McKay, F.G. Kinetic Models for the Sorption of Dye from Aqueous Solution by Wood. *Process Saf. Environ. Prot.* **1998**, *76*, 183–191. [CrossRef]
64. Ho, Y.S.; McKay, F.G. Pseudo-Second Order Model for Sorption Processes. *Process Biochem.* **1999**, *34*, 451–465. [CrossRef]

coatings

MDPI

Article

Enhancing Magnesium Phosphate Cement Paste for Efficient Fluoride Adsorption

Sana Gharsallah [1], Abdulrahman Alsawi [2,*], Abdulelah H. Alsulami [3], Clarence Charnay [4] and Mahmoud Chemingui [1]

1 Laboratory of Inorganic Chemistry, LR17-ES-07, Faculty of Science, University of Sfax, Sfax 3018, Tunisia; sana.gharsallah.etud@fss.usf.tn (S.G.); mahmoud.chmingui@fss.usf.tn (M.C.)
2 Department of Physics, College of Science, Qassim University, Buraidah 51452, Saudi Arabia
3 Chemistry Department, Faculty of Science and Arts in Baljurashi, Al-Baha University, Al-Baha 65527, Saudi Arabia; aalsulami@bu.edu.sa
4 Charles Gerhard Institut, UMR-5253 CNRS-UM-ENSCM, University of Montpellier, Place E, Bataillon, F-34095 Montpellier cedex 5, France; clarence.charnay@umontpellier.fr
* Correspondence: ansaoy@qu.edu.sa

Abstract: In the present study, we explore the synthesis and characterization of novel composite materials derived from magnesium phosphate cement by incorporating varying quantities of aluminum, iron oxide, or alumina. These composites demonstrate promising properties related to water resistance and significant specific surface areas. Furthermore, our investigations reveal that aluminum, iron oxide, and phosphate constituents exhibit an affinity for fluoride retention. Consequently, we apply these synthesized materials for fluoride adsorption. Our results indicate a noteworthy adsorption capacity, ranging from 2.35 mg/g for cement synthesized with 0.25 g of aluminum to 4.84 mg/g for materials synthesized with 1.5 g of aluminum. The influence of incorporating alumina or iron oxide into these matrices is thoroughly examined. Additionally, we investigated the optimal conditions utilizing a range of analytical techniques, including scanning electron microscopy (SEM) equipped with an energy dispersive spectrometer (EDS), the Brunauer–Emmett–Teller (BET) method, X-ray diffraction (XRD), Fourier-transform infrared spectroscopy (FTIR), and thermogravimetric analyses (TGA). To further elucidate this process, we perform equilibrium modeling and present experimental data in accordance with the Langmuir, Freundlich, Temkin, and Dubinin–Radushkevich isotherms.

Keywords: magnesium phosphate cement (MPC); X-ray diffraction; water purification; fluoride adsorption; isotherms; scanning electron microscopy

Citation: Gharsallah, S.; Alsawi, A.; Alsulami, A.H.; Charnay, C.; Chemingui, M. Enhancing Magnesium Phosphate Cement Paste for Efficient Fluoride Adsorption. *Coatings* **2024**, *14*, 9. https://doi.org/10.3390/coatings14010009

Academic Editors: Shoufeng Tang and Deling Yuan

Received: 24 November 2023
Revised: 17 December 2023
Accepted: 18 December 2023
Published: 20 December 2023

1. Introduction

Pollution presents an increasing threat to water, a basic resource that is essential to human survival. Pollution disturbs aquatic ecosystems and puts water safety at risk. Elevated amounts of harmful substances, either naturally occurring or due to human activity, are the cause of the contamination; the main contributors include agriculture, industrial operations, natural factors, and inadequate sewage treatment. In addition to affecting the health of aquatic ecosystems' fauna and flora, pollution also raises questions regarding the safety of water for human use and consumption [1,2].

Personalized and successful treatments are necessary to address these issues. Customized techniques are necessary due to the complexity of pollutants, which involves taking into account the many sources and types of contaminants found in water. Reducing water pollution requires strong sewage treatment facilities, stringent industrial discharge laws, and environmentally friendly farming methods. Furthermore, maintaining the long-term health of aquatic ecosystems and preserving the supply of safe water for the environment and human population depends heavily on public knowledge and active participation in water conservation initiatives [1–3]. Even at trace concentration levels in water, some toxic elements can harm human health, which is why it is imperative to perform accurate studies on water quality. Commonly found in underground water are ions such as fluoride, nitrate, phosphate, and arsenic, which can be harmful to humans if ingested in excess. The third most common element in nature is fluoride, which is also one of the essential trace elements that, when consumed in the right amounts, has positive effects on the human skeletal system and teeth [4]. Fluoride, a prevalent natural contaminant, is found in the air, plants, and water. Nevertheless, to avert fluorosis, which predominantly impacts teeth and bones and leads to bone weakening by influencing the assimilation of calcium and phosphorus over an extended period, it is imperative to ensure that its concentration does not surpass 1.5 mg/L, according to the World Health Organization [2,4]. Additionally, elevated levels of fluoride can lead to liver damage, bone cancer, Alzheimer's syndrome, alterations in DNA structure, brain damage, and a reduction in children's IQ [5,6]. Effective treatment techniques include adsorption, reverse osmosis, membrane separation, ion exchange, and chemical treatment [5,6]. Among these, adsorption stands out for its flexibility, high efficiency, ease of operation, and cost-effectiveness compared to alternative techniques [7,8]. The utilization of adsorbent materials can often be economically viable, offering a practical solution for water treatment needs [9]. Commonly used adsorbents for fluoride removal include bone char, chitosan, activated carbon, zirconium–carbon hybrids, aluminum-based material, activated alumina, and other alternative and economical natural materials [7,10,11]. Moreover, materials that include rare earth elements are regarded as prospective adsorbents owing to their strong affinity for fluoride. Various metals, such as Y (III), Zr (IV), La (III), Ce (IV), Al (III), Cu (II), and Fe (III), are incorporated into carbonates, oxides, and hydroxides and play a significant role in capturing fluoride in water. Multiple research studies have highlighted the efficacy of fluoride adsorption with materials such as alumina, aluminum, iron oxide, and phosphate. This is particularly notable as iron oxide, known for its cost-effectiveness, and composites, including aluminum oxide, exhibit significant fluoride adsorption capacity [3,7,8]. To incorporate these diverse adsorbents, we opted to synthesize composite materials using magnesium phosphate cement. This involved introducing varying amounts of aluminum, alumina, or iron oxide to investigate their respective adsorption capacities.

Magnesium phosphate cement (MPC), a type of chemical cement, was discovered and developed in the late 19th century [11]. Classified within the domain of construction materials, magnesium phosphate cements (MPCs) are recognized as a specific category of inorganic cementitious materials belonging to the phosphate cement family [12]. These cementitious materials are formulated at ambient room temperature through a neutralization reaction. This process involves combining basic magnesia with an acidic solution containing phosphates, commonly ammonium dihydrogen phosphate (ADP), supplemented with a setting retarder, typically from the boron family [10,13]. The chemical reaction yields the formation of the crystalline phase identified as struvite. This complex reaction can be accurately described by the following chemical equation:

$$MgO + NH_4H_2PO_4 + 5H_2O \rightarrow MgNH_4PO_4{\cdot}6H_2O \text{ (struvite)} \quad (1)$$

The uniqueness of this cement stems from its outstanding features: minimal permeability, accelerated strength gain, high initial strength, adhesive properties, and minimal drying shrinkage; moreover, MPCs demonstrate superior adaptability to changes in environmental temperature, a swift hydration process, solid durability, and rapid setting [14,15]. The distinctive characteristics of magnesium phosphate cement (MPC) pave the way for a wide range of applications, extending across civil engineering and the solidification and stabilization of nuclear waste [6]. Furthermore, within the medical domain, this cement demonstrates its value by serving applications in biomedical materials and contributing to bone restoration [13,16,17]. Despite the interesting characteristics of this type of cement, it presents disadvantages. It is essential to delve deeper into these persistent drawbacks. Its limited water resistance poses a challenge, and solving this problem requires innovative solutions. Its short, difficult-to-regulate setting period can hinder flexibility in various construction applications, requiring advanced techniques to extend working times without compromising performance. Additionally, reduced porosity, although characteristic of magnesium phosphate cement, requires research and development to improve its permeability for specific use cases. By addressing these aspects, progress can be made to alleviate the disadvantages and further optimize the performance of magnesium phosphate cement in practical applications [12]. In recent years, there has been a growing focus on the study of MPC-based repair materials. Research efforts primarily delve into the refinement of their preparation techniques, the exploration of modification mechanisms, and the assessment of their performance in conventional environmental settings.

The primary objective of this study is to enhance the resistance of magnesium phosphate cement when exposed to water by developing various composite materials. These composites will incorporate reinforcing elements such as aluminum, alumina, or iron oxide. Additionally, the study aims to investigate and optimize their capacity for adsorbing fluorine, a critical aspect of water treatment processes.

2. Material and Methods

2.1. Materials

In this study, cement pastes were meticulously prepared by blending weighted portions of powders, including high-purity magnesium oxide (MgO > 99%—Merck KGaA Frankfurter Str.25, Darmstadt, Germany), dihydrogen ammonium phosphate ($NH_4H_2PO_4$ > 99%—SIGMA-ALDRICH (Darmstadt, Germany)), borax, aluminum, alumina, and iron oxide. All raw materials were sourced from Sigma Aldrich (Darmstadt, Germany), ensuring consistency and quality. To reduce reactivity, magnesium oxide was calcined at 1500 °C for 6 h, resulting in dead-burned magnesia. Furthermore, in this research, we chose to exclude aggregates. The omission of aggregates is essential to preserving material purity during synthesis, as they have the potential to jeopardize the quality of the end product. The process entails regulating particle size, and the optimization of synthesis conditions involves fine-tuning parameters to minimize the likelihood of aggregates while maximizing yield [18].

2.2. Preparation

The synthetic protocol for creating the diverse composites starts with the mixing process, which involves carefully combining magnesium and acid in a precise ratio of Mg/P = 1, with the addition of distilled water. Following this, an optimal quantity of aluminum powder, aluminum oxide, or iron oxide is added, and rapid manual mixing ensues until a uniform paste is achieved. This step-by-step approach ensures a homogeneous distribution of components in the composites.

Various materials are formulated using a consistent cement matrix. The Al-cement is prepared by incorporating 1.5320 g of aluminum, while the sample S_1 includes 1.5640 g of aluminum and an additional 1.5320 g of Fe_2O_3. Moreover, the 0.25 g Al-cement is prepared using only 0.25 g of aluminum. The fourth synthesized material, named S_2, contains 0.2500 g of aluminum and 0.5470 g of alumina (Al_2O_3). These various synthesized materials are subsequently characterized for optimal quantities and utilized for the retention of fluorides.

Table 1 provides an overview of the chemical compositions of the various composites.

Table 1. Chemical compositions of synthesized samples.

	MgO (g)	$NH_4H_2PO_4$ (g)	Borax (g)	Al (g)	Al_2O_3 (g)	Fe_2O_3(g)	H_2O (mL)
Al-cement	1.0680	2.8580	0.3115	1.5320	-	-	2
S_1	1.0320	2.8628	0.3330	1.5640	-	1.5320	2
0.25 g Al-cement	1.0790	2.8800	0.3310	0.2560	-	-	2
S_2	1.0390	2.8580	0.3250	0.2500	0.5470	-	2

2.3. Analysis Method

The powder structure analysis was conducted using wide-angle X-ray diffraction. For every cement paste preparation, we used the X'per PRO PANalytical powder diffractometer (Philips, Farnborough, UK) with CuKα radiation (λKα = 1.54 Å) to identify the crystalline phases. Moreover, a comparative study of compounds synthesized using pure cement and Al-cement was provided.

Diffraction patterns were collected in the range of $10° < 2\theta < 80°$. The microstructures of the magnesium phosphate cement (MPC) pastes were examined using scanning electron microscopy (SEM) equipped with an energy dispersive spectrometer (EDS) (SEM, (JEOL)-Akishima, Japan) to analyze their elemental compositions. Thermal behavior was investigated through a thermogravimetric analysis (TGA, Perkin-Elmer, Waltham, MA, USA), tracking the mass evolution of samples subjected to temperature increments from 20 to 1150 °C at a heating rate of 20 °C/min^{-1}. Additionally, we determined the specific surface areas using the Brunauer–Emmett–Teller (BET) method (TriStar 3000 V6.06 A, Stevenage, UK) to examine the dehydration characteristics of struvite. Furthermore, the FTIR spectra were recorded in the 4000–400 cm^{-1} range using 32 scans at a nominal resolution of 4 cm^{-1} using a Perkin Elmer 100 FT spectrophotometer equipped (FTIR, Perkin-Elmer, USA) with an ATR Diamant unit and an MCT detector (liquid nitrogen).

2.4. Adsorption Study

The fluoride adsorption process was conducted at a consistent pH and room temperature (25 ± 2 °C) across all samples. The mass of solids used was standardized at 0.3 g for all materials. The solids were introduced into 20 mL tubes, each containing a fluoride solution with different initial concentrations ranging from 20 to 400 mg/L. The mixtures were stirred at 200 rev/min for a duration of one night (15 h). Subsequently, 0.2 µm filters were employed for filtration, and the filtrate was analyzed for its residual fluoride concentration using a fluorine-specific electrode.

The following equation (Equation (2)) was used to determine the amount of adsorbed fluoride [19], q_e (mg/g):

$$q_e = \frac{(C_0 - C_e) \times V}{m} \quad (2)$$

where m is the mass of the adsorbent (g), V is the volume of the solution (L), and C_0 and C_e are the initial and final concentrations of fluoride in the solution (mg/L).

Throughout the experiments, the pH is intentionally maintained constant within the range of 8 to 9. This choice is deliberate to minimize the potential influence of pH fluctuations on the adsorption process, thereby ensuring the reliability and consistency of our results.

3. Results and Discussion

3.1. Composites Analyses

The microstructures of the various synthesized cement pastes were examined and are presented in Figure 1. The SEM images are organized as follows: Figure 1(a1,a2) correspond to the material named S_1, Figure 1(b1,b2) relate to material S_2, and Figure 1(c1,c2) specifically represent Al-cement, a sample prepared with 1.5 g of aluminum. Additionally, the EDX analysis of materials S1 and S2 is presented in Figure 2.

As shown in Figure 1, the microstructures of the various materials exhibit a dense composition with tubular crystals predominantly in rod-like forms [6]. Additionally, a minor level of porosity is observed, primarily resulting from the significant gas evolution during the synthesis. Despite this, the structure demonstrates heightened strength, enhanced resistance to water contact, and remarkable stability. Notably, even after continuous stirring for several days, the solid structure remains intact.

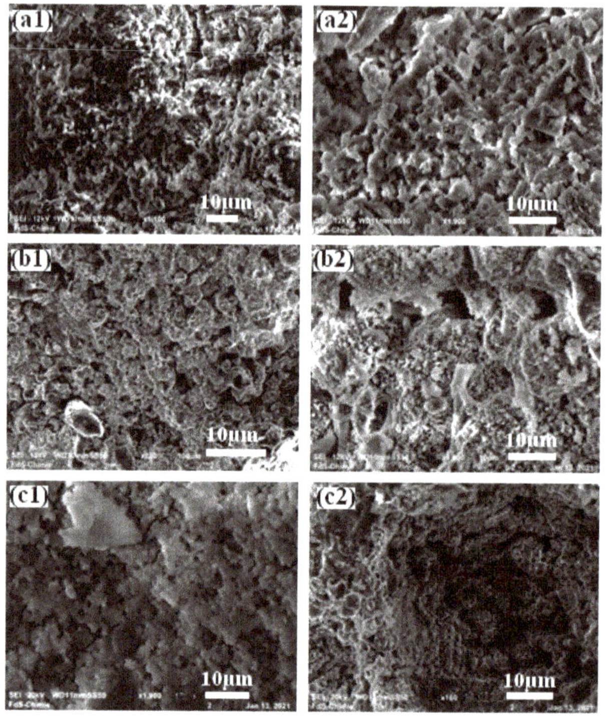

Figure 1. SEM of composites (**a1**,**a2**) S_1 and (**b1**,**b2**) S_2 and (**c1**,**c2**) Al-cement samples.

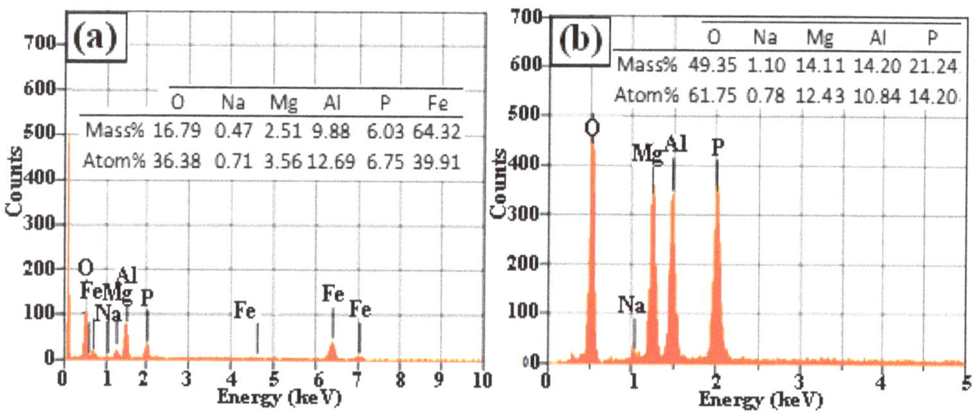

Figure 2. EDS analyses of (**a**) S_1 and (**b**) S_2 composites.

On the other hand, the incorporation of iron oxide yields a denser and more compact structure. The EDS analysis of this material validates the presence of trace amounts of phosphorus (P), sodium (Na), magnesium (Mg), and aluminum (Al) (Figure 2). This analysis confirms nearly complete acid dissolution and minimal unreacted MgO while underscoring the prominent presence of iron oxide. Moreover, the EDS analysis of the composite synthesized with an alumina addition confirms the presence of aluminum (Al), along with traces of magnesium (Mg), phosphorus (P), and sodium (Na).

The thermal gravimetric analysis (TGA) curves for the various composites are illustrated in Figure 3. Both synthesized materials display similar thermal behaviors, characterized by an initial weight loss starting at T = 50 °C and the process finishing near T = 700 °C. The total mass loss for S_1 is approximately 22.89%, whereas for S_2, it amounts to approximately 39.86%. This mass loss can be attributed to the decomposition of the material, occurring in two distinct stages [6,11].

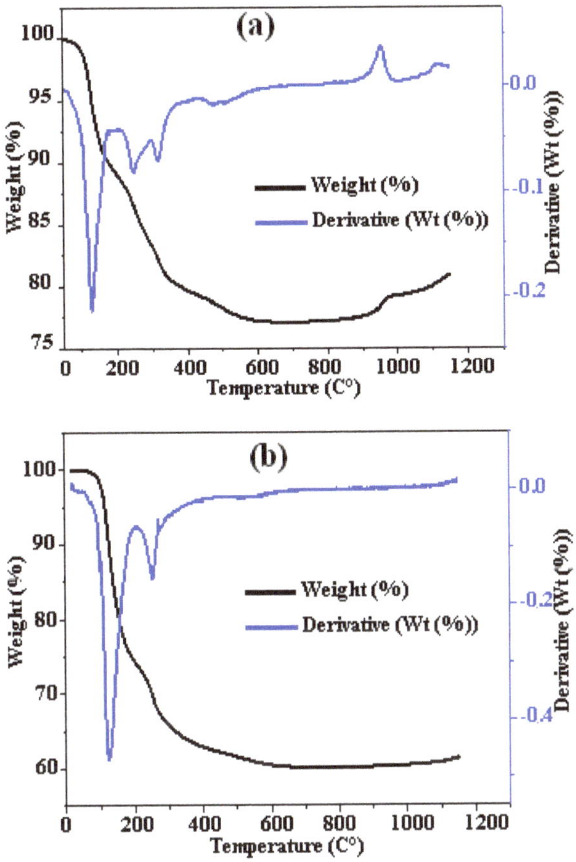

Figure 3. TGA curves of different composites (**a**) S_1 and (**b**) S_2.

The decomposition observed between 50 and 300 °C is attributed to dehydration [6,10], represented by the following reaction:

$$MgNH_4PO_4 \cdot 6H_2O \rightarrow MgNH_4PO_4 + 6H_2O_{(g)} \quad (3)$$

The decomposition reaction of the material occurs in the interval 300 to 700 °C, expressed in terms of:

$$MgNH_4\ PO_4 \rightarrow MgHPO_4 + NH_{3\,(g)} \quad (4)$$

For both composites synthesized with 1.5 g of aluminum, the specific surface of the aluminum-based cement is greater than that prepared by adding the same amount of iron oxide (Table 2). As shown, one can note that it is almost three times greater.

Table 2. Specific surface area, pore volume, and pore size values of different prepared samples.

	BET Surface	Pore Volume	Pore Size
1.5 g-Al-cement	64.0342 ± 3.7 m^2/g	0.060498 cm^3/g	37.7913 Å
S_1	21.8228 ± 1.2 m^2/g	0.030965 cm^3/g	56.7565 Å
0.25 g-Al-cement	83.524 ± 3.3 m^2/g	0.074313 cm^3/g	35.5886 Å
S_2	134.673 ± 2.3 m^2/g	0.15054 cm^3/g	44.715 Å

This increase is due to the strong release of gas during the synthesis of the cement. In contrast, the composites synthesized with only 0.25 g of aluminum or by adding 0.5 g of alumina exhibit a significantly greater surface area compared to the other two types of cement (Table 2). Additionally, the incorporation of alumina has a pronounced positive impact on the structural properties. According to Figure 4, it is evident that the specific surface area notably increases to approximately 134.6 m^2/g. For each of the synthesized materials, the hysteresis curve is type II, commonly observed in BET analyses, suggesting the appearance of multilayer adsorption and adsorption on low-porosity surfaces. These phenomena generally begin at higher pressures. The hysteresis loop is type H4, in particular, indicating that the sample includes divided pores with a size distribution mainly in the microporous range [6,11].

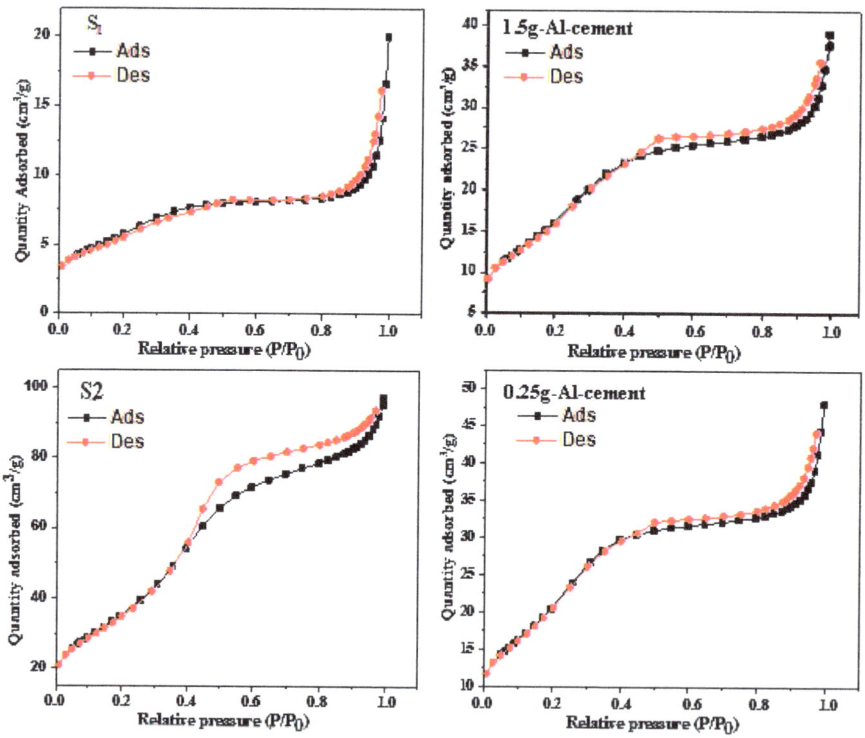

Figure 4. N_2 adsorption/desorption curves for different samples: 0.25-Al-cement, 1.5 g-Al-cement, S_1, and S_2.

An analysis of powder structure was conducted, and Figure 5 demonstrates a comparative study of the diffractograms of compounds synthesized using both pure cement and Al-cement.

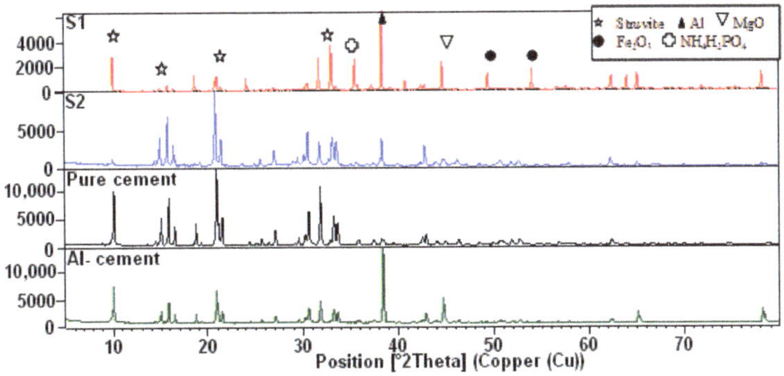

Figure 5. X-ray diffraction pattern of different composites: S_1, S_2, pure cement, and Al-cement.

The overlay of diffraction patterns confirms a distinct peak characteristic of struvite, which is the stable phase of cement, observed around 21°-2-theta. This peak signifies the crystallization of the material. Minor peaks attributed to unreacted magnesia were observed at approximately 43°, along with peaks indicating the presence of residual $NH_4H_2PO_4$ within the 30–33° 2-theta range. The main peaks associated with aluminum were identified in the 2θ range of 38–39°. Furthermore, iron oxide exhibited a prominent peak at an average of 50–54° 2-theta. It is noteworthy that no new phases were formed, indicating the material's ability to maintain its stable structure, attributed to the addition of aluminum, alumina, or iron oxide. Moreover, this enhanced composition resulted in improved cement hardness and water resistance [8,16].

Figure 6 presents the FTIR data for the composite cement. Both composites display similar bonding patterns, albeit with slight shifts towards shorter or longer wavelengths. The peak at 559 cm^{-1} is attributed to the bending modes of the P-O bonds in the phosphate groups. Distinctive Al-O stretching vibrations are observed in the range of 559 to 754 cm^{-1}, as well as at 1655 cm^{-1}, corresponding to the transient and stable phases of alumina (AlO_4 or AlO_6 vibrations). The adsorption bonds at 985 and 2346 cm^{-1} can be attributed to the asymmetry of P-O vibrations in PO_4^{3-} within the cement. The prominent peak at 2842 cm^{-1} is associated with the unsymmetrical stretching vibration of N–H in NH_4^+. Additionally, a smaller peak indicating the unsymmetrical bending vibration of the NH_4^+ group appears around 1432 cm^{-1}. Furthermore, the peak observed at approximately 3226 cm^{-1} confirms the presence of the –OH stretching bond in chemisorbed water [20–23]. For sample S1, a noticeable stretching vibration at 2940 and 2920 cm^{-1}, respectively, is specific to the CH2 and CH3 groups and is probably caused by the solvent used to clean the crystal. Also, the –OH stretching bond observed at 3560 cm^{-1} may be a further consequence of this cleaning solvent.

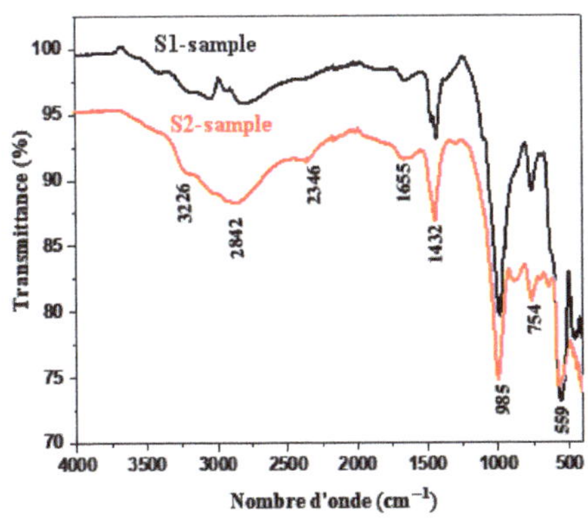

Figure 6. FTIR of different composites.

3.2. Adsorption Study

In order to assess the impact of various synthesized materials on fluoride removal efficiency, these solids were examined at varying concentrations. As depicted in Figure 7, it is evident that the rate of fluoride removal increased with higher initial concentrations.

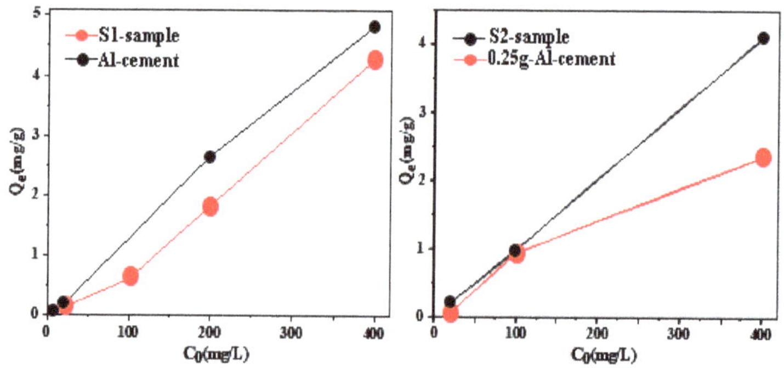

Figure 7. Influence of the initial concentration of fluoride on adsorption.

The adsorption capacity of various composite materials has been investigated and is presented in Figure 7. Notably, a comparison of the adsorption capacities between the two materials synthesized with 1.5 g of Al, one with and one without the addition of iron oxide, reveals that the Al-cement composite exhibits a higher adsorbed quantity compared to its iron oxide addition counterpart, measuring 4.84 mg/g and 4.29 mg/g, respectively. Furthermore, it is evident that the inclusion of alumina positively influences fluoride retention. For instance, the fluoride retention increases from 2.35 mg/g for the composite synthesized with 0.25 g of aluminum to 4.10 mg/g with the addition of only 0.5 g of alumina.

The adsorption isotherms illustrate the connection between the extent of adsorption and the concentration of the solute. By providing a detailed insight into the effectiveness of various materials, the isotherm enables the determination of the maximum quantity of adsorbed fluoride [22]. Equilibrium studies were conducted utilizing different isotherms. Various isotherms, such as the Langmuir, Freundlich, Temkin, and Dubinin–Radushkevich isotherms, are utilized to gain a comprehensive understanding of the adsorption process. The Langmuir isotherm effectively portrays monolayer adsorption, signifying that adsorbate molecules arrange themselves in a single layer on the adsorbent surface. This model assumes a uniform surface with a fixed number of identical adsorption sites, neglecting interactions between adsorbate molecules.

Conversely, the Freundlich isotherm characterizes multilayer adsorption, indicating that adsorbate molecules can form multiple layers on the surface. This model is suitable for heterogeneous surfaces with varying adsorption energies, providing flexibility in representing the adsorption process.

The Temkin isotherm takes into account the indirect interactions between the adsorbate and the adsorbent, introducing the concept of a uniform distribution of binding energies. It assumes a linear decrease in adsorption energy as the surface coverage increases.

Lastly, the Dubinin–Radushkevich isotherm is employed for microporous adsorbents and describes the adsorption process in terms of the volume filling of micropores.

By employing these diverse isotherms, we can unravel the complex dynamics of adsorption, capturing various aspects of the interaction between adsorbate molecules and the adsorbent surface [19,24]. Each isotherm is characterized by an equation that posits specific assumptions and identifies the prevailing adsorption mechanism, as presented in Table 3 [25,26]. Referring to Figure 8 and Table 3 and considering the correlation coefficient, R^2, it is evident that both materials adhere to Langmuir's isothermal model, indicating monolayer adsorption. Furthermore, the determined R_L separation factor affirms the favorable nature of the adsorption.

Table 3. Values of different parameters of Langmuir, Freundlich, Temkin, and Dubinin–Radushkevich isotherms for both composites S_1 and S_2.

Isotherm	Equation	S_1	S_2
Langmuir	$\frac{1}{q_e} = \frac{1}{q_m} + \frac{1}{q_m b C_e}$	b = 0.205 L/g q_0 = 4.87 mg/g R_l = 0.806 K_{ap} = −0.0010 R^2 = 0.998	b = 0.023 L/g q_0 = 43.47 mg/g R_l = 0.993 K_{ap} = 0.0016 R^2 = 0.988
Freundlich	$\log q_e = \log K_f + \frac{1}{n} \log C_e$	1/n = 1.197 K_f = 0.0038 R^2 = 0.999	1/n = 0.968 K_f = 0.0131 R^2 = 0.918
Temkin	$q_e = \frac{RT}{b} \ln A + \frac{RT}{b} \ln C_e$	A = 0.040 B = 1.281 R^2 = 0.740	A = 0.185 B =0.705 R^2 = 0.658
Dubinin–Radushkevich	$\ln q\,\varepsilon = \ln q_m - \beta\,\varepsilon^2$	q_d = 0.951 B = −135.2 R^2 = 0.750	q_d = 1.799 B = 101.7 R^2 = 0.729

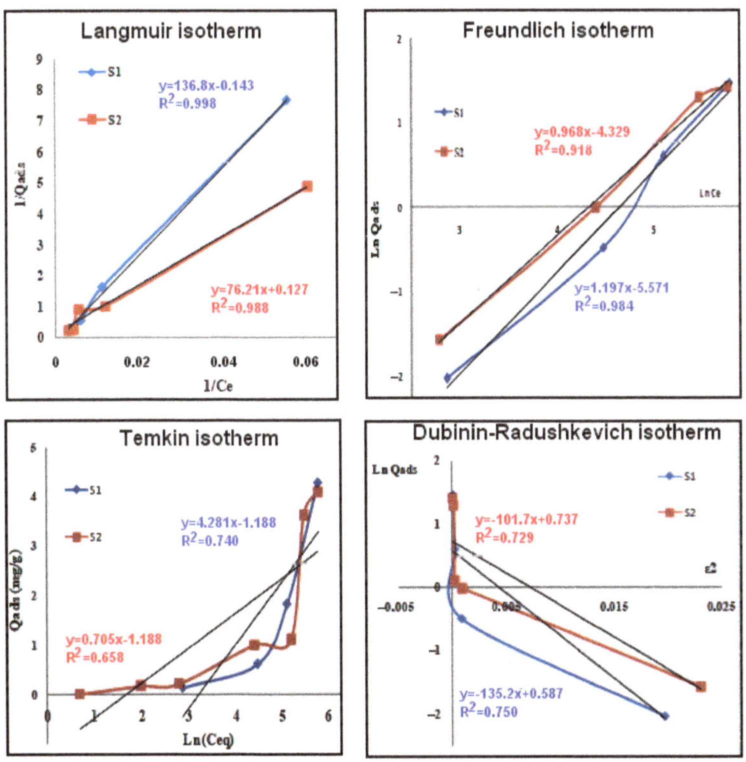

Figure 8. Modeling of composites according to Langmuir, Freundlich, Temkin, and Dubinin–Radushkevich isotherms.

3.3. Comparison with Other Used Adsorbents

A comparison of the two synthesized materials with other composites studied in previous articles [5,9], as well as with various adsorbents, reveals significant information (Table 4). The introduction of iron oxide into the cement matrix alongside aluminum slightly reduced the adsorption amount from 4.48 mg/g to 4.29 mg/g. Conversely, the incorporation of alumina results in an improved adsorption capacity, increasing from 2.35 mg/g to 4.10 mg/g with a simple addition of 0.5 g of alumina. It should be noted that the inclusion of alumina has a favorable impact on the retention of fluoride within the aluminum-based cement matrix compared to peroxide. The adsorbed quantity increases from 1.61 mg/g to 4.10 mg/g with the addition of alumina.

Table 4. Comparison of removal capacity of the present composites with others previously studied.

Adsorbent	Removal Capacity (mg/g)	pH	Refs.
Portland Pozzolana Cement	0.25	6–11.8	[27]
Calcite	0.39	6	[28]
Mg-HAP	1.40	9–10	[29]
Alumina cement	1.61	8–10	[10]
0.25 g Al-cement	2.35	8–9	Present study
1.5 g Al-cement	4.84	8–9	Present study
Composite S_1 cement	4.29	8–9	Present study
Composite S_2 cement	4.10	8–9	Present study

4. Conclusions

A novel composite material based on magnesium phosphate cement was successfully synthesized by incorporating varying amounts of aluminum, alumina, or iron oxide. The primary objective was to enhance the properties of this cement in both water and dry environments, as well as assess its capability to adsorb fluoride. The analyses conducted led to several significant conclusions:

- The addition of iron oxide, aluminum, and alumina significantly enhances the strength of the cement matrix, improving its resistance to water and dry conditions even after several days of direct contact. Notably, the addition of alumina has the additional benefit of reducing the exothermicity of the reaction, improving the cement's setting time.
- The addition of aluminum or alumina to the cement matrix enables an improvement in the specific surface area of the material.
- The highest adsorption capacity for fluoride observed was 4.84 mg/g, achieved with the material synthesized using 1.5 g of aluminum.
- Incorporating iron oxide into a matrix containing 1.5 g of aluminum slightly decreases the adsorbed quantity, shifting from 4.84 mg/g to 4.29 mg/g.
- An alumina addition markedly improves adsorption, especially for the composite initially containing only 0.25 g of aluminum, increasing from 2.35 mg/g to 4.10 mg/g with the addition of a mere 0.5 g of alumina.
- The Langmuir isotherm aptly describes the adsorption process, indicating monolayer adsorption, where adsorbate molecules form a single layer on the adsorbent surface. This model assumes a homogeneous surface with a fixed number of identical adsorption sites and does not account for interactions between adsorbate molecules. Additionally, the calculated separation factor, R_L, further confirms the favorable nature of the adsorption process.

The modification of magnesium phosphate cement with aluminum, iron oxide, or alumina demonstrates promising potential as an effective adsorbent for fluoride removal, showcasing remarkable fluoride adsorption capabilities.

Author Contributions: Conceptualization, A.H.A., C.C. and M.C.; Formal analysis, S.G. and M.C.; Investigation, A.A.; Resources, A.A.; Data curation, M.C.; Writing—original draft, S.G.; Writing—review & editing, A.A., A.H.A. and C.C.; Supervision, M.C. All authors have read and agreed to the published version of the manuscript.

Funding: This research received no external funding.

Institutional Review Board Statement: Not applicable.

Informed Consent Statement: Not applicable.

Data Availability Statement: Data can be requested from the authors.

Acknowledgments: Researchers would like to thank the Deanship of Scientific Research, Qassim University for funding the publication of this project.

Conflicts of Interest: The authors declare no conflict of interest.

References

1. Schweitzer, L.; Noblet, J. Chapter 3.6—Water Contamination and Pollution. *Green Chem.* **2018**, 261–290. [CrossRef]
2. Ben Mbarek, W.; Pineda, E.; Escoda, L.; Suñol, J.J.; Khitouni, M. High-efficiency decolorization of azo dye Reactive Black 5 by Ca-Al particles. *J. Environ. Chem. Eng.* **2017**, *5*, 6107–6113. [CrossRef]
3. Hafshejani, L.D.; Tangsir, S.E.; Daneshvar, M.; Maljanen, A.; Lähde, J.; Jokiniemi, M.; Naushad, M.; Bhatnagar, A. Optimization of fluoride removal from aqueous solution by Al_2O_3 nanoparticles. *J. Mol. Liq.* **2017**, *238*, 254–262. [CrossRef]
4. Borgohain, X.; Boruah, A.; Sarma, G.K.; Rashid, M.H. Rapid and extremely high adsorption performance of porous MgO nanostructures for fluoride removal from water. *J. Mol. Liq.* **2020**, *305*, 112799. [CrossRef]
5. Karkar, S.; Debnath, S.; De, P.; Parashar, K.; Pillay, K.; Sashikumar, P.; Ghosh, U.C. Preparation, characterization and evaluation of fluoride adsorption efficiency from water of iron-aluminium oxide-graphene oxide composite material. *Chem. Eng. J.* **2016**, *306*, 269–279. [CrossRef]
6. Gharsallah, S.; Alsawi, A.; Hammami, B.; Khitouni, M.; Charnay, C.; Chemingui, M. Synthesis and Characterization of New Composite Materials Based on Magnesium Phosphate Cement for Fluoride Retention. *Materials* **2023**, *16*, 718. [CrossRef] [PubMed]
7. Mondal, P.; Purkait, M.K. Preparation and characterization of novel green synthesized iron aluminum nanocomposite and studying its efficiency in fluoride removal. *Chemosphere* **2019**, *235*, 391–402. [CrossRef] [PubMed]
8. Gasparotto, J.M.; Roth, D.; Perilli, D.O.; Franco, D.S.P.; Carissimi, E.; Foletto, E.L.; Jahn, S.L.; Dotto, G.L. A novel Fe-Al-La trioxide composite: Synthesis, characterization, and application for fluoride ions removal from the water supply. *J. Environ. Chem. Eng.* **2021**, *9*, 106350. [CrossRef]
9. Liu, X.; Xiao, M.; Li, Y.; Chen, Z.; Yang, H.; Wang, X. Advanced porous materials and emerging technologies for radionuclides removal from Fukushima radioactive water. *Eco-Environ. Health* **2023**, *2*, 252–256. [CrossRef]
10. Li, Y.; Zhang, G.; Hou, D.; Wang, Z. Fluoride adsorption properties of three modified forms of activated alumina in drinking water. *J. Water Health* **2014**, *12*, 715–721. [CrossRef]
11. Gharsallah, S.; Mallah, A.; Alsawi, A.; Hammami, B.; Khitouni, M.; Charnay, C.; Chemingui, M. Study of Modified Magnesium Phosphate Cement for Fluoride Removal. *Materials* **2023**, *16*, 5749. [CrossRef]
12. Li, Y.; Zhang, G.; Hou, D.; Wang, Z. Nanoscale insight on the initial hydration mechanism of magnesium phosphate cement. *Constr. Build. Mater.* **2021**, *276*, 122213. [CrossRef]
13. Ribeiro, D.V.; Morelli, M.R. Influence of the addition of grinding dust to a magnesium phosphate cement matrix. *Constr. Build. Mater.* **2009**, *23*, 3094–3102. [CrossRef]
14. Le Rouzic, M.; Chaussadent, T.; Stefan, L.; Saillio, M. On the influence of Mg/P ratio on the properties and durability of magnesium potassium phosphate cement pastes. *Cem. Concr. Res.* **2017**, *96*, 27–41. [CrossRef]
15. Tansel, B.; Lunn, G.; Monje, O. Struvite formation and decomposition characteristics for ammonia and phosphorus recovery: A review of magnesium-ammonia-phosphate interactions. *Chemosphere* **2018**, *194*, 504–514. [CrossRef]
16. Zhang, Q.; Cao, X.; Ma, R.; Sun, S.; Fang, L.; Lin, J.; Luo, J. Solid waste-based magnesium phosphate cements: Preparation, performance and solidification/stabilization mechanism. *Constr. Build. Mater.* **2021**, *297*, 123761. [CrossRef]
17. Lu, X.; Chen, B. Experimental study of magnesium phosphate cements modified by metakaolin. *Constr. Build. Mater.* **2016**, *123*, 719–726. [CrossRef]
18. Bouaoun, I.; Hammi, H.; M'nif, A. Box-Behnken design optimization of magnesium potassium phosphate cement properties using sodium chloride as retarder. *J. Tunis. Chem. Soc.* **2016**, *18*, 152–159.

19. Limousin, G.; Gaudet, J.P.; Charlet, L.; Szenknect, S.; Barthès, V.; Krimissa, M. Sorption isotherms: A review on physical bases, modeling and measurement. *Appl. Geochem.* **2007**, *22*, 249–275. [CrossRef]
20. Ding, Z.; Dong, B.; Xing, F.; Han, N.; Li, Z. Cementing mechanism of potassium phosphate based magnesium phosphate cement. *Ceram. Int.* **2012**, *38*, 6281–6288. [CrossRef]
21. Sotiriadis, K.; Mácová, P.; Mazur, A.S.; Tolstoy, P.M.; Viani, A. A solid state NMR and in-situ infrared spectroscopy study on the setting reaction of magnesium sodium phosphate cement. *J. Non-Cryst. Solids* **2018**, *498*, 49–59. [CrossRef]
22. Spirovski, F.; Kuzmanovski, I.; Lutz, H.D.; Engelen, B. Infrared and Raman spectra of magnesium ammonium phosphate hexahydrate (struvite) and its isomorphous analogs. I. Spectra of protiated and partially deuterated magnesium potassium phosphate hexahydrate. *Mol. Struct.* **2004**, *689*, 110. [CrossRef]
23. Chu, Y.; Khan, M.A.; Zhu, S.; Xia, M.; Lei, W.; Wang, F.; Xu, Y. Microstructural modification of organo-montmorillonite with Gemini surfactant containing four ammonium cations: Molecular dynamics (MD) simulations and adsorption capacity for copper ions. *J. Chem. Technol. Biotechnol.* **2019**, *94*, 3585–3594. [CrossRef]
24. Foo, K.Y.; Hameed, B.H. Insights into the modeling of adsorption isotherm systems. *Chem. Eng. J.* **2010**, *156*, 2–10. [CrossRef]
25. Zhu, S.; Khan, M.A.; Wang, F.; Bano, Z.; Xia, M. Exploration of adsorption mechanism of 2-phosphonobutane-1,2,4-tricarboxylic acid onto kaolinite and montmorillonite via batch experiment and theoretical studies. *J. Hazard. Mater.* **2021**, *403*, 123810. [CrossRef] [PubMed]
26. Chaudhary, M.; Jain, N.; Maiti, A. A comparative adsorption kinetic modeling of fluoride adsorption by nanoparticles and its polymeric nanocomposite. *J. Environ. Chem. Eng.* **2021**, *9*, 105595. [CrossRef]
27. Shyamal, D.S.; Ghosh, P.K. Efficiency of Portland Pozzolana Cement as an adsorbent in removing excess fluoride from groundwater. *Groundw. Sustain. Dev.* **2019**, *9*, 100248. [CrossRef]
28. Fan, X.; Parker, D.J.; Smith, M.D. Adsorption kinetics of fluoride on low-cost materials. *Water Res.* **2003**, *37*, 4929–4937. [CrossRef]
29. Mondal, P.; George, S. Removal of Fluoride from Drinking Water Using Novel Adsorbent Magnesia-Hydroxyapatite. *Water. Air. Soil Pollut.* **2015**, *226*, 241. [CrossRef]

MDPI
St. Alban-Anlage 66
4052 Basel
Switzerland
www.mdpi.com

Coatings Editorial Office
E-mail: coatings@mdpi.com
www.mdpi.com/journal/coatings

www.ingramcontent.com/pod-product-compliance
Lightning Source LLC
Chambersburg PA
CBHW061611210926
51850CB00009B/2289